"Information through Innovation"

Management of Information Technology

Carroll W. Frenzel

University of Colorado at Boulder

boyd & fraser

bf

boyd & fraser publishing company

DEDICATION

This book is dedicated to my wife, Barbara, with love.

Publisher: Thomas Walker
Acquisitions Editor: James H. Edwards
Developmental Editor: Arthur Weisbach
Production Editor: Barbara Worth
Interior Design and Composition: Gex, Inc.
Cover Design: Anne L. Craig

boyd & fraser

 ©1992 boyd & fraser publishing company
A Division of South-Western Publishing Company
Boston, Massachusetts 02116

Manufactured in the United States of America
Dust jacket photograph © Peggy Skycraft/West Light

Library of Congress Cataloging-in-Publication Data

Frenzel, Carroll W.
    Management of information technology / Carroll W. Frenzel.
        p.   cm.
    Includes index.
    ISBN 0-87835-508-1
    1. Information technology--Management.   I. Title.
T58.64.F74   1991
658.4'038--dc20   90-40254
                                                          CIP

2 3 4 5 6 7 8 9 D 4 3 2

Contents

PREFACE xix

PART ONE *THE NATURE OF INFORMATION TECHNOLOGY*
MANAGEMENT

Chapter 1 Management In The Information Age 2

Introduction 3
Expectations 4
Organizations and Information 6
    The IT Organization Within the Firm 6
    Structures and Information 8
Information Technology Management 14
The Difficulties of IT Management 15
    Technological Complexity 16
    Pervasiveness of Technology 17
    Applications and Data 17
    Production Operations 17
    Business Controls 18
    Environmental Changes 18
    Strategic Considerations 19
    People and Organizations 19
Technology Assimilation 19
    Stages of Growth 19
Current Issues 21
Skills That Favor Success 27
    Critical Success Factors 27
A Model for the Study of IT Management 29
Summary 29
Review Questions 31
Discussion Questions 31
Assignments 32
References and Readings 34

Chapter 2 The Strategic Importance of Information Technology . . 36

   A Business Vignette 37
   Introduction 40
   Strategic Issues for Senior Executives 40
      Why Strategic Issues Are Important 40
      Relation to Other Critical Issues 42
   Strategic Information Systems Defined 43
      Major Types of Systems 43
      The Character of SIS 44
   A Framework for Visualizing Competitive Forces 44
      A Model of Forces Governing Competition 44
      Strategic Thrusts 45
   Strategic Systems in Action 47
      Airline Reservation Systems 47
      Stock Brokerage 50
      Order Entry Systems 53
      Delivery Services 55
   Where Are the Leverage Points? 56
      Opportunities 56
      The Importance of Technology 57
      The Time Dimension 57
   The Characteristics of SIS 59
      Organization and Environment 59
      Financial Implications 59
      Additional Considerations 59
   The Strategist Looks Inward 60
   External Strategic Thrusts 60
   Integrating the Strategic Vision 61
   Summary 63
   Review Questions 63
   Discussion Questions 64
   Assignments 65
   References and Readings 67

Chapter 3 Developing an Information Technology Strategy 68

   A Business Vignette 69
   Introduction 70

Considerations in Strategy Development 71
Strategies and Plans 73
Types of Strategy 74
    Stand-Alone Strategies 76
    Business and Functional Strategies...................... 77
Requirements of a Strategy Statement 78
Strategy Document Outline.................................. 78
Strategy Development Process 83
    The Strategic Time Horizon 83
    Steps in Strategy Development 84
Strategy as a Guide To Action.............................. 85
The Strategy Maintenance Process 86
The IT Strategy Statement 86
    The Business Aspects 87
    Technical Issues....................................... 88
    Organizational Concerns 88
    Financial Matters...................................... 89
    Personnel Considerations............................... 89
Summary 89
Review Questions 91
Discussion Questions 92
Assignments 93
References and Readings 95

Chapter 4 Information Technology Planning................. 96

A Business Vignette 97
Introduction... 98
The Planning Horizon 99
Strategic Planning.. 102
Tactical Plans 102
Operational Plans and Controls........................... 103
    Planning Schedules 103
A Planning Model for IT Management 105
    Application Considerations 107
    Production Operations 107
    Resource Planning.................................... 109
    People Plans... 110
    Financial Plans 110

Administrative Actions 111
Technology Planning 112
Other Approaches to Planning 113
Stages of Growth 113
Critical Success Factors 114
Business System Planning 115
The Integrated Approach 115
Management Feedback Mechanisms 117
Summary 118
Review Questions 119
Discussion Questions 120
Assignments 121
References and Readings 123

PART TWO *INFORMATION TECHNOLOGY TRENDS*

Chapter 5 Hardware and Software Trends 126
A Business Vignette 127
Introduction 128
The Semiconductor Industry 130
Semiconductor Technology Trends 131
Advances in Recording Technology 133
Magnetic Recording 134
Alternative Technologies 134
From Microcomputers to Mainframes 136
The Microcomputer Revolution 138
What's Happening with Supercomputers? 138
Trends in Systems Architecture 141
Current Trends in Programming 141
Operating Systems Considerations 142
The Evolution of Operating Systems 142
Contemporary Operating Systems 144
Communications Technology 144
Technology Trends 145
The Meaning for Management 145
The Meaning for Organizations 146

Summary 147
Review Questions 148
Discussion Questions 149
Assignments 150
References and Readings 152
Appendix 153

Chapter 6   Advances in Telecommunication Systems 154

A Business Vignette 155
Introduction 157
The Telecommunications Services Industry 158
Telecommunications Equipment Suppliers 159
Telecommunications Systems 159
    Cluster Controller 161
    Line Adapters 161
    Telecommunications Lines or Links 162
    Communication Controller 163
A More Sophisticated Telecommunications System 163
Service Offerings 164
Some Current Developments 166
    Integrated Services Digital Network....................166
    The T-1 Service 168
Network and Communications Standards 171
Open Systems Interconnect 172
Vendor Network Architectures 174
    IBM Systems Network Architecture.......................174
    DEC Digital Network Architecture 174
    Transmission Control Protocol/Internet Protocol 175
Additional Types of Networks 176
    Local Area Networks 176
    Wide Area Networks 177
Summary 179
A Business Vignette 180
Review Questions 182
Discussion Questions 183
Assignments 184
References and Readings 185

Appendix A 186
Appendix B 186

PART THREE *THE APPLICATION PORTFOLIO RESOURCE*

Chapter 7 The Existing Portfolio Resource....................190

A Business Vignette 191
Introduction 192
The Application Resource 193
   Depreciation and Obsolescence 193
   Maintenance and Enhancement 193
The Data Resource 194
Applications as Depreciating Assets 194
Spending More May Not Be the Answer 195
   The Programming Backlog 195
   The Need to Prioritize 196
Enhancement and Maintenance Considerations 197
   The High Cost of Enhancement 198
   Trends in Resource Application 200
Spending Wisely 202
A Process for Portfolio Management 203
   Satisfaction Analysis..................................204
   Strategic and Operational Factors 206
   Costs and Benefits 208
   Some Additional Important Factors 209
Managing the Data Resource 211
Why the System is Valuable 212
Summary 212
Review Questions 213
Discussion Questions 214
Assignments 215
References and Readings 217

Chapter 8 Managing Application Development 218

A Business Vignette 219
Introduction 220

The Application Development Problem 221

    Reasons for Development Difficulties 221

The Traditional Life Cycle Approach 223

Application Project Management 224

Business Case Development 225

The Phase Review Process 228

    Phase Review Objectives 230

    Timing of Phase Reviews.............................230

    Phase Review Contents 230

    Who Are the Participants? 234

    Phases for Large Projects 234

Managing the Review Process 235

Resource Allocation and Control........................236

Risk Analysis 237

A Business Vignette 242

Risk Reduction Actions 243

Successful Application Management 243

Summary 244

A Business Vignette 246

Review Questions 247

Discussion Questions 248

Assignments 249

References and Readings 251

Chapter 9 Alternatives to Traditional Development 252

A Business Vignette 253

Introduction 255

Advances in Tools and Techniques 255

    Fourth-Generation Languages 255

    CASE Methodology 258

    The Object Paradigm 261

    Prototyping.......................................262

Subcontract Development 265

Purchased Applications 266

    Advantages of Purchased Software 267

    Disadvantages of Purchased Applications 268

Additional Alternatives 271

A Business Vignette 273

Managing the Alternatives 273
Summary 275
Review Questions 275
Discussion Questions 276
Assignments 277
References and Readings 278
Appendix 279

Chapter 10 Successful End-User Computing 280

A Business Vignette 281
Introduction 282
Why Adopt End-User Computing? 283
What are the Issues? 285
Organizational Changes 289
    The Workstation Store 289
    The Information Center 291
Policy Considerations.....................................292
Downsizing...293
    The Attributes of Downsizing 294
    What to Downsize...................................296
Office Automation 297
Planning for Office Automation 299
    Stages of Growth 299
Implementation Considerations 300
    Coordination 301
    Some Possible Difficulties 302
People Considerations 302
    Change Management 303
Summary 304
Review Questions 305
Discussion Questions 307
Assignment: A Case Discussion 308
Case Discussion Questions 310
References and Readings 312

PART FOUR *TACTICAL AND OPERATIONAL CONSIDERATIONS*

Chapter 11 Developing and Managing Customer Expectations . .316

A Business Vignette 317
Introduction 319
Tactical and Operational Concerns 320
Expectations 321
The Disciplined Approach 321
Service-Level Agreements 323
What the SLA Includes 325
      Schedule and Availability 327
      Timing 328
      Workload Forecasts 329
      Measurements of Satisfaction 331
The Role of User Satisfaction Surveys 332
Additional Considerations 332
Congruence of Expectations and Performance 333
Summary 334
Review Questions 334
Discussion Questions 335
Assignments 336
References and Readings 337
Appendix 337

Chapter 12 Problem, Change, and Recovery Management 338

A Business Vignette 339
Introduction 340
Problem Definition 342
What Is Problem Management? 342
      The Scope of Problem Management 343
      The Process of Problem Management 345
      The Tools of Problem Management 345
      Problem Management Implementation 346
Problem Management Reports 349
What Is Change Management? 349
      The Scope of Change Management 350
      The Change Management Process 351
Reporting Change Management Results 353

What Is Recovery Management? 354
A Business Vignette 356
    Emergency Planning 357
Contingency Plans 358
    Crucial Applications 358
    Environment 359
    Strategies 359
Recovery Plans 362
Summary 363
A Business Vignette 364
Review Questions 365
Discussion Questions 366
Assignments 367
References and Readings 368
Appendix A 368
Appendix B 369
Appendix C 369

Chapter 13 Managing Production Operations 370
A Business Vignette 371
Introduction 372
Managing Systems 374
    Batch Systems Management 374
    On-Line Systems Management 376
Performance Management 379
    Defining Performance 380
    Performance Planning 380
    Measuring Performance 382
    Analyzing Measurements 383
    Reporting Results 383
    System Tuning 384
Capacity Management 384
    Capacity Analysis 385
    Capacity Planning 386
    Additional Planning Factors 387
The Link to Service Levels 389
Management Information Reporting 389

Summary 390
Review Questions 391
Discussion Questions 392
Assignments 394
References and Readings 395

Chapter 14 Network Management 396
A Business Vignette 397
Introduction 398
The Importance of Network Management 399
    Networks Are Strategic Systems 399
    Networks Are International 400
    Networks Facilitate Restructuring 401
The Scope of Network Management 403
Management Expectations of Networks 404
The Disciplines Revisited 405
    Network Service Levels 407
    Configuration Management 408
    Network Problems and Changes 410
    Network Recovery Management 411
    Network Performance Assessment 412
    Capacity Assessment and Planning 413
    Management Reporting 414
Network Management Systems 414
International Considerations 415
The Network Manager 417
Summary 418
Review Questions 419
Discussion Questions 421
Assignments 422
References and Readings 425

PART FIVE *CONTROLLING THE INFORMATION RESOURCE*

Chapter 15 Accounting for Information Technology Resources . . 428
A Business Vignette 429
Introduction 431

Why Account for IT Resources? 432
The Objectives of Resource Accountability 435
Should IT Recover Costs from Users? 437
    The Goals of a Chargeback System 439
Alternative Methodologies................................ 441
    The Profit Center Method 441
    The Cost Center Method 443
Some Additional Considerations in Cost Recovery 445
    Funding Application Development and Maintenance 446
    Cost Recovery in Production Operations 447
    Network Accounting and Cost Recovery 449
Relationship to Client Behavior 450
Some Compromises to Consider 451
Expectations.. 452
Summary 453
Review Questions 454
Discussion Questions 455
Assignments 456
References and Readings 457

Chapter 16   Information Technology Controls and
                  Asset Protection 458

A Business Vignette 459
Introduction.. 461
The Meaning of Control 462
Why Controls Are Important 462
Some Principles of Business Control 464
    Asset Identification and Classification 464
    Separation of Duties.................................. 466
    Efficiency and Effectiveness of Controls 466
Control Responsibilities.................................. 467
Application Controls 469
    Application Processing Controls........................ 469
    Application Program Audits 475
    Controls in Production Operations 476
Network Controls and Security 476
Additional Control and Protection Considerations 480
The Keys to Effective Control 482

Summary 483
Review Questions 484
Discussion Questions 486
Assignments 488
References and Readings 489

PART SIX *PREPARING FOR ADVANCES IN INFORMATION TECHNOLOGY*

Chapter 17 People, Organizations, and Management Systems . . 492
A Business Vignette 493
Introduction 494
Technology Shapes Organizations 495
Organizational Transitions 495
Centralized Control—Decentralized Management 496
The Impact of Telecommunications 498
The Span of Communication 498
People Are the Enabling Resource 500
People and Information Technology 501
Essential People Management Skills 502
Effective People Management 503
Attitude and Beliefs of Good People Managers 503
Achieving High Morale 505
The Collection of Management Processes 506
Strategizing and Planning 507
Portfolio Asset Management 509
The Disciplines of Production Operations 512
The Management of Networks 515
Financial and Business Controls 516
Business Management 517
The IT Management System 519
Summary 519
Review Questions 521
Discussion Questions 522
Assignments 524
References and Readings 525

Chapter 18 The Chief Information Officer's Role 526

    A Business Vignette 527

    Introduction 528

    Corporate Challenges for the Senior IT Executive 529

    The Chief Information Officer 530

        Organizational Position 530

        Career Paths 531

        Performance Measurement 532

    Challenges Within the Organization 533

    The Chief Information Officer's Role .534

        CIOs Manage Technology 534

        CIOs Introduce New Technology 536

        CIOs Facilitate Organizational Change 538

    A Business Vignette 541

        CIOs Must Find New Ways of Doing Business 541

    Successful CIOs Are General Managers 543

    What CIOs Must Do for Success 543

    A Business Vignette 544

    Summary 546

    Review Questions 547

    Discussion Questions 549

    Assignments 550

    References and Readings 552

INDEX 553

Preface

The intense pace of technological innovation fosters rapid advances in information handling. These advances, in turn, are leading to the pervasive deployment of information technology throughout the industrialized world. The complexity of this technology; its widespread application; its vast potential for generating value; and its widespread adoption are altering the nature of management work in many firms today. The technological capability currently available for pursuing a firm's objectives and the projected growth place a premium on management's ability to capitalize on the wide array of opportunities and to anticipate the areas of potential pitfalls. Colleges, universities, and corporations are recognizing these facts; they are increasing their attention to the management of information technology as a result.

Management of Information Technology focuses on this complex subject for advanced undergraduates or graduate students and for current practioners. To take full advantage of this material, students studying this subject are expected to have a background in information-processing systems gained through course work in Computer Science, Telecommunications, Management Information Systems, or Engineering. However, students from fields as diverse as Economics, Law, and Military Science will also find the management of information technology important to their future.

SCOPE OF THIS TEXT

This text is directly concerned with the management issues surrounding information technology and with the ingredients of management knowledge necessary for success for information technology managers. *Management of Information Technology* describes what managers must know, and it also teaches what managers must do. The text outlines proven techniques, processes, and procedures through which management knowledge is applied. Successful IT managers must know management principles and must know when and how to apply them; this text teaches both.

The view of information technology taken by this text is from the perspective of management at several levels—from the first line manager to the chief executive officer. The text material provides frameworks and principles that managers or aspiring managers can utilize as they reap the benefits of the challenges of rapidly advancing technology and its ramifications.

Scope of the Subject

Information technology embraces the large mainframes usually owned and operated by the central information systems organization; the information-processing capabilities dispersed throughout the firm; and the network and telecommunications systems used for voice, video, and data transmission. (Separate telecommunications and information system organizations are now being consolidated in many firms, because their common technology eliminates distinctions previously existing between voice and data functions.)

Management of Information Technology integrates the subjects of telecommunications and information systems management into one entity, information technology management. By integrating these topics at the outset, the text presents a coherent view of the coalescence of the technology and its management and organizational consequences.

Management Implications

Executives in charge of the consolidated organizations acquire significant responsibilities; they need a broad array of finely tuned management skills. Information technology managers also have substantial line responsibility in most firms. Furthermore, they may acquire considerable staff responsibility arising from the rapid dispersion of computing resources. This dispersion may be driven by downsizing, use of personal workstations, and employment of end-user computing. As both line and staff responsibilities grow, the nature of the senior IT manager's job at many firms will expand in scope and responsibility, giving rise to the position of Chief Information Officer.

These trends and others demand that IT managers have skills more closely resembling those of general managers. General management skills are important, but a keen awareness and knowledge of technical issues, particularly technology trends, is vital. This text dissects and displays the many issues and examines the complex task of preparing individuals for their IT responsibilities.

Workplace Perspective

Graduates entering the workplace are confronted by a wide variety of situations for which they need preparation. The deployment of information technology in their firms may alter each organization, changing individual work patterns and demanding skills and management practices that are frequently unique to information technology. *Management of Information Technology* focuses attention on these practices and a myriad of other important issues, bringing the subject matter together in a logical, rational, and easy-to-understand manner. Students learn the major underlying issues and grasp the management principles required to deal with them. They are also exposed to the processes needed for implementing the principles. The themes of information asset management, people management, long- and short-range planning, disciplined management processes, and managing expectations unify the subject matter.

Discussion of Management Issues

Another text feature consists of a disciplined methodology applied to the significant management issues surrounding information technology. The student learns management processes, tools and techniques, and organizational adjustments which provide important leverage in dealing with the issues. These practices surface in the areas of strategy and planning; technology assessment; applications portfolio management; operational activities, and business controls.

Realistic Business Examples

Students need an appreciation of the information industry. They should be familiar with some of the ways in which firms utilize the technology for competitive advantage. The text addresses this subject in the chapter entitled "The Strategic Importance of Information Technology" and in business vignettes at the beginning of each chapter. For example, the vignette "Who's On Second?" exposes the reader to some facets of semiconductor development and manufacturing and illustrates some business aspects of second-sourcing; some international ramifications of networking are developed in the vignette "Volvo's Net Gains."

Organization of the Text

Management of Information Technology is organized in a modular fashion. This gives instructors the ability to adjust both the volume and emphasis of material to suit their purposes. Some instructors may prefer to concentrate on business controls, emphasizing Part Five, while reducing the focus on operational topics found in Part Four. Likewise, some instructors may focus less on technology trends found in Part Two and cover the remaining topics in detail.

CONTENT OVERVIEW

Part One: The Nature of Information Technology Management

This section describes the nature of information technology and the issues related to its management. The reader is presented with frameworks and examples through which a perspective on this subject can be gained. Part One also outlines the strategic importance of information technology. It presents the essential ingredients for developing strategic statements and translating these statements into strategic plans. The important topic of strategic planning for information technology is covered in detail. The concepts of tactical and operational planning are developed as logical extensions to the strategic planning process. The planning process is explored and developed via models which relate tools, information, and organizations interacting through time.

Part Two: Information Technology Trends

Information delivery systems and technology trends are explored in Part Two. Trends in semiconductor logic and memory are related to advances in microcomputers, mainframes, and supercomputers. Advances in programming, secondary storage, telecommunications, and personal workstations are covered in this section. The student will gain a perspective on the rapid pace and direction of this innovative activity and will see these trends related to management opportunities. The text focuses extensively on the importance of telecommunications and how it enables the integration of voice, data, and video information. Open System Interconnect, Integrated Services Digital Network, and other standards activities important to information technology are presented. The relationship between these developments and industry implementation give students a realistic view of the potential and opportunities facing current and future IT managers.

Personal computers have had a revolutionary impact on information systems in government, business, and industry. The nature of this impact and the accompanying business and sociological consequences are related to the concomitant management challenges and opportunities. The importance of personal workstations and their relationship to the broader field of information technology are highlighted.

Part Three: The Applications Portfolio Resource

The applications portfolio is a major source of benefits for the firm. It also represents the most profound source of difficulty for the firm's managers. Development backlogs remain high; schedules for completion seem inordinately long; and the investment of increasing amounts of money seems not to eliminate the problems. The enormous amounts of input and output data supporting any applications set add significant complications to the applications environment. The management of application development and acquisition is given special attention. Discussion of the successful implementation of end-user computing and office systems in the firm follows application management.

Part Four: Tactical And Operational Considerations

The fourth section of this text focuses on tactical and operational concerns. The approach in this section is to develop descriptions of disciplined processes for managing the tactical and operational situations encountered in IT operations. The development of customer and client expectations, along with IT's response to user requirements, introduces this section. The management of problems and changes and the development of plans for recovering from disastrous or catastrophic failures follow. Techniques for managing daily production operations, including the disciplines required to manage large networks, are explored next. Management reporting systems respond to established customer expectations and tie the operational disciplines together. This section deals with the successful management of information and telecommunications systems within tactical and operational time frames.

Part Five: Controlling The Information Resource

Rapid and pervasive implementation of information technology is accompanied by the requirement for increased management control. The growing complexity of the technology and its application, the speed with which it is progressing, and heightened legal implications make this subject increasingly relevant.

IT accounting systems are discussed in relation to financial control and to customer and client relations and expectations. Applications resources demand internal controls and audits. The reader is given insights into the need for internal controls and audits in the applications portfolio resource and gains insights regarding how to accomplish these goals. These topics set the scene for the broader subject of protection of information assets. The widespread deployment of information resources needs to be accomplished within a framework of sound business controls. Information technology managers need to provide these controls and must be able to establish a program which ensures senior executives that the controls are effective and efficient. *Management of Information Technology* develops this subject matter in detail.

Part Six: Preparing for Advances in Information Technology

This final section considers the relationship of people and management systems to technology and organizations. The impact of technology advances on people and organizations is explored from the viewpoint of first-line managers and senior managers. Personnel management issues and the need for people-oriented managers in high-technology organizations form an important ingredient of this section. Interpersonal skills employed by most successful managers are reviewed in this chapter. This section summarizes the management principles and practices developed throughout the text and molds them into a cohesive management system designed to serve both CIOs and their subordinates.

The role of chief information officers is presented in this section with the perspective of management principles and practices found throughout this book. This presentation serves to illuminate their role and to summarize the approaches which permit them to achieve success. The final chapter of this section links together the main threads developed from the beginning and rounds out the management system central to this book's theme.

THE THEME

The theme of this text is to illustrate the challenges and opportunities found in information technology management through the exploration of trends. These trends shed light on future issues and concerns. Models of successful behavior are presented, and tools and techniques designed to capitalize on opportunities and minimize the effects of pitfalls are developed. Students of this text achieve an appreciation for the value of management; they are prepared for events certain to face future managers dealing with technology issues.

A NOTE TO THE READER

The subject matter in this text is broad, and the material is comprehensive. Students who take the necessary time to relate the vignettes, the text material, the questions and exercises, and the references will be rewarded with significant new knowledge. This knowledge will be of substantial value to the aspiring IT manager. This book, however, does not pretend to answer all the questions that could possibly be raised about this complex subject, nor does it intend to portray management as responding to stimuli with conditioned responses. People and technology are incredibly complex; it's a wise person who, after extensive study and prolonged experience, maintains a high level of humility. Management can be a highly satisfying career, and it is my hope this book will bring increased satisfaction to current and future members of this profession.

ACKNOWLEDGEMENTS

I derived much of the material in this book from more than 20 years of experience as a manager for the IBM Company in the area of information technology. IBM gave me a great deal of formal training for the responsibilities inherent in my positions, and I am grateful for these educational experiences. Finally, my management experience was enriched by the associations with many wonderful people; I was taught by superiors, peers, and subordinates alike. I hereby acknowledge the contribution to my education by all of my colleagues. The contents of this book are my responsibility alone, however. They do not imply any endorsement by IBM.

I would like to thank 13 special people for their thorough reviews of the manuscript for this textbook. Almost all comments offered by these reviewers were incorporated in the final text. The following reviewers receive my grateful acknowledgement of their assistance:

Carol V. Brown
Purdue University

Paul H. Cheney
Texas Tech University

J. Daniel Cougar
University of Colorado
at Colorado Springs

Albert L. Harris
Appalachian State
University

C. Brian Honess
University of
South Carolina

Mehdi Khosrowpour
Penn State University

Lewis E. Leeburg
UCLA

Jane Mackay
Texas Christian
University

Rodney Pearson
Mississippi State
University

John V. Quigley
East Tennessee State
University

John Sviokla
Harvard University

James T. C. Teng
University of
Pittsburgh

Robert Trent
University of Virginia

Part
One

1 Management in the Information Age

2 The Strategic Importance of Information Technology

3 Developing an Information Technology Strategy

4 Information Technology Planning

The Nature of Information Technology Management

Rapid advances in information and telecommunications systems have created enormous opportunities for skillful managers. The management of information technology (IT) demands a high level of general management expertise focusing on strategy development, long-range planning, and business controls. Successful information technology managers exhibit outstanding resource and people management skills. This Part includes chapters on Management in the Information Age, The Strategic Importance of Information Technology, Developing an Information Technology Strategy, and Information Technology Planning.

The management of information technology is an exceptionally difficult task. The ability to develop IT strategic directions and plan effective implementations is a vital first step toward successful IT management.

1

Management in the Information Age

Introduction
Expectations
Organizations and Information
  The IT Organization Within the Firm
  Structures and Information
Information Technology Management
The Difficulties of IT Management
  Technological Complexity
  Pervasiveness of Technology
  Applications and Data
  Production Operations
  Business Controls
  Environmental Changes
  Strategic Considerations
  People and Organizations
Technology Assimilation
  Stages of Growth
Current Issues
Skills That Favor Success
  Critical Success Factors
A Model for the Study of IT Management
Summary
Review Questions
Discussion Questions
Assignments
References and Readings

The typical large business 20 years hence will have fewer than half the levels of management of its counterpart today, and no more than a third of the managers. In its structure, and in its management problems and concerns, it will bear little resemblance to the typical manufacturing company, circa 1950, which our textbooks still consider the norm. Instead it is far more likely to resemble organizations that neither the practicing manager nor the management scholar pays much attention to today: the hospital, the university, the symphony orchestra. For like them, the typical business will be knowledge-based, an organization composed largely of specialists who direct or discipline their own performance through organized feedback from colleagues, customers, and headquarters. For this reason, it will be what I call an information-based organization.[1]

INTRODUCTION

The introduction of new technology to process and transport data and information has proceeded at exceptional rates for more than three decades. This innovation introduction has significantly affected employees, managers, and their organizations. New technology has created countless opportunities and challenges for millions of individuals. In particular, the challenges to managers responsible for introducing this technology have been exceptionally high. In our information-based society, management must attempt to capture the advantages offered by information technology, yet must also avoid the pitfalls along the way toward increasing automation. As information technology has altered the way many people do their jobs and has changed the nature of work in industrialized nations, the practice of management has been greatly affected. The management of many firms will require substantial future readjustment.

The revolution that we are experiencing has occurred largely during the past 35 years. Many individuals entering the workplace today have directly experienced this technological phenomenon, and perhaps they take it for granted. Employees completing their careers have seen the whole spectrum of events unfold during their lifetimes. For many people, the growth of information technology has been an unmitigated blessing. The technology was a major part of their formal education; it forms the basis of their employment. It is a platform on which their future depends. For others, information technology has been a complicating factor, perhaps something to be feared, or at least to be viewed with apprehension. For nearly everyone it has brought change.

Information technology (IT) has become increasingly vital for creating and delivering the products and services in industrialized nations. For those who manage the information technology function within a larger enterprise, the rapid pace of innovation has meant an unprecedented growth of job opportunities, fueled by an ever-increasing need for skilled managers. Senior executives who can manage these complex tasks are in great demand. This demand has been driven by the growth of computer applications such as decision support systems, expert systems, computer-aided design, and computer-aided manufacturing. Demand for skilled executives has also been created by the rapid dispersion of technology via advances in telecommunications and personal computing hardware and software. The outlook for skilled IT managers remains bright.

DuWayne Peterson, executive vice president and general manager of information systems for Merrill Lynch, is an example of a successful IT manager. Merrill Lynch recently decentralized its $1.2 billion MIS organization and now the majority of its 14,000 information systems staff reports directly to the business units they support. "MIS's role is to give business units the structure they need so they can be turned loose to do what they do best—make money," said Peterson. He was hired to accomplish the introduction of new products and to reduce Merrill's back office expenses. It is reported that he was able to reduce expenses by $300 million.[2]

This introductory chapter sets the scene for the management task in the field of information technology. It analyzes the nature of information technology and relates it to the challenges of management. Executives leading organizations and individuals working within them have visions or expectations of what the technology can do for them, but they may have less appreciation of the impact the technology will have on them. The impacts are very significant, however, and must be well managed for the firm to capture the opportunities. A discussion of the issues facing senior IT executives is presented in this chapter, and frameworks for dealing effectively with these issues are found throughout the text. Leading IT managers of the future will need to have strong foundations in general management principles. They will perceive how these principles apply to the complex task of managing technology introduction.

EXPECTATIONS

Although the opportunities available to highly skilled managers are bright indeed, the challenges of the profession remain exceptionally high as well. These challenges show no evidence of decreasing dramatically in the future.

The challenges arise from the volatility of the technology and from the relationship of the evolving technology to the structure of the firm. Other sources of challenge are the expectations held by members of the firm at many levels.[3] IT managers must anticipate these challenges, and they also need thoughtful plans to contend with the issues. Well-prepared managers will seek sound understanding of the issues surrounding these challenges, and they will prepare to deal with their consequences.

For IT managers to be successful within their organizations, they must not only be prepared to cope with the issues as they arise; they must also take a leadership position in formulating and shaping them. Their vision of future technology is valuable in enabling executives to develop strategies that permit the firm to gain competitive advantage. IT managers will need to prepare executives in the firm with their technological vision so that the firm can anticipate future structural changes, and prepare for them well in advance. IT managers must accomplish this task from the perspective of the CEO. Their vision must be that of a general manager. Their vision must be presented; but above all, it must inspire the executive team with a realistic and practical view of the IT future.

IT management must be at its best when dealing with expectations. Senior executives have expectations that include using information technology for competitive advantage and attaining bottom-line results for the firm. Given that corporations spend anywhere from 1 to 5 percent of revenue on information technology, these expectations are entirely reasonable.[4] CEOs have every reason to require the IT organization to conduct its affairs in a businesslike manner and to conform to business practices common to their corporations. It is reasonable that CEOs require general management skills on the part of their senior IT managers.

In addition, there are numerous sources of information, many of which originate outside the firm, that influence CEOs and their management teams in particular. These information sources include trade associations, government agencies, and informal communication with peers, in addition to sources internal to the firm. Information is a basis for expectations, and senior IT managers need to respond to these expectations in a disciplined manner.

IT managers must have a good understanding of the corporate culture. Corporate culture or corporate philosophy consists of the basic beliefs or basic ideas that guide members of the organization in their behavior within the organization. These behavior patterns describe "how we do things around here."[5] IT managers must also have a clear and realistic view of technology futures and an appreciation for the degree of technological maturity of the firm. This knowledge is essential for providing the CEO and others with information upon which expectations can be built.

Expectations held by the firm's senior executives constitute a yard-stick against which IT managers will ultimately be measured. It doesn't seem to matter where the expectations originated or whether they were realistic. But completely fulfilled expectations usually lead to a satisfactory performance appraisal in most organizations. Skillful IT managers, those with a general management view of the business in which they are operating, are likely to position themselves and their organizations to maximize this eventuality. Superior IT managers understand the importance of expectations and cope with them effectively.

Less-skilled managers interact with the process of setting expectations less effectively. They find themselves and their organizations more frequently overcommitted or operating reactively. Unskilled managers and the organizations they lead create expectations they are unable to fulfill. Through lack of discipline or excess enthusiasm, such managers sow the seeds of their demise. Setting and managing expectations relate directly to the perceived performance of executives and are vital to their personal success and to the success of their organizations. The general management skills needed to cope with expectations and to operate effectively in this arena will be discussed at appropriate points throughout this text.

Skills alone, however, are not enough. Tools and processes, along with a management system in which these processes can operate effectively within the corporate culture, are a necessary condition for success. These processes involve various members of the firm engaged in activities ranging from long-term strategic considerations at one extreme to very short-term considerations at the other. In short, successful IT management can flourish only in a supportive environment. All players engaged in this activity bear some responsibility for the success of the processes.

ORGANIZATIONS AND INFORMATION

The IT Organization Within the Firm

The IT organization performs a vital function for the firm and usually enjoys functional status within it. IT holds a position similar to that of marketing, accounting, or manufacturing. The IT organization is structured with departments that have specific responsibilities, reporting through department managers to the senior IT manager. In many firms, the senior manager reports within the finance organization, but there are

current trends toward alternate reporting relations. Organizational alignment and the rise of the chief information officer, or CIO, will be discussed later in this text.[6]

The IT organization operates as a business within a business,
supporting many other functional units in a variety of ways. Functions
usually supported and some typical applications in support of these functions are illustrated in Table 1.1.

TABLE 1.1   Functions and Applications Supporting Them

Functions	Applications Supported
Product development	Design automation, parts catalog
Manufacturing	Materials logistics, factory automation
Distribution	Warehouse automation, shipping and receiving
Sales	Order entry, sales analysis, commission accounting
Service	Call dispatching, parts logistics, failure analysis
Finance and accounting	Ledger, planning, accounts payable
Administration	Office systems, personnel records

The typical large firm uses information technology extensively
throughout its business functions. The examples indicated in Table 1.1
are rather common and easily recognized. But they are only a small part
of the total portfolio of computerized applications. The application portfolio of a medium-size corporation may contain several thousand computer programs.[7]

A typical structure for the IT organization and some main activities
for the elements within the structure are depicted in Table 1.2. But not
all IT functions follow this pattern. The industry within which the firm
operates and the culture of the firm are governing factors. For example,
in high-tech firms the role of technical support may be enlarged or
expanded, while in other firms this activity may be delegated to the
development group or to computer operations. As will be discussed in
Chapter 10, the organizational positioning of the information center has
a number of possible variants. Structures can vary, and they are evolving, but what is shown is typical of many firms today.

This portrayal indicates the line management organization of the IT
manager. The operating units report directly to the IT manager; this
arrangement is called line management. In many firms the CIO has more
responsibility than is shown in Table 1.2. Typically the CIO position

includes responsibility for information activities dispersed throughout the firm in organizations not directly reporting to the CIO. The CIO position may have staff organizations to assist in discharging these responsibilities. The responsibility for activity not directly reporting to the CIO is called staff responsibility.[8] Subsequent portions of this text will refer to this structure.

TABLE 1.2   Information Technology Organization

Applications Development	Computer Operations	Technical Support	Systems Planning
Develops and maintains applications in support of the firm's business units	Operates and manages the computer and communication equipment and runs application systems	Installs and maintains operating systems and communication software	Develops and maintains plans and strategy and manages the information center

Structures and Information

Organization structures have been actively studied for many years, and an inevitable conclusion is that some sort of structure is necessary for the effective operation of human enterprise. The pyramid structure with the nonmanagerial employees forming the base and the chief executive at the top is one familiar structure. Discussions of organizational structure include, among other topics, the concepts of levels of management and span of control.

The concept of levels of management implies differences in responsibility and degrees of authority, both increasing from the base of the pyramid to the senior executive at the top. The responsibility differences can be distinguished in part by the characteristics of the data and information utilized by the incumbents and the manner in which they are used. In particular, three facets help to differentiate the levels of management and their relation to and use of information:

1. The sources of information applicable to the position
2. The degree of judgment involved in the application of the information
3. The time span in which the information is considered pertinent for making decisions.

There are other characteristics as well, such as level of aggregation, currency, and frequency of use.[9]

These differentiating characteristics are useful for understanding the manner and degree to which information technology has been applied within the structure. Keep in mind, however, that the introduction of information technology within a structure acts to change that structure and, as Drucker[10] points out, these alterations may be large and very important indeed.

Information employed by individuals at the base of the pyramid is likely to originate within the firm. The information will have value over a relatively short time measured in hours or days, and will require relatively little judgment in its application. The expression most applicable to this activity is data processing because the commodity in use is data rather than information. Systems to do this type of work are generally called transaction processing systems or structured decision systems. In contrast, information of value to the CEO is more likely to originate outside the firm. This information will have value or meaning for a much longer period—one to five years or more—and will require a high degree of experience and judgment in its application. Executive activity involves information and knowledge. The systems that assist in this work are called executive information systems.

As one moves up the pyramid, the data or information is applied in differing ways. At lower levels in the firm, data and information are applied operationally, but senior managers use data and information tactically or strategically. The time frame of value of the commodity increases from perhaps one day to several years. To utilize this commodity fully requires increasing amounts of judgment as one ascends the pyramid. In the ranks of middle management, the situation falls somewhere between the two extremes. Middle managers utilize information originating inside and outside the firm which requires modest amounts of judgment for application. The period applicable to this information is normally one to two years.

The text will say more about organizational structure, but keep in mind the transformations that are taking place. For a number of reasons, including the use of information technology, organizational structures are changing. Peter Drucker, Tom Peters, and others have pointed out the tendency for levels of management to decrease and for the organization to become "flatter." For example, Nucor, a billion-dollar steel company, has only three layers between the millworker and the chairman of the board. Wal-Mart, with sales of $25 billion, also has three layers in its hierarchy. Wal-Mart operates a satellite network with more than 1100 terminals. More than 1800 of its 5000 suppliers use electronic data interchange, EDI, and its on-line storage is growing at 100% annually. Wal-Mart invested more than $500 million in IT over the past five years.[11]

We know from the flatter structure that these very successful companies operate with a style of management entirely different from conventionally organized companies. We also realize that the responsibilities of managers in the flattened organizations differ from those in conventional organizations. In either type of organization, however, the functions performed at the top of the organization differ markedly from those performed at the lower levels. The next few paragraphs discuss some of the characteristics of these positions and describe how they vary with level in the organization.

In what follows, this text adopts Drucker's definition of information: "Information is data endowed with relevance and purpose. Converting data into information thus requires knowledge. Knowledge, by definition, is specialized."[12] Knowledge is information that has been distilled via study or research and augmented by judgment and experience. Actually, these distinctions become blurred because one person's information may become someone else's data as communication occurs within the firm.

Figure 1.1 illustrates the relationships among the management levels in an organization and the time frames within which managers or executives normally operate. In this traditional view of organizational structure, the first-line managers are responsible for the firm's activities on a day-to-day basis, while higher levels of management maintain a longer-term perspective. First-line managers plan tomorrow's activities; the senior executives plan for the next several years or the next decade. These perspectives require vastly different types of information. The information delivery system also varies considerably among the managerial levels.

The terms *strategic* and *tactical* are used in this text to describe the kind of activity and the periods involved at the CEO level and the middle-management level, respectively. There is also a relationship among these terms, the organizational levels, and the quantity of judgment and experience that may be required. For example, the application of data to the operational activities of loading a supertanker may require little or no judgment. But the use of data or information by middle managers to schedule the tanker fleet requires more sophistication and experience. The CEO, who must decide whether to expand or contract the tanker fleet, requires an exquisite blend of knowledge, experience, intuition, and judgment. Management control or middle-management activities combine some attributes from each end of the management spectrum. This structure or taxonomy was first described by Anthony.[13]

Another useful framework for understanding these concepts was developed by Gorry and Scott-Morton.[14] Their model relates levels of management to structured or unstructured decision making. Structured decisions are those based on a relatively firm understanding of the underlying parameters. Determining how much inventory is on hand is structured because all the factors in the determination can be quantified. Deciding how many products to schedule in the production line tomorrow is an example of a semistructured decision because customer demand, work in process, and other necessary factors are not usually completely known. Deciding how much money to spend on research is unstructured because the factors on which to base this decision are not crisply known. The results of the research activity are an uncertain but important factor. Table 1.3 illustrates the Gorry and Scott-Morton model.

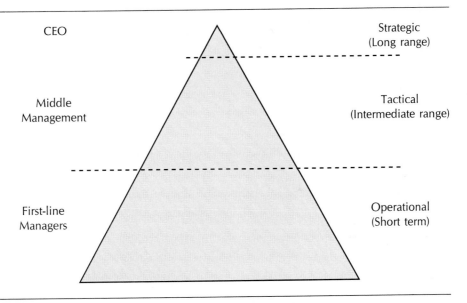

FIGURE 1.1   The Traditional Managerial Hierarchy

These characteristics can be depicted in yet another manner by developing an orthogonal relationship among levels of management, the source of information, and the temporal considerations. Figures 1.2 and 1.3 show these relationships. In each case the vertical scale represents 100 percent of the item under consideration.

As one moves horizontally from the first-line management position toward the CEO position, the source of information changes from mostly internal to mostly external. First-line managers have little need to seek information from external sources in most cases. Their major sources of information are the databases supporting the transaction processing systems at their command. The CEO, on the other hand, utilizes and relies on external sources of information to a far greater extent. For the most part, this external information cannot be found in the databases internal to the firm.

TABLE 1.3   The Gorry and Scott-Morton Model

	Operational Control	Tactical Control	Strategic Planning
Structured	Accounts receivable	Budget analysis	Tanker fleet mix
	Order entry	Short-term forecasting	Factory location
	Inventory control		
Semistructured	Production scheduling	Variance analysis	Mergers and acquisitions
	Cash management	Budget preparation	New product planning
Unstructured	PERT/COST systems	Sales and production	R&D planning

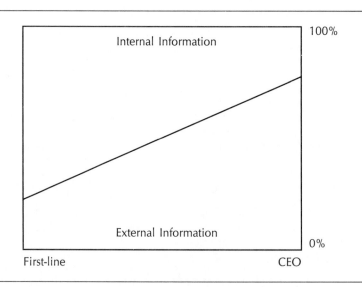

FIGURE 1.2   Sources of Information *vs.* Level of Management

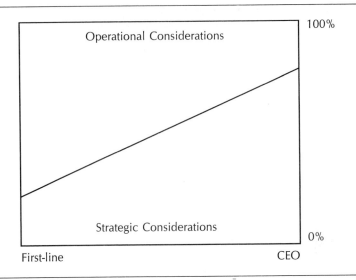

100%

Operational Considerations

0%

Strategic Considerations

First-line   CEO

FIGURE 1.3   Type of Information *vs.* Level of Management

First-line managers are concerned mostly with operational considerations and data. Their view of the firm is operational and near-term. CEOs
must take the long-term or strategic view. They are not dealing with operational issues in most of their activities.

A useful way to combine these concepts is shown in Figure 1.4.

This portrayal does not imply a smooth continuum in the variables
as one moves from the first-line position to the CEO post. Additionally,
the relative values are not the same for all firms. They may also differ
for various functions within the same firm.

For example, in some firms the CEO has developed very capable
subordinates who assume nearly complete responsibility for the operational activities of the firm. In these firms, the CEO develops and manages
external, long-range relationships with customers, suppliers, governments,
and others. In other firms, particularly those in distress, the CEO and
other senior executives concentrate their energy and talents on the immediate, mostly internal actions that must be accomplished successfully for
survival. Likewise, a smoothly operating manufacturing function will
receive little executive attention, while a marketing organization facing
stiff competition will undoubtedly attract an abundance of attention.

As Drucker stated, future organizations will experience major structural changes as they transform themselves into knowledge-based organizations through the pervasive employment of information technology.
Technology will permit CEOs to interact directly and efficiently with more

members of the management team. Their effective span of control will increase. The span of control of the other managers in the firm will also increase. Consequently, the firm will require fewer levels of management and fewer managers.

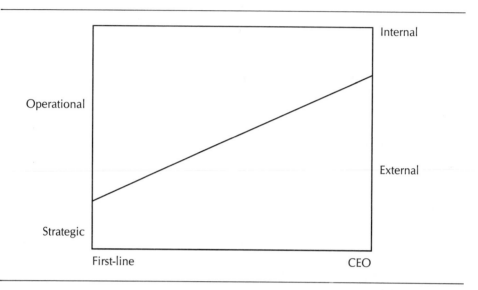

FIGURE 1.4   Types and Sources of Information *vs.* Level of Management

Information technology will have profound impacts on business enterprises and their internal structures. The technology will affect the firm's relationships with the external competitive environment and will alter the environment for employees and managers. Nonhierarchical structures and strategic alliances are two of the many key forces that leaders of organizations must contend with in the future. Noel Tichy, of the School of Business Administration, University of Michigan, uses the words *corporate transformation* to describe the organizational restructuring taking place. This restructuring "is like trying to change the fan belt while the motor is running," he claims.[15] The challenges imposed on managers at all levels and in all functions will be enormous. These impacts and challenges will be explored at appropriate points throughout the text.

INFORMATION TECHNOLOGY MANAGEMENT

For reasons that are beginning to emerge, information technology managers have a difficult task to perform. Although many IT managers have established brilliant performance records, others have done less well. Mediocre performance and a lack of success has dominated the careers of many. The literature chronicles some of the difficulties of IT management over the years.

Firings and terminations of EDP managers are common and both top management and EDP managers must reassess the function of information systems and must focus on the management dimensions of the EDP manager's job. Top management often fails to recognize the dual managerial and technical role of the EDP manager's position and the EDP manager is frequently the scapegoat for things that go wrong. EDP managers need to recognize that applications cause change and they must view themselves as change agents. Active collaboration between the EDP function and the user organizations is required.[16]

The information systems manager's job is changing. The technical content of the work is declining and the managerial content is rising. Organizations now have many technical experts and specialists on which they can rely. Many IS managers have surprisingly little contact with users; a study by Ives and Olson was unable to determine the IS manager's effectiveness.[17]

Reorganization, decentralization, and budget cuts are causing many top-level IS executives to seek other employment opportunities. Many IS executives have done a poor job of educating their bosses on the importance of information systems, and some have not managed their budgets properly. Some top executives have not developed sound plans, and some are being pressured by organizational changes due to mergers or acquisitions.[18]

Joseph Izzo believes IS organizations are in trouble because they have not lifted productivity for the firm and are not contributing to the firm's business goals. The IS organization is isolated from the remainder of the firm and resists change. IS organizations must make computer technology more accessible in a logical and businesslike manner.[19]

THE DIFFICULTIES OF IT MANAGEMENT

Many information technology managers find themselves and their organizations in untenable positions. They are implementing new technologies that have high potential value to the organization but that carry high risk. They are supporting clients who are requesting increased services while senior executives are questioning the increasing expense levels. The demand for services is manifested in expanding backlogs of work. Expectations in many parts of the firm are rising, and senior IT managers in many firms are unable to cope effectively with those challenges.

As some of the articles discussed above suggest, IT managers have been trained in technology, but they lack the general management skills demanded by their organizational role. They and their organizations are under continuing scrutiny. Their jobs demand knowledge of people management and organizational considerations rather than programming or hardware expertise. Their jobs also require them to demonstrate productivity improvements for the organization in return for the resources they consume. In addition, mergers, acquisitions, and reorganizations of all kinds are threatening their positions.

The root causes of the difficulties of information technology management are ill-prepared managers in charge of complex, rapidly changing technology. The technology causes structural and personal dislocations and generates high expectations within the organization and among its managers. To address this situation, future IT managers must acquire general management skills that will allow them to contend successfully with these complex issues. Their skills must include managing expectations and coping with personal and structural changes. In short, the IT manager of the future must be the organization's technological leader and a superb generalist as well.

In addition, it is imperative that IT managers have extensive management systems and processes to assist them in discharging their responsibilities. (The development of these systems and processes forms a considerable portion of this text.) Understanding these management systems and processes and knowing their application and use within the firm are necessary conditions for success. The text devotes considerable attention to these important topics.

Technological Complexity

Semiconductor computer chips less than one-fourth square inch in area are commercially available for several hundred dollars each. These chips carry millions of electronic circuits and cost $100 million or more to develop. The peoples of the world are linked electronically through strands of fiber-optic cable capable of carrying 8000 simultaneous telephone conversations. More than 130 satellites relay enormous amounts of data of all kinds. High-tech storage devices maintain huge databases at modest costs which serve as sources and sinks of data for telecommunications and computer systems around the globe. Not only is the complexity of this technology increasing, but the rate of change is accelerating as well. The development of modern computer systems is the most complex human activity ever undertaken. The application of computer systems is also incredibly complex.

Modern computer and communication systems bring promise of great benefit to firms adopting this technology. However, the complexities inherent in technology adoption and utilization raise the management challenges to new heights. The opportunities for individuals well prepared to manage the new challenges are also unprecedented.

Pervasiveness of Technology

Computer and telecommunications technology is pervasive in the modern firm. Advanced business users of information technology are installing intelligent workstations at a rapid pace; in some firms, most professional employees have their own individual workstations.[20] In a few years, most office and professional workers will have workstations for their sole use. Soon these workstations will be interconnected via local area networks that access large mainframe computers. The mainframes will function as message-switching systems and large data repositories. Sophisticated telecommunication networks will interconnect a firm's sites nationally and internationally and will provide data highways linking firms to their customers and suppliers wherever they may be located.

The Gartner Group, an information systems consulting firm located in Stamford, Connecticut, reported worldwide microcomputer sales of $21 billion in 1988 and forecasts sales of $40 billion in 1995. The sales of mainframe computers are expected to rise from $36 billion to $86 billion during this same period.[21] Likewise, the telecommunications industry is one of the most rapidly growing segments of the U.S. economy. The penetration of electronic information processing into the fabric of human activity will continue unabated into the foreseeable future.

Applications and Data

Rapidly growing advanced computer and telecommunication systems capability will facilitate large, sophisticated, and very valuable applications programs for the firm. The firm's databases will grow in size and importance. The acquisition and maintenance of this vast program and data resource demands careful attention from many members of the firm's senior management team. The application program and database resource, largely intangible and not recorded on the balance sheet, is the cornerstone of the information-based organization of the future.

Production Operations

Skillfully managed production operations are vital for the efficient performance of the firm. Today, some firms process hundreds of revenue-producing transactions per second. Each must be executed flawlessly and in a timely manner. Loss of service for even a few seconds in this type of system has serious consequences to the firm's financial health and to its reputation. Management performance demanded by this kind of operation is extremely challenging.

Production operations usually consist of one or more large mainframe computers running routine batch and on-line applications. In some cases minicomputers and large microcomputers are being used for part of the firm's production work. IT activities supporting production processing on small and large systems are growing in importance.

Significant growth in large systems can be expected too. The Gartner Group forecasts a 45% increase in the annual number of mainframe computers shipped between 1988 and 1995 and indicates the average mainframe growing in capability from 30 MIPS (million instructions per second) to 150 MIPS. During this period, the number of new mainframe MIPS shipped annually will increase by a factor of seven.[22] The task of managing this expanding computer power will increase greatly. And it will bring new and exciting challenges to a larger group of IT managers.

Business Controls

Knowledge-based organizations must operate under control. Weak and ineffective controls or loss of control in highly valuable and highly automated operations allow very rapid error propagation. Additional threats, both external and internal, to sophisticated human and machine business operations must be neutralized through careful attention to business controls of all types. The controls issue grows in importance with increasing automation.

Environmental Changes

The social and political environment surrounding the technological evolution is also undergoing significant change. Internationalization of business and growing international competition, partly the result of technology advances, is altering the way firms conduct their affairs. The role of governments is changing the shape of business enterprise worldwide, and the pace of change in the business sector is increasing with time. This moving backdrop against which business strategy and planning takes place greatly increases the management task for the firm's executives and for the IT executive.

Strategic Considerations

Information technology has great strategic value and significance for most organizations today. Information technologists and their organizations are expected to provide the tools with which the firm can capture strategic competitive advantage. Many information technology organizations and their managers are on their firm's critical path to success. The IT organization may be limiting the firm's long-term performance. Obviously this poses a very special challenge for the firm and its executives.

People and Organizations

Information technology introduction alters the nature of work in our industrialized society. The technology impacts organizations, managers, and workers at all levels. Not all the consequences are perceived favorably; many are considered threatening or intimidating to managers and workers alike. These personal and organizational dislocations further heighten the management challenges accompanying the technology. Tom Peters speaks very forcefully about these changes:

> Hierarchies are merely machines that process and agglomerate information, each level adding a further degree of synthesis. So, as we develop technology-based information-processing (and add twists such as expert systems), and especially as we link the systems in networks, inside and outside the corporation, hierarchy's reason for being recedes. It must go.[23]

Effective managers cope with these challenges via increased awareness and keen appreciation of the phenomena and with finely tuned people-management skills. People functioning within organizations are the enabling resource. People enable firms and their executives to capture the benefits of the advanced technology of the information age. Effective executives understand this extremely well.

TECHNOLOGY ASSIMILATION

Stages of Growth

Organizations go through predictable stages of growth as they adopt and implement technology. The stage hypothesis was first developed in 1974 by Nolan and Gibson; the number of identifiable stages was later expanded from four to six.[24] These important writings form a significant basis for understanding the introduction and assimilation of new technology. Table 1.4 identifies the six stages.

TABLE 1.4   Six Stages of Growth

1. Initiation
2. Contagion
3. Control
4. Integration
5. Data administration
6. Maturity

Stage 1.   *Initiation.* In this stage, the technology is initially introduced into the organization and some users begin to find
applications. The use grows slowly as people become
familiar with the technology and its applications.

Stage 2.   *Contagion.* As more individuals and departments become
acquainted with it, demand increases and use of the technology proliferates. Enthusiasm for the new technology
builds rapidly during this stage.

Stage 3.   *Control.* During the control stage, the issue of costs versus
benefits intensifies and management becomes increasingly
concerned about the economics of the technology.

Stage 4.   *Integration.* As systems proliferate within the organization
and databases continue to grow, the notion of systems
integration becomes dominant. Management becomes interested in leveraging integrated systems and their databases.

Stage 5.   *Data administration.* During the data-administration stage,
management is concerned with the valuable data resources.
Functions are created to manage and control the databases and to ensure that they are utilized effectively.

Stage 6.   *Maturity.* In this stage, if it ever occurs, the technology
and the management process are integrated into an efficiently functioning entity.

The stages-of-growth concept is useful because this phenomenon can
be observed at several different levels within an industry or within an
organization and because the concept provides predictability. Not all firms
within an industry are in the same stage of growth, and not all functions
within a corporation are at the same stage. Executives can observe where
their organizations are in the stage theory. They can make reasonable

assertions concerning the future behavior of their organizations. Thoughtful managers can prepare themselves and their organizations for the future according to their observations of the present.

The stages-of-growth concept is also useful in information technology planning. It forms the basis for one of the frameworks of IT planning. The stage hypothesis can be employed to provide guidance in planning future activities.

As new technology becomes available and is introduced into the organization, the stage theory comes into play again. Managers can observe the technology adoption process occurring. They can implement management practices required to capitalize on the technology's potential while minimizing the effects of inevitable pitfalls. This text explores these notions in later chapters in connection with IT planning, the introduction of personal workstations, and other topics.

The stage hypothesis is an important management concept because it provides insight into the technology adoption process. Skillful IT managers are cognizant of these processes and they utilize this knowledge to improve their performance and their organization's performance.

CURRENT ISSUES

The task of managing the information technology function remains difficult for a variety of important reasons. The issues concern some basic concepts relating to the inner workings of the firm and the utilization of information technology within the organization. For the most part, these issues are management oriented and are not primarily technical. Management skills are much more valuable in coping with the difficulties than are technical skills.

Several individuals and organizations have developed portrayals of the issues facing information technology managers. These analyses categorize the concerns facing IT managers, their peers, and their superiors. Researchers at the University of Minnesota and the University of Colorado conducted key issue surveys of corporate members of the Society of Information Management periodically. These surveys identify current issues and also reveal trend information. The Index Group, Inc., an international consulting group based in Cambridge, Massachusetts specializing in information technology, has developed issue lists and performed analyses of the trends and considerations behind the issues. The consulting firm Arthur Andersen and Company also surveys IT managers to identify their major concerns.

The important findings from the survey data by these three organizations is presented in Tables 1.5, 1.6, and 1.7. Collectively, these studies represent a good cross section of the industry and the results are important to IT managers and their superiors.

Table 1.5 presents the current findings of Neiderman, Brancheau, and Wetherbe.[25] The analyses have been compiled repeatedly so that comparisons over time can be derived. The table displays the most recent results and those from three years previously along with trend information over this period.

Strategic planning is very important because it is crucial that the firm align its business plan with the IT strategic plan. The business environment is changing rapidly, making strategic planning skills critical. Information systems can provide competitive advantage. They are essential to survival in many firms. Opportunities for innovative systems must be pursued vigorously. Organizations must learn to make effective use of information technology in their operations. A structure or architecture to guide the use of information within the firm and to relate data resources to business processes is very important to the entire organization.

After topping the list in 1980, 1983, and 1986, strategic planning dropped slightly to third position. This topic is an important, long-term issue for IT executives. Executives are concerned about data resources. The firm's structure or architecture for information flow has risen dramatically as an issue. Management issues dominate the top third of the list, and new items comprise a significant portion of the 1989 list. Some issues are important because of their longevity: data resources, strategic planning, IS human resources, organizational learning, and IS organization alignment were in the top third of the 1980 list. Senior managers in many firms expect to gain competitive advantage from their investments in information technology. Attaining competitive advantage requires solid strategic planning. Of particular significance is the extent to which the management issues dominate the list. These ideas will receive continued attention throughout the text.

Table 1.6 presents the results of studies by the Index Group entitled Top Ten 1991 Management Issues. The ranking of 1991 issues for the three preceeding years is shown in the righthand columns.

The issues in Table 1.6 were obtained by surveying 394 IS executives at large North American corporations. Most of the management issues are enterprise-wide and some of them are persistent as indicated by their ranks in previous years. In particular, aligning IS and corporate goals and IS strategic planning have maintained high ranks in this study and that conducted by Neiderman, Brancheau, and Wetherbe.

TABLE 1.5   1986 and 1989 Issues

IS Rank 1989	IS Rank 1986	Three-Year Change	Description of Issue	Issue Type
1	8	+7	Information architecture	T
2	7	+5	Data resource	M
3	1	-2	Strategic planning	M
4	12	+8	IS human resource	M
5	3	-2	Organizational learning	M
6	NR	new	Technology infrastructure	T
7	5	-2	IS organization alignment	M
8	2	-6	Competitive advantage	M
9	13	+4	Software development	T
10	11	+1	Telecommunications	T
11	4	-7	IS role & contribution	M
12	14	+2	Electronic data interchange	T
13	NR	new	Distributed systems	T
14	NR	new	CASE technology	T
15	16	+1	Applications portfolio	T
16	9	-7	IS effectiveness measures	T
17	NR	New	Executive/decision support	M
18	6	-12	End-user computing	M
19	18	-1	Security & control	T
20	NR	new	Disaster recovery	T
21	NR	new	Organizational structure	M
22	10	-12	Technology integration	T
23	NR	new	Global systems	M
24	NR	new	Image technology	T
25	NR	new	IS asset accounting	M

Notes: NR Issue was not ranked in 1986 study
          M Concerned primarily with management problems
          T Concerned primarily with technology and application-related problems

TABLE 1.6   Top Ten 1991 Management Issues

		1990	1989	1988
1.	Reshaping business processes through IT	1	11	NR
2.	Aligning IS and corporate goals	4	2	1
3.	Instituting cross-functional systems	3	7	NR
4.	Boosting software development productivity	6	13	12
5.	Utilizing data	7	6	7
6.	Developing an IS strategic plan	5	4	2
7.	Improving software development quality	14	NR	NR
8.	Creating an information architecture	9	5	5
9.	Integrating information systems	16	12	6
10.	Improving leadership skills in IS	NR	NR	NR

NR = Not Ranked

Survey conducted by the Index Group, Inc., a Cambridge, Massachusetts-based management consulting firm. Reprinted with permission.

Recently, IS executives are increasingly concerned with using IT to reshape and improve fundamental business processes. They desire to use IT as a change agent, institute cross-functional systems for the firm, and ensure congruence between IS and corporate goals. IS strategic planning and creating an information architecture are recognized by IS executives as important issues. New, strategically important systems for their firms must be developed with high productivity and quality.

In addition, the Index Group study reveals that less than half of the companies' executives understand the importance of using IT to reshape business processes according to the respondents. Less than one-third of the IS executives are involved in strategic planning activities.[26] These figures are troublesome because they indicate that the top issues will remain prominent for some time. Earlier studies by the Index Group revealed that only about one-fifth of the surveyed IS managers believed their executives had a strong desire to learn about information management.[27]

A recent survey of 125 senior information systems executives from large public and private companies in North America by Andersen Consulting reveals additional detail on key IS issues.[28] Top issues from a study conducted in 1988 were included in the 1990 study and many respondents still considered them very important. The issues and the percentage of respondents considering them important are:

1. Communicating with top management, functional managers, and end users (93%),
2. Developing a quick response capability to handle changing business conditions (76%),
3. Managing information resources such as computer files and databases (83%),
4. Training and educating the organization's workforce in the effective use of applications (79%), and
5. Improving the productivity of application development (73%).

But the 1990 study identified new issues involving broader business and human resources questions. The new issues surfaced in the 1990 study are:

1. Getting functional managers involved in using IT to reshape business process (88%),
2. Training and educating the workforce in the use of IT to expand their capabilities (87%),
3. Integrating IT into corporate strategy (82%),
4. Managing and mastering change (81%), and
5. Developing a corporate-wide IT strategy (79%).

The Andersen Consulting study also examined the extent to which the respondents believed their firms were effective in managing each issue. They defined the leadership gap for each issue as the difference in percentage rating on importance versus effectiveness. Issues with a leadership gap are found in three primary areas: human resources, business strategy, and technology. This analysis is presented in Table 1.7.

According to Andersen Consulting, the number and scope of issues facing IS executives are growing and firms are experiencing difficulties in mastering them. Thus it is not surprising that a leadership gap exists in many areas. The gaps represent important challenges for IT and other executives as they strive to improve their firm's effectiveness in dealing with the key issues.

Key issue studies have been conducted outside the U.S., with some variance in results. Management issues dominate technical issues around the world and the topic of IT strategic planning is at or near the top of most issue lists worldwide. The order of the key issues reflects country- or continentwide circumstances or cultural differences.

The typical IT organization is struggling in its relationships with senior executives and is experiencing frustration in long-range planning. The firm is affected by the increasing pace of change and the growing complexities wrought by the technology. IT executives are expected to contribute to the firm's financial results and are spending more money on strategic systems. Meanwhile, they are experiencing growing backlogs of work in other areas. The IT executive needs a keen awareness of business issues and refined general management skills to contend successfully with this environment.

TABLE 1.7   The Leadership Gap on Key Issues

Business/Strategy Issues	Importance	Effectiveness	Gap
Getting functional managers involved in using IT to reshape business process	88%	30%	58%
Integrating IT into corporate strategy	82%	31%	51%
Developing a corporate-wide strategy	79%	28%	51%
Human Resource Issues			
Training and educating the workforce in the use of IT	87%	34%	53%
Managing and mastering change	81%	38%	43%
Training and educating IS staff about the business	76%	36%	40%
Defining the role and structure of IS in the organization	69%	36%	33%
Technology Issues			
Developing a quick response capability to handle changing business conditions	76%	24%	52%
Improving application development productivity	73%	23%	50%
Defining an architecture that will enable integration of all information systems	69%	33%	36%
Integrating systems across diverse organizational structures inside the company	57%	21%	36%

©1990 Andersen Consulting. Reprinted with permission.

Having knowledge of the important problems facing IT executives and their organizations, managers are positioned to describe the factors necessary for success. What actions must IT executives carry out successfully? What management systems and processes are vital for their personal success and for the success of the IT organization?

Critical Success Factors

The answers to some of these questions are found in the notion of critical success factors (CSF). The idea of critical success factors was developed by Rockart in the late 1970s to help executives define their information needs.[29] The methodology is useful in information technology planning. It will be discussed further in succeeding chapters. Critical success factors identify those few areas where things must go right; they are the executive's necessary conditions for success. They apply to IT executives, to their subordinate managers, and to senior executives in the firm.

Rockart identified four sources or areas where executives should search for critical factors: the industry in which the firm operates; the company itself; the environment; and time-dependent organizational areas. The last source accounts for the possibility that some organizational activity may be outside the bounds of normal operations and require intense executive attention for a short period. In addition, Rockart identified two types of CSFs: the monitoring type and the building type. The monitoring type keeps track of the ongoing operation; the building type initiates activity designed to change the functions of the organization in some way.

Critical success factors apply to the IT manager. What are the conditions necessary for the IT manager's success? What tasks must be carried out very well in order for the manager to be successful? The critical issues studied previously provide a great deal of information needed to answer these questions. The factors necessary for success can be grouped into several classes, as shown in Table 1.8.

TABLE 1.8   Critical Areas for IT Managers

1. Strategic and competitive issues

2. Planning and implementation concerns

3. Operational items

4. Business issues

The firm's IT management team must take action in these critical areas. The IT manager must have goals and objectives to solve problems, if there are problems, and to prevent the development of issues. Outstanding managers may use a roadmap such as the following to assess their posture on these vital topics.

1. Strategic and Competitive Issues (long range)
   a. Develop IT strategies supporting the firm's strategic goals and objectives.
   b. Provide leadership in the use of technology to attain advantage for the firm.
   c. Take the lead in educating the management team on the opportunities and the problems surrounding technology introduction.
   d. Ensure realism in long-term expectations.

2. Planning and Implementation Concerns (intermediate range)
   a. Develop plans in support of the firm's goals and objectives.
   b. Provide effective communication channels so that plans and variances to them are widely understood.
   c. Establish a partnership between the IT organization and its clients during planning and implementation.
   d. Maintain a realistic perspective within the organization regarding intermediate-term expectations.

3. Operational Items (short range)
   a. Provide customer service with high reliability and availability.
   b. Deliver service of all kinds on schedule and within planned costs.
   c. Respond to unusual customer demands and to emergencies.
   d. Maintain a management process that aligns operational expectations of users with IT capabilities.

4. Business Issues
   a. Improve the productivity of the IT organization.
   b. Attract and retain highly skilled people.
   c. Practice good people management skills.
   d. Operate the IT function within the norms of the parent organization.
   e. Position the IT function to provide technical and business leadership to the firm.

Not all of these items will be on every IT manager's list of critical success factors. In many cases, some of the critical areas will be operating smoothly and routine attention will maintain high-quality operation. In other instances, managers must add temporal organizational factors or company-specific factors to the list. In all cases, superior managers remain attentive to those factors necessary for their success.

Information technology managers who can accomplish the tasks outlined above have a good chance of becoming highly successful. They have developed general management skills that will prepare them for increased future responsibility. Throughout the remainder of this text, management tools, techniques, and processes will be developed to enable the IT manager to accomplish these critical tasks successfully. Achieving success in these critical areas is a high-priority goal for the aspiring information technology manager.

A MODEL FOR THE STUDY OF IT MANAGEMENT

The study of information technology management concentrates on business results; attaining efficiency and effectiveness; and achieving competitiveness with the external environment. The intended result is improved profitability for the firm. The discussion in this text is related to business results and is portrayed in Figure 1.5.

Each topic on the right side of Figure 1.5 contributes to the success of the firm in an essential manner. Each represents a tangible or intangible asset to be deployed effectively for the firm's benefit. In every area, the IT executive needs to cooperate with and needs the cooperation of managers throughout the organization to gain full value from these assets. Each section in this text addresses one of these vital assets.

SUMMARY

Information technology is a powerful force for change in our industrialized world. The change is fundamental in nature, profoundly affecting people, organizations, industries, and nations. Peter Drucker addresses the emerging organizational upheaval and Tom Peters amplifies the theme.

This chapter concentrated on some of the basic ideas surrounding the management of organizations and introduced topics and issues important for the managers of information technology in particular. The subjects of expectations, stages of growth, important current issues, and critical success factors are interwoven through the remaining text material. The text focuses on attaining business success through the application of management principles to information processing technology.[30]

Information technology is an important ingredient in the strategies of nearly all firms today. The IT manager plays a vital role in achieving long-range success for the firm. *Management of Information Technology* illuminates the general management principles that are necessary for success in this endeavor.

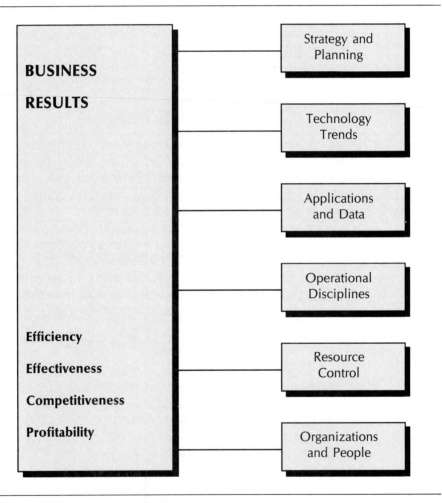

FIGURE 1.5   A Model for the Study of IT Management

Review Questions

1. According to Drucker's thesis, how will the introduction of information technology alter the future structure of the firm?

2. What causes the challenges facing IT managers to remain high?

3. Why are expectations so important to the IT manager?

4. Relate levels of management to sources of information and to type of decision making.

5. What appear to be the causes of the difficulties experienced by IT managers, as highlighted in the section entitled "Information Technology Management"?

6. What kind of preparation improves the IT manager's chances of success?

7. During which stage of growth does senior management take a serious interest in the technology and why?

8. Why is it unlikely that a firm will reach the stage of maturity?

9. Why is an understanding of stages of growth useful to management?

10. What important concepts do we glean from the 25 issues presented in Table 1.5?

11. What is the difference between necessary conditions for success and sufficient conditions for success?

12. Why doesn't the text present a set of sufficient conditions for success?

13. Why does the IT manager's list of critical success factors contain statements regarding expectations?

14. What are the causes of the dramatic structural changes predicted by Drucker and Peters?

Discussion Questions

1. How does Drucker's hypothesis bear on the concepts displayed in Tables 1.2 and 1.3 and Figures 1.1 and 1.2?

2. Information technology has high potential value for most firms. Why does this make the IT manager's job so demanding?

3. What personal characteristics would you expect to find in the individuals responsible for the initiation of a technological thrust?

4. Why is it possible that not all parts of a corporation will be at the same stage at the same time? What opportunities does this present to management?

5. Some individuals in an organization may be extremely reluctant to embrace a new technology. What special problems does this raise? What tools are available to management to solve this problem?

6. One can conclude from reviewing the lists of issues that executive management in most firms is undereducated in the subject of information technology. Why might this be happening, and what can and should be done about it?

7. Discuss the relationship between the idea of corporate culture and that of critical success factor 4d.

8. Why is it important that IT managers prepare themselves for increased future responsibility?

Assignments

1. Read the referenced article by Peter Drucker and prepare a written summary of his thesis, concentrating on the meaning for future organizations and managers. Specifically, how do you think this will affect the managers of the IT organization?

2. Using library reference material, find an IT manager's success story and prepare a synopsis for class presentation. Identify the chief factors contributing to success.

3. Which of Nolan's stages of growth best describes your school's level of information processing maturity? Prepare your answer by discussing this subject with a knowledgeable individual in the information processing section of your school's administration.

[1] Peter F. Drucker, "The Coming of the New Organization," *Harvard Business Review* (January-February 1988): 45.

[2] Jeanne Iida, "Merrill Lynch Decentralizes MIS," *MIS Week*, November 13, 1989, 1.

[3] See Business Vignette, Chapter 3.

[4] "The Premier 100," Supplement to *Computerworld*, September 12, 1988. This article presents expense-to-revenue ratios in information systems for 100 companies. The ratio ranges from a low of .77 percent to a high of 9.14 percent. The estimated budget and estimated value of installed equipment, the percent of budget spent on staff and the percent spent on training, the number of PCs and terminals installed and the number of these per 100 employees are also given.

[5] Marvin Bower, *The Will to Manage* (New York: McGraw-Hill Book Company, 1966), 22. Also see Terrence E. Deal, and Allen A. Kennedy, *Corporate Culture* (Reading, MA: Addison-Wesley Publishing Co., 1982).

[6] The term *chief information officer* was coined in 1981 by William Synott. See Chapter 17 for trends in IT organizations and Chapter 18 for discussions of the concept and role of the chief information officer.

[7] The total number of applications is increasing rapidly as end users develop programs at their individual workstations. The increasing availability of applications for sale is also contributing to the growth in application portfolio size. In some firms the real costs and benefits of the complete portfolio cannot be estimated at this time.

[8] Harold Koontz, and Cyril O'Donnell, *Principles of Management* (New York: McGraw-Hill, 1964), 262.

[9] Gordon Davis, and Margrethe Olson, *Management Information Systems* (New York: McGraw-Hill, 1985). In particular, refer to Chapter 7, "Concepts of Information" and Chapter 11, "Organizational Structure and Management Concepts," for a detailed presentation of these topics.

[10] Peter F. Drucker, "The Coming of the New Organization," *Harvard Business Review* (January-February 1988): 45.

[11] Tom Peters, "Tomorrow's Companies," *The Economist*, March 4, 1989, 19. Also see Ellis Booker, "IS Trailblazing Puts Retailer on Top," *Computerworld*, February 12, 1990, 69.

[12] Drucker, "Coming of the New Organization," 46.

[13] Robert N. Anthony, *Planning and Control Systems: A Framework for Analysis* (Cambridge, MA: Harvard University Press, 1965). For a clear perspective on how this model relates to management information systems (MIS), see David Kronke, *Management Information Systems* (Santa Cruz, CA: Mitchell Publishing, Inc.), 101-102.

[14] G. A. Gorry, and Michael Scott-Morton, "A Framework for Management Information Systems," *Sloan Management Review* (Spring 1989): 49. This article updates the classic article with the same title by these authors in *Sloan Management Review* (Fall 1971): 55.

[15] *Executive Summary*, The 1990 Society for Information Management Institutional Member Conference, March 14-16, 1990, 1.

[16] Richard Nolan, "Plight of the EDP Manager," *Harvard Business Review* (May-June 1973): 143.

[17] Blake Ives and Margrethe Olson, "Manager or Technician? The Nature of the Information Systems Manager's Job," *MIS Quarterly* (March 1981): 49.

[18] Jeff Moad and Ralph Carlyle, "A Rash of Top-Level Departures Erupts at IS Shops Across the Nation," *Datamation*, July 15, 1988, 21.

[19] Joseph E. Izzo, *The Embattled Fortress: Strategies for Restoring Information Systems Productivity* (San Francisco, CA: Jossey-Bass Publishers, 1989).

[20] "The Premier 100," *Computerworld*, September 12, 1988.

[21] *Solutions*, Gartner Group, Inc., 1989.

[22] *Solutions*, Gartner Group, Inc., 1989.

[23] Tom Peters, "Tomorrow's Companies," *The Economist*, March 4, 1989, 19.

[24] Richard L. Nolan, and Cyrus F. Gibson, "Managing the Four Stages of EDP Growth," *Harvard Business Review* (January-February 1974): 76. Also see Richard L. Nolan, "Managing the Crisis in Data Processing," *Harvard Business Review* (March-April 1979): 115.

[25] Fred Neiderman, James C. Brancheau, and James C. Wetherbe, "Information Systems Management Issues in the 1990s," Faculty Working Paper 90-16, University of Colorado, Graduate School of Business Administration, Boulder, CO. Reprinted with permission. Also see James C. Brancheau and James C. Wetherbe, "Key Issues in Information Systems Management," *MIS Quarterly* (March 1987): 23.

[26] James Champy, "Organizational Revisionism," *CIO Magazine* (December 1990): 20.

[27] Thomas H. Davenport, "Directions for 1988," *CIO Magazine* (January-February 1988): 6.

[28] *The Changing Shape of IS: Redefining Technology Leadership*, 3rd Ed., Andersen Consulting, 1990.

[29] John F. Rockart, "Chief Executives Define Their Own Data Needs," *Harvard Business Review* (March-April 1979): 81.

[30] The concepts in this text are applicable to institutions, agencies, government entities, and not-for-profit organizations. The use of terms such as bottom-line results and business results is intended to encompass the results of all forms of organizations, not just profit-making firms.

REFERENCES AND READINGS

Ackoff, Russell L. "Management Misinformation Systems." *Management Science* (December 1967): B147.

Allen, Brandt. "An Unmanaged Computer System Can Stop You Dead." *Harvard Business Review* (November-December 1982): 77.

Anthony, R. N. *Planning and Control Systems: A Framework for Analysis*. Cambridge, MA: Harvard University Press, 1965.

Applegate, Lynda M., James I. Cash, Jr., and D. Quinn Mills. "Information Technology and Tomorrow's Manager." *Harvard Business Review* (November-December 1988): 128.

Davenport, T., and R. Buday. *Critical Issues in Information Systems Management in 1988*. Index Group, 1988.

Dickson, Gary W., and James C. Wetherbe. *The Management of Information Systems*. New York: McGraw-Hill Book Company, 1985.

Drucker, Peter F. "The Coming of the New Organization." *Harvard Business Review* (January-February 1988): 45.

Ives, Blake, and Margrethe Olson. "Manager or Technician? The Nature of the Information Systems Manager's Job." *MIS Quarterly* (1981): 49.

Izzo, Joseph E. *The Embattled Fortress*, San Francisco: Jossey-Bass Publishers, 1989.

Keen, Peter. "Vision and Revision," *CIO* (January/February 1989): 9.

Nolan, Richard L. "Plight of the EDP Manager." *Harvard Business Review* (May-June 1973): 143.

Rappaport, Alfred. "Management Misinformation Systems—Another Perspective." *Management Science* (December 1968): B133.

Stromer, Richard. "Management Issues Top Concerns for Information Systems Execs." *PC Week*, March 13, 1989, 154.

2

The Strategic Importance of Information Technology

A Business Vignette

Introduction

Strategic Issues for Senior Executives
   Why Strategic Issues Are Important
   Relation to Other Critical Issues

Strategic Information Systems Defined
   Major Types of Systems
   The Character of SIS

A Framework for Visualizing Competitive Forces
   A Model of Forces Governing Competition
   Strategic Thrusts

Strategic Systems in Action
   Airline Reservation Systems
   Stock Brokerage
   Order Entry Systems
   Delivery Services

Where Are the Leverage Points?
   Opportunities
   The Importance of Technology
   The Time Dimension

The Characteristics of SIS
   Organization and Environment
   Financial Implications
   Additional Considerations

The Strategist Looks Inward

External Strategic Thrusts

Integrating the Strategic Vision

Summary

Review Questions

Discussion Questions

Assignments

References and Readings

A Business Vignette

Strategy: Avon Products

Ranier Paul believes. Mr. Paul, vice president of corporate MIS, sits with restless energy in his office at Avon, high above 57th Street in New York. He talks with evangelistic zeal about the company's new Executive Information System (EIS), and his enthusiasm is catching. So far, the top 15 executives at Avon Products Inc. have caught it; the feeling's going all the way up to the executive suite.

Avon's EIS uses PCs to deliver information that helps executives find potential business problems and spot trends much more quickly than before. The system not only represents a new way of working at Avon, it also reflects changes at the firm. While Avon is best known for its door-to-door beauty business, the company, which posted sales of almost $3 billion last year, also includes thriving health-care and direct-response divisions. Recent acquisitions include such high-end fragrance manufacturers as Giorgio Inc. and Parfums Stern, which sells the Oscar de la Renta, Perry Ellis and Valentino designer-fragrance lines.

Avon's corporate MIS department was set up just two years ago; its charter was to manage better the company's continuing metamorphosis into a new, multifaceted corporation.

Mr. Paul and others in the MIS group engaged in an interview process to uncover the information that would help the executives run the business. "We asked, for example, 'If you came back from a two-week vacation, what information would you want to see on your desk?'" Mr. Paul said. The MIS group interviewed executives and staff, gathered information needs from both the top down and the bottom up. Based on the interviews, the group identified over 70 applications of importance, from human-resource information to business indicators.

With the luxury of the new corporate computing architecture, Mr. Paul urged top management to deliver some of this information electronically, as an Executive Information System. The idea caught on. "People really

---

[†] Excerpted from *PC WEEK*, February 9, 1988, 48. The author, Jon Pepper, is a journalist and consultant based in Sunderland, Massachusetts.

---

understood the need for this system," Mr. Paul explained. The entire development process went quickly. "We interviewed in April and May, presented the concept in June, got the OK and set up the first prototype in September of 1986."

The prototype met with instant approval. "The CEO saw it and immediately said, 'Let's implement it,'" Mr. Paul said. An unexpected and brief setback in the health-care division in 1986 helped seal the system's fate. "Our COO (Chief Operating Officer) Jack Chamberlin said that if he had Avon EIS in place prior to that, it would have been possible to see the problem and act to head it off before it developed into anything serious," Mr. Paul noted.

The programming team assembled by Mr. Paul and his managers was a unique aspect of the project. Rather than recruit programmers who understood business, they looked for M.B.A.s who knew or wanted to learn programming. "I think the quality of our product is better because we hired business people and what we have today reflects that," Mr. Paul explained. "If you bring in good people, they will challenge the functional departments to think differently, and I think it helped us add a lot of value to the project." The EIS resulting from this effort is a marvel of conciseness, extreme ease of use, and usefulness.

All selections from the menu-driven product are made by pointing and clicking with a mouse. This means that busy executives don't have to bother using the keyboard at all. When they log onto the system, they see a menu divided in half; the left side contains internal Avon information, such as financial indicators, while the right side contains external information, such as news and competitive analyses. Under the theory that less is more, the Avon EIS is heavily graphics oriented, presenting more information concisely. Using a mouse, the user simply moves the cursor over a menu choice, and if there is another level of information behind that selection, it changes color. Clicking the mouse brings up the next level of detail.

Individuals who want to take further action based on the information presented can choose a utility icon on the screen, and can either lay out a report, edit notes, download the figures to Lotus 1-2-3 or print out a hard copy. An electronic-mail function is being added so users can add notes and send the entire screen.

From the beginning, part of the plan was presenting information not only graphically, but also in full color. Consistent with that intent, all hard copy is delivered in color as well. Centrally located output devices give users the option of creating hard-copy reports on a color printer, making 35mm slides or full-color transparencies. Using color graphics "allows the executives to see the trends and see what has happened from a business point of view much faster than they would by just looking at numbers," according to Mr. Paul.

All analysis on the Avon EIS is done dynamically. Users can customize important views, or set upper and lower limits or variances they want to examine. Other main-menu choices include 30 days worth of business and financial news from the Dow Jones News service and competitive information. The competitive analysis lets users compare the performance of any of Avon's divisions against those of selected competitors. Executives can also customize which ratios or companies they want to compare with Avon based on quarterly statements, sales growth, inventory turnover or a number of other measures. Up to 15 companies can be compared at one time, letting an executive see instantly how Avon stacks up against the competition.

Perhaps the most concise indicator on the system is the Hot Buttons, an addition suggested by Avon CEO Hicks Waldron. He explained that after looking at a demonstration of the system, he wanted to be able to identify major business trends easily so Hot Buttons, which are colored flags on the screen, were added. The top part of the flag represents the current month, and the bottom represents the trend for the last two months. If the top part of the flag is green, it means the trend is above plan. If the color is blue, the trend is flat. If it is red, the trend or month is behind. It is easy to compare the actual month to the business plan and to the two-month trend.

Response to the system has been extremely positive, and training time is virtually non-existent but help screens are provided.

Mr. Chamberlin related how the system let Avon executives quickly see a business downturn in the Avon division. "One of the indicators helped to show us that we had eliminated some low-priced items from our Christmas line, and that cut our profitability," he said. "So we included those items in our 1987 Christmas line and the results were much more satisfactory." Mr. Chamberlin felt that without the system, it might not have been possible to correct them for the next season.

Despite examples of how the system has affected profits, there was never a demand to cost-justify it, because upper management believed it would help executives make better decisions. This is a trend emerging as corporate computing evolves, according to Mr. Paul. "The higher you get in this movement, the less you can cost-justify things," Mr. Paul said. "It's no longer a case of cranking up inventory level, but more in the area of highlighting opportunities or problems in the organization."

Clearly, Avon is counting on EIS, for in the aggressive world of beauty and health-care products, simply looking good is not enough.[1]

INTRODUCTION

The preceding chapter presented information-processing technology and its relation to the firm's workers, from nonmanagerial employees to the CEO. Individuals employed by the firm are served by a wide variety of information technology capabilities. However, the technology has penetrated most deeply at the lower levels of the employment hierarchy and has had relatively less impact on the more senior positions. This is a natural development considering the nature of work performed by individuals in the firm and the evolution of the technology. Although routine transaction processing is well served by modern information systems, considerable future potential for the technology lies in giving more senior decision makers assistance in discharging their responsibilities. Only 30 percent of *Fortune 500* executive had terminals in their offices in 1987 but International Data Corporation estimates companies will spend $115 million on executive information systems in 1991.[2]

Senior executives expect their IT organizations to provide more support to executive-level workers in the future than is currently available. Although their expectations include additional automation and cost reduction in the more routine processes in the organization, they clearly have larger expectations for themselves and for their organization as a whole. Their vision includes significant additional support for the activities they conduct personally and, more importantly, they are increasingly questioning whether information technology is providing sufficient leverage to their firms in the competitive arena.[3]

This chapter discusses the importance of strategic information systems to organizations and describes the character of these systems. Some examples of important strategic systems are discussed to illustrate their features and to portray their enormous value. Some useful techniques for identifying strategic applications within the firm are developed in this chapter and some mechanisms for integrating these potential applications into the fabric of the application portfolio are discussed.

STRATEGIC ISSUES FOR SENIOR EXECUTIVES

Why Strategic Issues Are Important

Senior managers in many firms have been increasingly directing their attention toward the employment of information systems for strategic purposes. There are several reasons why this is happening. These are tabulated in Table 2.1. For the last several decades, executives have authorized expenditures for the development of systems directed toward automating the relatively routine transaction-processing activities of the firm. But executives also note that significant resources are being deployed on more sophisticated applications. The penetration of information technology in the transaction-processing realm is relatively deep, and only minimal future gains are expected for most firms.

TABLE 2.1   **Forces Driving the Strategic Vision**

1. Transaction processing has matured.
2. IT budgets continue to rise.
3. IT offers great opportunity for innovation.
4. The technology is pervasive.
5. Precedents are becoming widespread.

Many firms have reached the state of rather complete automation of transaction processing. If an organization is to attain further gains over competition from information technology, more than cost reductions from further automation of routine activities is required. In a sense the playing field has been leveled as far as transaction processing is concerned.

Total expenditures for information processing have continued to rise for most firms. The annual budget for most IT organizations has reached a level where senior executives are expecting a more significant contribution to bottom-line business results. Firms are spending anywhere from 1 to 5 percent of revenue on information-processing activities, and CEOs are demanding visible contributions to the firm's financial health from this investment.

Increasingly, managers view information technology as an investment to be deployed in pursuit of the goals of the firm and not just as an activity to reduce costs via routine automation. For example, some firms in the insurance industry spend nearly 10 percent of revenue on information technology.[4] IBM spends 17 percent of revenue on IT and consumes 10 percent of its own output.[5] This is not to say that cost reductions are not important; they are important, but senior managers have greater expectations for their IT investments.

Senior executives believe that information technology holds great promise for innovation and corporate or organizational leadership. They desire to capitalize on technological innovation and strive for a leadership position for the firms they head. Executives believe that past and future investments in information technology should result in leadership via technological innovation. This view by senior executives has great importance for IT managers. It must direct their thinking if they are to achieve success.

Over the years, but particularly in the recent past, CEOs in many firms have observed the productive use of information technology in virtually every aspect of their operations. The funds deployed for this technology have been invested across the board and nearly every facet of their organizations is now being affected significantly. The pervasiveness of the technology demands strategic thinking about its future use.

Finally, the belief that information systems can have a very significant impact on a firm's strategic direction and its long-range position in the industry is not the result of wild speculation. This belief is validated by many examples from current experience. Senior executives have specific, concrete precedents on which to base their desires and aspirations. They are familiar with the success enjoyed by firms in the airline industry that own important and vital reservation systems, and others as well. They are aware of significant systems in the brokerage industry and of highly valuable order entry systems. These precedents validate their desire to see their firms enjoy comparable successes.

For these reasons, senior executives, IT managers, controllers, and others are focusing their attention on the long-range strategic implications of the technology and are concentrating their energies on attaining competitive advantage from their IT investments. Therefore, utilizing information technology for competitive advantage is a high-priority issue for business managers in most corporations today.

Relation to Other Critical Issues

Strategic concerns have occupied the attention of senior IT executives for many years, but these issues are not fully appreciated if considered in isolation. Strategic issues are frequently accompanied by or occur in conjunction with other important questions, such as: "Is there an effective planning process in the firm and in the IT organization?" "Does the firm have a well-defined information architecture?" "Do the data resources support the functions and goals of the firm in an effective manner?" Missions, goals, and organizational alignment among and between the functional units within the firm all have a bearing on the strategic concern. In particular, the general effectiveness of the IT organization internally and its effectiveness in relating to other functions are very important to strategic concerns.

The task of making improvements in any of these issues generally involves making improvements in all of them. The mutuality of these topics demands that management make progress across a broad front. It is misleading for organizations to believe that competition can be overtaken or beaten by the development and implementation of one new strategic system. A significant leap forward using information technology appears unlikely if the organization has major weaknesses in planning, alignment, or other critical areas.

STRATEGIC INFORMATION SYSTEMS DEFINED

According to Charles Wiseman,

> Strategic information systems are information systems in which the primary function of the system is either to process predefined transactions and produce fixed-format reports on schedule or to provide query and analysis capabilities. The primary use of SIS is to support or shape the competitive strategy of the enterprise, its plan for gaining or maintaining competitive advantage or reducing the advantage of its rivals.[6]

This definition mostly addresses the business activities of a firm but, for completeness, it must also include technical systems that generate competitive advantage for the firm. For instance, support to product development and manufacturing with sophisticated, unique electronic design or automated production systems must be included within the scope of SIS. Strategic systems shape the competitive posture and strategy of the firms that own them whether they are administrative or technical in nature.

Major Types of Systems

There are many types of systems defined in the IT literature that are prominent in the discussions and the thinking of information systems professionals. These include transaction processing systems (TPS), decision support systems (DSS), office automation systems (OAS), management information systems (MIS), and end-user computing systems (EUC). Transaction processing systems handle routine information items, usually manipulating the data in some useful way as it enters or leaves the firm's database. Many transaction processing systems are on-line, which means that many users interact with the database, performing updates or retrievals. Decision support systems are computer programs that provide support to the user in reaching a decision. They support the personal decision-making style of individual managers.[7] Management information systems provide a focused view of information flow as it develops during the course of business activities. Report generation is prominent in management information systems.

Office automation systems provide electronic mail, word processing, electronic filing, scheduling, calendaring capability, and other support to office workers. End-user computing places computational capability into the hands of users to initiate the execution of programs or to develop programs for later execution.

Some of these systems are operational in nature; they support the firm in the short term. Others aid managers in making intermediate or long-term decisions. All of these systems have been designed to improve the manner in which employees and managers carry out the functions of the firm.

The Character of SIS

To the extent that any system gains or maintains competitive advantage for its owner, it forms part of the spectrum of strategic information systems. The attribute of competitive advantage is what distinguishes a strategic system from all others.

A FRAMEWORK FOR VISUALIZING COMPETITIVE FORCES

A Model of Forces Governing Competition

To understand the possible roles that can be played by information systems in shaping or altering the competitive posture of a firm, managers must visualize business competition in its broadest terms. A succinct and lucid view of the forces shaping competition was first presented by Michael Porter in 1979.[8] Porter views an industry as consisting of firms jockeying for preferred positions while being impacted by the bargaining power of suppliers, the bargaining power of customers, the threat of new entrants, and the threat of substitute products or services. These forces, affecting all competitors, must be contended with strategically by the firm if it is to grow and prosper.

Porter suggests that companies need to address strategic actions based on the factors identified in the model. These actions consist of diminishing customer or supplier power, lowering the possibility of substitute products entering the marketplace, discouraging new entrants, or gaining a competitive edge within the existing industry.

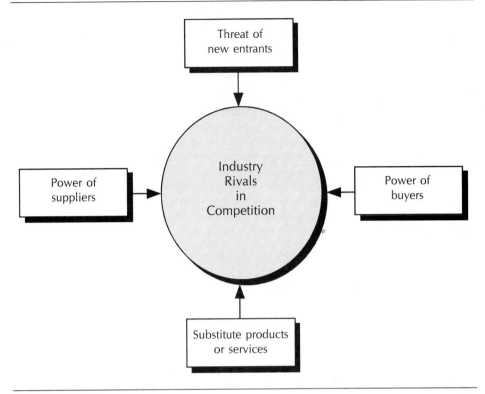

FIGURE 2.1   Forces Governing Competition

Porter's model, shown as Figure 2.1, is very helpful in thinking about competition in its broadest terms, and it suggests areas in which a firm may need to examine its competitive posture. The model is a framework for judging the firm's position versus its competitors and for analyzing various strategies the firm may elect to employ.

Strategic Thrusts

An addition to the framework of strategy development is presented in detail by Wiseman.[9] His discussion gives insight into the theory of strategic thrusts and is complete with numerous examples of strategic information systems. The ingredients of these strategic moves are contained in five basic thrusts. I have added "time" as a strategic thrust to the five defined by Wiseman. Time, specifically the reduction in time or the increase in responsiveness, is an important factor in business competition in today's environment. Time must be considered a strategic thrust.[10] Its inclusion as a basic strategic thrust is mandated by the increasing importance of telecommunications in strategic systems and by the growing number of firms that consider themselves to be engaged in time-based competition. The six basic thrusts discussed in this chapter are presented in Table 2.2.

TABLE 2.2   **Strategic Thrusts**

1.   Differentiation

2.   Cost

3.   Innovation

4.   Growth

5.   Alliance

6.   Time

The following paragraphs define the essential characteristics of these basic strategic thrusts.

1.   *Differentiation.* The firm's products or services are distinguished from competitors' products or services, or a rival's differentiation is reduced. For instance, automated teller machines distinguish the services of some financial institutions from others.

2.   *Cost.* Advantage is attained by reducing costs to the firm, the firm's suppliers, or its customers, or by increasing costs to competing firms. Advanced order entry systems in the health care industry reduce ordering costs to both the customer and supplier, for example.

3.   *Innovation.* Advantage is gained by introducing changes to the product or process, yielding fundamental changes to the manner in which the industry conducts its business. For example, brokerage firms have used information technology innovations, including telecommunications, to provide new products and services for their clients.

4.   *Growth.* Advantage is secured by expansion, forward or backward integration, or by diversification in products or services. Telecommunications technology is employed for expansion by several daily newspapers that seek broad national markets. The *Wall Street Journal* and *USA Today* are two examples.

5. *Alliance*. Advantage is attained by achieving agreements, forming joint ventures, or making strategic acquisitions. The Business Vignette in Chapter 9 describes an interesting joint venture. Even large firms are using agreements and turning to joint ventures as a strategic thrust.

6. *Time*. Competitive advantage is secured by rapid response to changing market conditions or by supplying a more timely flow of products or services. Electronic design automation tools, computer-aided manufacturing systems, and the integration of the CAD/CAM systems and production logistics systems are thrusts aimed at increasing manufacturing's response to the marketplace.[11] "Innovation via technology requires resources of capital, technology, management and time. Firms cannot buy time off the shelf. They need a catalyst to integrate the other resources in such a way as to 'buy' them the time they need."[12]

Information technology is an important ingredient in the competitive strategies of many firms. It is commonly used to shape strategic thrusts. The desire for strategic information systems stems from business managers' convictions that properly conceived systems can provide enormous advantages to their owners.

STRATEGIC SYSTEMS IN ACTION

Airline Reservation Systems

In the late 1960s, American Airlines developed a rudimentary computerized reservation system called the Semi-Automated Business Research Environment, or Sabre. Sabre capitalized on the then advanced third-generation computing equipment available from IBM. American made an initial investment of $350 million in Sabre but did not achieve profitability on this investment until 1983.[13] Today, Sabre is the world's largest travel agency reservation system, serving 14,000 agents. The system enables the agents to provide their customers with airline, hotel, and automobile reservations and other services. It is one of the most widely known and most valuable strategic systems in the world.

The Sabre reservation system combines advanced computerized data processing and telecommunications technology. Reservations are taken by 8000 reservations operators accepting calls at five regional sites and from 85,500 terminals in travel agency locations in 47 countries. The Sabre network consists of 45 high-capacity digital lines over which 470,000 airline reservations are booked daily.

During the 1970s, Sabre became an important marketing force in the industry and a valuable strategic asset for American Airlines. Travel agencies were considering the development of a system for their own uses, and faced with this competitive threat, American decided to make Sabre available to them. Because of the high cost of reservation system development, many airlines joined the Sabre system and purchased reservation services from American.

To attain dominance, American has continually updated and expanded the system and introduced new products and services. Sabre provides a basis for travel agent office automation, serves corporate customer's reservation needs, and makes available a host of travel-related services. American also utilizes the Sabre system internally in its advanced yield management system. Yield management is the process of allocating airplane seats to various fare classes to maximize the profit on each flight. Sabre databases are used to predict demand for each flight and to observe the rate at which seats are booked prior to flight departure. Based on these data, seats are dynamically allocated to full fare or discount fare passengers, maximizing seat utilization and flight profitability.[14]

The introduction of EAASY Sabre for individual home use and Commercial Sabre for corporate use have been successful areas of growth for American. Only the schedules are available to these users; customers must make reservations through agents, thus preserving commissions for the agents. These new products stem the threat that new low-cost technology such as personal computers may provide competition and undermine Sabre's dominant position.

American has continually upgraded the system to permit more flexibility with its end users. It implemented a feature named Calltrack which records all incoming calls to Sabre central. The calls are recorded by agency, caller name, and type of problem, thus providing American with valuable market and customer relations data.[15] Other innovations such as the Frequent Flier program have helped Sabre grow and given American a competitive advantage by developing bargaining power and comparative efficiency.

Some companies have elected to develop their own reservation systems. Delta Airlines has invested more than $120 million in its reservation system, DatasII, and its software update, DataStar, yet, after six years, has only 11 percent of travel agents nationwide enrolled. DatasII accounts for only 8.2 percent of revenues generated by all such reservation systems. (United has captured 24.7 percent of the market, and American has 32 percent of the market on its system).[16] Delta is pushing its system by offering back-office capability for the travel agents through small networks of PCs and by providing stored video images of cruise ships, rental cars, and other travel products on hard disks. But the company admits that it is playing catch-up.

Recently Delta purchased 40 percent of a firm known as Worldspan Travel Agency Information Services. Northwest owns 33.3 percent and TWA owns 26.6 percent of the firm, which combines Delta's DatasII with PARS, previously owned by Northwest and TWA. The new firm, which combines the fourth and fifth largest reservation systems, expects to capture 26 percent of the U.S. market.[17]

Many companies have chosen not to compete with Sabre but prefer to lease the service. Air France, KLM, and others are airline customers of American's reservation system. The system generates 5 percent of the gross revenue of the AMR Corporation, American's parent company, and accounts for 15 percent of its profits.

In the early 1970s United Airlines automated its in-house reservation system, and in 1976 the system was introduced as a travel agency system called Apollo. United spent $250 million to develop Apollo. It gives agents access to listings for all major airlines as well as for hotels, car rentals, and other travel-related services. The system was transferred to a separate business unit of United Airlines called the Apollo Services Division in 1976.

The objectives of the division are to add new, marketable services to the reservation system. One such service is an office automation system for travel agents which provides accounting, reporting, and other office features. This service has since been enhanced with a new modular Enterprise Agency Management System containing advanced features and functions. Clearly United intended to invest in Apollo to improve the quality of its information-processing assets.

In 1987 the UAL Corporation decided to spin off a subsidiary centered around its computerized reservation system. This subsidiary became an independent affiliate of United Airlines known as COVIA Corporation. COVIA's mission is to enter new data processing business ventures and to capitalize on opportunities outside the travel industry through the use of its large worldwide network.

There have been many legal actions taken against United in connection with Apollo, ranging from charges of monopolization of the reservation system market to charges of bias in the display and presentation of flight information. Many travel agencies have tried to break contracts with United to join other systems. For the most part, United has been successful in obtaining relief through the courts in these contract disputes. Recently, competing firms have been offering to pay legal fees for agents who elect to break contracts with United in favor of their systems. Competitors are using every means available to combat the advantages of United and its system.

A market value of $1 billion was placed on the Apollo reservation system in 1988 when UAL Corp. sold 49.9 percent of COVIA for $499 million. The buyers were USAir, British Airways, and Swissair at 11.3 percent each, KLM Royal Dutch at 10 percent, and Alitalia at 6 percent. U.S. airlines sold $42 billion in tickets in the last year, 90 percent through travel agents. It is estimated that Sabre and Apollo have annual returns on investment of 50–100 percent.[18]

In Swindon, England, a joint business venture composed of 10 airlines is developing an internationalized reservation system to offer a wide array of services. The joint venture, called Galileo, is owned by Aer Lingus, Alitalia, Austrian Airlines, British Airways, KLM, Olympic Airways, Sabena, Swissair, TAP Air Portugal, and the U.S. company, COVIA.[19] The company plans to include reservation services for most airlines worldwide, tens of car rental companies, thousands of hotels, and major theater and sporting events. Office management services for travel agents will also be offered. A goal of the system is to give travel agents in Europe one user-friendly terminal to handle all their customers' needs.

Computerized reservation systems are increasingly being treated as saleable assets by large corporations. Delta has discussed merging its reservation system with those of American and Trans World Airlines; Texas Air has discussed the sale of its reservation subsidiary to Electronic Data Systems, a unit of General Motors Corp.[20] There are a variety of reasons behind these moves, but clearly computerized reservation systems make up a large, dynamic business today.

Stock Brokerage

In 1977 the brokerage firm of Merrill Lynch announced the development of its Cash Management Account (CMA) at a New York press conference and began test marketing the product in Atlanta, Denver, and Columbus, Ohio. The CMA product consists of a combination checking account, debit card, and brokerage margin account supported by a computerized cash management system. The system provides customers with current information via phone and detailed printed reports at month end. Customer net cash balances are invested in one or more money market funds generating interest income for the client. Subscribers' expenditures are applied first against their net cash balance and, when depleted, against their lendable equity in the margin account. This innovative product involved an alliance with Banc One of Columbus, Ohio, which processes the checking activity for the CMA, thus preserving the separation of banking and brokerage as required by law.

In 1978 Merrill Lynch expanded the CMA product to 38 offices in five states, and by 1980 the CMA became available in 39 states. By this time the number of accounts had grown to 186,000. To expand account growth, Merrill Lynch launched its first specialized version of the CMA designed for estate administrators. The CMA became available in all 50 states in 1981, by which time the number of accounts had passed the half million mark. The Professional Golfers Association adopted the CMA automatic transfer system in 1981 to manage funds and pay tournament winners. Additional features were under development to expand the range of services to CMA clients.

The international CMA was launched in 1982, and the Working Capital Management Account, serving the needs of businesses and professional corporations, debuted in 1983. Accessing CMA reserves through cash machines became widespread in 1984. The Capital Builder Account, tailored for the needs of individual investors, became popular in 1985 and 1986. Additional features were under development—and required large investments at that time.

By 1987, the tenth anniversary of the CMA, 1.3 million accounts with $150 billion in assets were being served. Merrill Lynch introduced CMA enhancements including increased ATM access (24 hours a day at more than 22,000 locations) and the new CMA Premier Visa program. The Premier Visa program provides financial benefits and administrative features to the client and an additional $25 per year fee to Merrill Lynch per client. Additional functions are under development.

The CMA has been enormously successful, and Merrill Lynch has derived a major competitive advantage from the program. The minimum balance required to open a CMA is $20,000, but the average balance is close to $100,000 for the 1.3 million clients. The minimum fee is $65 per client per year. Fee income is substantially augmented by commissions on securities transactions, interest charges on debit balances, and service fees on the more than $28 billion in Merrill Lynch–managed money market funds.

It was not until 1984 that the competition begin to employ similar information technology. Merrill Lynch carefully protected its position by securing a U.S. patent on the computer program and by defending its turf in court. In 1983 Merrill Lynch won a $1 million settlement from Dean Witter, its closest rival at that time.

The CMA patent application was filed on July 29, 1980 and was granted on August 24, 1982. The patent, number 4,346,442, lists Thomas E. Musmanno as the inventor and is assigned to Merrill Lynch. The patent contains four drawings (flow charts) and six claims and is displayed in total on 11 pages. It represents an uncommon but important form of protection for a valuable strategic asset.

Additional enhancements continue to be announced for the CMA, implying continued investments by Merrill Lynch. *Fortune* reports that Merrill Lynch is spending $160 million each year on software.[21] Although Merrill Lynch's market share is declining slowly, it holds approximately 50 percent of this important and valuable market. With the CMA and other services, Merrill Lynch manages a total of $300 billion in client assets. The CMA exemplifies a brilliant combination of information technology and financial services employed for strategic competitive advantage.

But is Merrill Lynch secure in its position? Is it possible for a competitor to provide a more attractive offering through the use of information technology?

The answers to these questions must await the market response to SMARTVEST, an innovative offering from OLDE & Co., a discount brokerage firm located in Detroit. SMARTVEST appeals to Merrill's CMA clients by offering a wide range of services to investors through their PCs at home. From a PC connected to the system through the phone network, investors are able to obtain the status of their accounts, access market reports or research information, and place orders to buy or sell securities. Some of the many features of the system are summarized in Table 2.3.

TABLE 2.3   SMARTVEST Features

1.   Real-time securities quotations

2.   Research reports on 4800 stocks

3.   Current financial and business news

4.   Order entry

5.   Account retrieval

6.   Portfolio management

7.   Insider and institutional activity

8.   *USA Today* summary news reports

Clients of this service have the option of opening a Credit Access Account (CAA) with a credit union in Toledo, Ohio, which provides a free checking account and credit card services. SMARTVEST allows investors to view their transactions, which are updated daily, on their PCs at home or at any other location convenient to the customer.

Using network, database, and personal computer technology now widely available to investors, OLDE & Co. combines a series of strategic thrusts. It offers a differentiated product; utilizes innovative information technology; allies itself with other institutions; and provides services in a timely manner. Surely these are strong ingredients for success, but the implementation process and customer acceptance are the final determinants.

Order Entry Systems

In 1976, Baxter Health-Care Corporation, then American Hospital Supply Corporation (AHSC), introduced the first automated order entry system for health care products. Since that time, the company has continued to provide customers with the most efficient system in the industry by incorporating the latest technological advances.

Since 1976, order entry systems have become commonplace. They span a range of industries including grocery chains, mail-order catalog firms, pharmaceutical supply firms, and many others. Today, if order entry is a required function, the firm without an automated system is at a serious disadvantage. It wasn't always that way. In 1976, when AHSC pioneered with its ASAP (Analytic Systems Automatic Purchasing) program, it was well on the way to capturing important advantages in the hospital supply industry.

The system began life in the early 1960s as a conglomerate of punched-card and telephone technology, and through continued infusions of money and technology it blossomed into one of the most prominent strategic systems. Between 1976 and 1984, the ASAP program advanced through five versions, each building on and increasing the functions offered by its predecessors.

Since 1985 there has been a proliferation of vendor-initiated order entry systems, and individual hospitals have spent duplicate resources attempting to cope with them. The health care industry is seeking a strategy to achieve an all-vendor system for order entry. Baxter is shaping its system to meet these needs using the standardized data formats spelled out in the X.12 standard from the American National Standards Institute. Customer needs are the driving force behind Baxter's strategy.

Perhaps the most significant indication of the value of ASAP is that American Hospital Supply was purchased by Baxter Travenol Laboratories in 1985 in large measure because of its very successful and innovative order entry system. The new firm, Baxter International, of which Baxter Health-Care is a part, is consolidating its distribution business around ASAP. It should be noted that capturing the benefits of the consolidated organization has not been easy.[22]

Baxter is taking a more global view of its ASAP order entry system. It considers the system to be automated communications, not just automated order entry, and it is shaping the system accordingly. Baxter is following the lead of the airlines. It sees its system more as a general-purpose vehicle to provide service offerings to its customers. It is responding to client requirements in the broadest possible manner.

Baxter is providing the first electronic invoicing/electronic funds transfer system in the industry. It recently introduced ASAP Express, an all-vendor, all-transaction order-entry system implemented with ANSI X.12. Baxter is also active in electronic data interchange (EDI). The company takes full advantage of EDI technology. In a further effort to make life easier for its customers, Baxter is looking into automated purchase orders. This feature, to be implemented in ASAP Express, will provide the hospitals an easy, automated way to get data into the system.

The system is so successful that Baxter allows customers to order supplies from competitors. Customers use the system for competitive orders because the features of ASAP materially reduce their costs, and this use provides increased revenue and profit for Baxter. This kind of activity illustrates the tremendous value of the ASAP program and confirms its importance as a strategic asset to Baxter.

Baxter's systems development department excels in other ways. Their strategy is to provide a leadership vision for the firm through outstanding client service, adding information resource value, providing organizational development, and adding value to the business. Client service is attained by performing to or bettering requirements and meeting system service levels in response time, availability, and report delivery performance. The department manages development projects well, attaining on-schedule delivery within budget. Information-resource value additions are measured by reductions in expenditures for administration, programming efforts, and problem resolution time, as well as through improved project management.

Baxter's thrust toward organizational development centers around quality recruiting, career development, and sound people management practices. The system development department measures its value to the business in terms of operational cost reductions and reduced inventory levels. Other measures include sales increases due to its systems and increases in strategic value as defined by the business units they support. The development department firmly believes in adopting new technology when applicable, and it uses advanced tools to reduce development time, to increase quality, and to attain productivity enhancements in the development process.[23]

Baxter Health-Care has an IS budget of about $160 million, which represents 2.57 percent of revenue. It spends 5 percent of its IS budget on training for its more than 750 IS personnel. The company uses more than 25,000 terminals internally, and the network serves more than 12,000 customer terminals accessing various Baxter services.[24]

Delivery Services

Federal Express began the overnight package delivery business in 1973, processing 100 percent of the packages manually. Federal Express processes more packages overnight than any of its competitors, yet charges more for its services. Despite significant competition, Federal Express maintains its 40 percent share of the market, due in large part to Cosmos, a database system that tracks all letters and packages handled by the company. With Cosmos it can tell customers where their package is in 30 minutes or less, thereby easing customer fears of late delivery or lost parcels.

Much of Federal Express's growth is due to innovation in the systems area.[25] Nine thousand of its 14,000 couriers are on-line to local dispatching centers through console-mounted CRTs in their vans. The system supporting this capability is called Courier Dispatching. It helps drivers track packages and keeps them informed of schedules and locations, and it allows the couriers to pick up new orders quickly.

Bar codes are used extensively by Federal Express to track packages throughout the delivery system. The scanned data is uploaded within two minutes via telecommunications links into a central database where it is available to answer company or customer questions. Misdirected parcels average only one in 22,000, and this error rate is expected to decline as couriers are issued hand-held scanning devices. Packages are sorted by a scan-recognition system as they move at 500 feet per minute on conveyer belts. Federal Express processes more than 400,000 items in a two-hour window, and they believe these innovations are necessary to maintain their own service standards. Federal Express sorts packages and prepares bills in a paperless office where terminals display all necessary information.

The strategic advantage of Cosmos is that it differentiates Federal Express from its competitors, thereby providing substantial advantage to the company. Additional resources applied toward advanced features in Cosmos will enable the company to handle more volume and offer superior service in the future.

The examples we have studied relate directly to Porter's model of competition and to the theory of competitive thrusts. Telecommunications is a major ingredient of many important strategic systems. Many firms are using this technology to advantage. Some very successful strategic systems are composed mostly of telecommunications elements.[26]

WHERE ARE THE LEVERAGE POINTS?

Information systems theory and the examples of strategic systems implementations that we explored in this chapter serve to validate both the theory and the practice of strategic systems. Some strategic systems are proprietary and are not discussed in the literature. Many other valuable systems, smaller and less obvious than those we discussed, are installed and operating successfully in firms worldwide. Hundreds, perhaps thousands of firms own and operate information systems that provide advantage for them in their competitive spheres of operation. Hundreds, perhaps thousands more are making investments in hopes of capturing similar advantages for their firms.

Opportunities

As the examples illustrate, there are numerous opportunities for leveraging the investment in information systems. These opportunities frequently consist of product or service offerings that have been differentiated from their competitors by the application of information technology. All the systems we studied exemplify this principle. The technology provides elements of superiority in the minds of the firm's customers through perceptions of high quality in the product or service. The application of technology inspires confidence and promotes customer loyalty.

Customer loyalty is achieved in other ways as well. Effective application of information technology can lower costs to the producer and to the consumer, thus improving the cost effectiveness of both. Significantly enriched services command higher prices; customers are willing to pay for higher quality. The leverage points are exploited for growth of revenue and profit for the producer of the product or service. This is true for the order entry system and the delivery system we studied.

The examples illustrate how innovative application of information technology can be exploited to create value for the consumer. New markets can be developed through the creation of innovative new products.

Demand is developed and satisfied to the benefit of producer and consumer. Some new products or services require the collaboration of several firms. Alliances must sometimes be formed to develop, distribute, or market the product or service and to capitalize on the technology most effectively. The Cash Management Account created new products and new markets, and the airline systems led to alliances.

Strategic systems are focused not only on the end customer but also on supplier targets and competitors. Organizations that supply raw materials to the firm are affected by sophisticated systems that optimize buying strategies. And competitors feel the influence of systems used to design, develop, or manufacture superior competing products. Information technology is used by many firms to support or mold strategic thrusts. Effective utilization of information technology in the strategic competitive environment demonstrates high potential in many contemporary organizations.

The Importance of Technology

Technological advances and their introduction into business and industry worldwide constitute two of the principal drivers of competition. Advances in technology shape the products and services of the future and offer opportunity for innovative organizations to increase their value to the stream of economic activity. We have only to consider the advances in communication or transportation, or observe the revolution in chemicals or pharmaceuticals, or fathom the significance of information processing, to realize the importance of technological advances on international competition.

Information technology is particularly important because it is pervasive in the processes leading to advances in most other activities. All activities create, transport, disseminate, or use information. Advances in IT have a compounding effect on technological advances everywhere. Accordingly, information technology is exerting a remarkable and profound effect on worldwide competition.[27]

Advancing technology is important because it alters industry structure, shapes and molds competitive forces within and between companies and industries, and thereby changes forever the behavior patterns of billions of individuals. Technology for its own sake is not important. What is important is its positive impact on society as a whole.

The Time Dimension

Time is an asset and a source of competitive advantage. Firms must think about time resources much as they do about capital, facilities, materials, technology, and management resources. The firm that views time as a valuable asset will strive to use this asset for advantage both externally and internally.

The driving force behind the notion of time as a competitive factor is simply response to customers, markets, and changing market conditions. Responsiveness cuts across all functions of the firm, from product requirements definition to installation and service. Timeliness involves suppliers and customers and impacts competitors. Thus time provides opportunities for competitive advantage. Within the firm, time is important in planning, implementing, and controlling, and it is highly susceptible to manipulation with information technology. Although there are many examples, the notion of just-in-time manufacturing, or JIT, illustrates the point.

The JIT approach was implemented at the Irvine, California, factory of the McDonnell Douglas Computer Systems Co., a division of McDonnell Douglas Corp. It required 111 new programs and 97 modifications to installed programs, for a total of 1900 personhours of programming changes and additions.[28] Additional time was expended in planning, team meetings, and training and some expenses were incurred for tags, labels, and other items. A thorough understanding of the flow of parts, products, and information through the company formed the basis for constructing the JIT approach. Benefits are achieved in many areas and the results are impressive:

Inventory reduction	38%
Work in process inventory reduction	40%
Printed circuit board assembly cycle time	80%
Rework reduction	40%
Improvement in inventory turns	100%
Quality-control process yield (now 99% perfect)	80%
Setup time reductions	50%

Using information technology in conjunction with parts-logistics control, attacks on setup time, design process improvements, and supplier logistics yielded substantial savings in time and money for this manufacturer of computer products. Other computer manufacturers agree. "We've ignored a critical success factor: speed," says John Young, CEO of Hewlett-Packard Corporation. "Our competitors abroad have turned new technologies into new products and processes more rapidly. And they've reaped the commercial rewards of the time-to-market race."[29]

Telecommunications technology speeds the flow of information between organizational entities bridging the gap in space and time. Given the pervasive role of information in business, telecommunications offers great potential for competitive advantage through reductions in time and mitigation of distance barriers. Multinational firms must employ this technology to remain in business, and they are continually searching for innovative applications.

THE CHARACTERISTICS OF SIS

The strategic systems studied in this text, and many that were not analyzed, have some common characteristics worth noting.

Organization and Environment

Strategic systems may alter the market environment and change the field in which the competitors are engaged. In the process, these systems will probably alter the internal environment or change the form of organizational structure. As the strategic system becomes effective, it influences the environment. In turn, the firm modifies its structure to accommodate the changed environment and the system itself. The metamorphosis of the competitive situation and the firm's response continues unless some form of stability becomes established among the competitors.

Financial Implications

Strategic systems require continued investments of resources to sustain their advantage. The owners of these systems find that competitors are always looking for ways to build systems for themselves in order to negate the owner's advantage. As the examples revealed, it is a challenging and never-ending task to stay ahead of the competition. Relatively secure positions in the brokerage industry, for example, are subject to threat by smaller but innovative competitors.

Some strategic systems produce revenue by themselves and become profit centers within the corporation that owns them. In the extreme case, the profit center grows into a major international business entity in its own right, as illustrated by the evolution of COVIA Corporation.

Additional Considerations

The owners of some successful strategic systems eventually become engaged in legal struggles with the competition. The legal questions involve competitive issues surrounding the appropriate use of the business advantage held by the owner. Some legal issues relate to protection of the competitive advantage. The owners of major airline reservation systems have been in court numerous times over the years litigating issues arising from the advantages derived from their systems.

THE STRATEGIST LOOKS INWARD

Many important systems originated from the idea of one individual who visualized a way to capitalize on emerging technology or improve the functioning of some business aspect of the firm. This person's insight became the catalyst that spawned the business opportunity.

Most strategic systems developed from the process of looking at the firm's internal functions. Efforts to automate the internal activities of the company more fully paid off later in competitive advantage. (The Sabre system began life this way.) A logical starting point to search for strategic opportunities is the firm's portfolio of application systems.

Many firms own application portfolios comprised of several thousand programs. Some of these applications may be candidates for strategic development. The thrusts these applications may potentially contain range widely from cost reduction to innovative methods for attaining product or process superiority. These internal systems are found in marketing, development, manufacturing, sales, service, and administration. Superior market analysis tools coupled with automated design systems can significantly reduce the time and cost required to respond to changing customer requirements and market conditions. Automated manufacturing processes and sophisticated distribution systems speed the new products to the customer efficiently and at reduced cost.

Systems to handle these basic tasks exist in most firms today. How can these systems be augmented or enhanced to improve the firm's posture in the marketplace? What new technology can be employed in these processes? What innovative actions will permit the firm to utilize internal resources in order to maximize its competitive position? These and other questions directed toward current applications form a basis from which to search for strategic opportunities internally.

EXTERNAL STRATEGIC THRUSTS

Another view which is useful in identifying potential strategic opportunities is to consider external factors. These factors include changing industry environment, recent actions of competitors, changing relations among suppliers, potential business combinations, and many others. This view of the firm asks the questions: "What is happening external to the firm that may influence our opportunities to gain competitive advantage?" and "How can we capitalize on external factors through the use of information technology?"

The process of answering these questions is different and more complicated than that used to view the firm introspectively. The individuals best suited to this task are found in the top positions in the company. They may be located in the marketing function that has responsibility for competitive analysis or industry analysis, or they may be in product distribution with responsibility for ensuring timely and accurate dissemination of the firm's goods or services. The key senior people responsible for guiding the firm's long-term direction will have valuable insights. Their visions must be influenced by the information technologists, who have their own vision of what is possible and feasible technologically. Potential opportunities will emerge from the intersection of these two visions.[30]

Several cautions about strategic information systems are in order. Strategic systems usually develop from deliberate attempts to improve or enhance current management information systems. They do not begin life as separate and distinct systems from those in the applications portfolio. Most strategic systems do not emerge from specific attempts to meet corporate strategic objectives. They are the result of many incremental enhancements and sustained improvements to current systems. They do not derive from radical changes to operational systems or from totally new systems.[31]

Successful executives focus on improving corporate performance through constant attention to many details of their business. They search for improvement through new technology, enhancements to systems and and operational procedures, and modifications to the organization and its culture. The task of getting ahead of competition and staying ahead is difficult; it requires adaptation to environmental changes and response to competitive forces. Successful executives know they cannot attain lasting competitive advantage from a few grand strokes.

INTEGRATING THE STRATEGIC VISION

Building on a model presented by Wiseman and MacMillan,[32] the interrelationships of the strategic variables can be depicted graphically. Figure 2.2 displays these relationships and illustrates the internal and external sources and uses of strategic systems. Time, the highly important strategic thrust, complements the other five thrusts in this model.

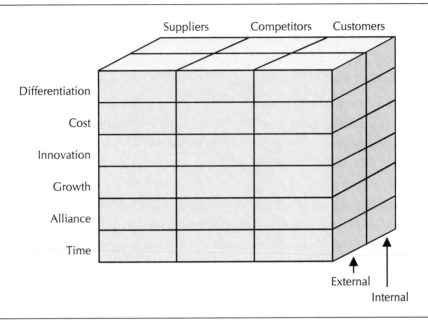

FIGURE 2.2   Integrated Model of Strategic Influences

Most strategic systems are found to cover several areas in Figure 2.2. A system may utilize several of the strategic thrusts itemized down the left side of the diagram, or it may have an impact on more than one of the groups illustrated across the top. For example, cost reductions and time savings obtained internally from the application of innovative technology may lead to growth in both revenue and profits, affecting customers and competitors. Computer-aided design/computer-aided manufacturing systems have this potential. In the computer industry itself, innovative use of advanced technology in sophisticated electronic design automation systems enables firms to spawn new, high-quality products in ever shorter development life cycles. Competitors and customers are impacted significantly.

Systems change character over time as their potential is exploited by the firm. Some systems start life as an internal thrust and become important to the firm externally. The Sabre system is an example. Throughout the life cycle of a strategic system, it may migrate through the model. It may play an important role for the firm in several arenas, as its potential is developed via redirection and further investment. Change is a fact of competitive life, and successful strategic systems management involves anticipating, initiating, or reacting to change to obtain or sustain competitive advantage for the firm. Management of strategic systems is a major challenge for senior executives in nearly all organizations today.

SUMMARY

Information technology is becoming increasingly important and, in many instances, vital to the success of modern-day firms in the industrialized world. Senior executives have many valid reasons to expect that information technology can provide advantage for their firms. They believe that innovative use of information technology gives promise for corporate or organizational leadership. They are prepared to invest in the technology and they expect their firms to attain a substantial return on this investment. Many realize this investment is not optional, but is required for the future health of the organization.

There are numerous examples to sustain an executive's belief in the importance of strategic systems. This chapter examined several systems to establish their features and to relate them to models of competitive behavior. It also presented models that help firms search for strategic opportunities. However, it is far easier to recognize strategic systems in action than it is to identify opportunities and capitalize on them.

Review Questions

1. Why are strategic concerns regarding information systems increasingly important today?

2. What distinguishes strategic information systems from other kinds of systems?

3. Explain Porter's model of forces governing competition.

4. What are the six basic strategic thrusts?

5. What are some examples of transaction-processing systems that have strategic value to their owners?

6. What thrusts are employed by American Airlines in its reservation system?

7. As a competitive weapon, what distinguishes United's Apollo system from American's Sabre system?

8. In what ways did Merrill Lynch's CMA alter the environment in which brokerage firms were operating in 1977?

9. What protection is afforded Merrill Lynch as a result of the patent on the CMA program?

10. With what current trends in the health care industry is Baxter contending in its development of the ASAP system? What innovations is it introducing?

11. How would you describe the Cosmos system in terms of the Porter model?

12. What leverage can information technology provide to business organizations?

13. What are some common characteristics of strategic systems?

14. What are the advantages and disadvantages of searching internally for sources of strategic systems?

15. What insights can the IT organization bring to the process of searching for strategic systems?

16. Using the information provided in the text and the model in Figure 2.2, trace the evolution of the Apollo system.

Discussion Questions

1. What strategy has Avon been pursuing with its executive information system?

2. What appear to be the strengths of this strategy? Do you think the Avon strategy is proactive or reactive, and why?

3. Where on the grid in Figure 2.2 does the strategy adopted by Avon fall?

4. Discuss the connection between the Avon information system and the Drucker quotation that begins Chapter 1.

5. Research reveals that the issue of office automation has been declining in importance for IT executives. What might this mean in terms of strategic systems?

6. Discuss Porter's model as it applies to the computer industry today. How is the telecommunications industry different from the computer industry as viewed from the model?

7. The six strategic thrusts presented in this chapter are not mutually exclusive. Discuss the implications of this fact.

8. Deregulation of the airline industry has encouraged new entrants into the industry. Given the enormous advantage to the competitor who owns a reservation system, how can new entrants overcome the barriers to entry?

9. Discuss the role played within the firm by IT managers as the firm seeks to improve its competitive posture. What contributions can they make, and in what areas must they take the lead?

10. What additional actions might be taken by OLDE & Co. to capitalize further on its current position?

11. Discuss the importance of telecommunications to the strategy of Federal Express.

12. How might firms seeking international competitive advantage rely on information technology? What thrusts might they employ?

Assignments

1. Using the Wiseman book or another reference, select an additional example of a strategic system and prepare an analysis of its characteristics. Be prepared to summarize your findings for the class.

2. Study the *Harvard Business Review* article by Stalk and prepare a report on why he thinks time is the next source of competitive advantage.

ENDNOTES

[1] Avon earned $195 M on sales of $3.5 B in 1990. Avon employs 30,000 but has 1.5 million sales representatives worldwide according to *Value Line*, April 19, 1991.

[2] Kathryn Kukula, "An Interface Even an Executive Could Love," *Solutions*, special issue, 1989, 18.

[3] James I. Cash, F. Warren McFarlan, James L. McKenney, and Michael R. Vitale, *Corporate Information Systems Management: Text And Cases*, 2nd ed. (Homewood, IL: Irwin), 7.

[4] "The Premier 100," *Computerworld*, September 12, 1988.

[5] Jack Kuehler, president of IBM, reported this data to the Institutional Member Conference of the Society for Information Management, April 12-14, 1989.

[6] Charles Wiseman, *Strategic Information Systems* (Homewood, IL: Irwin, 1988), 98.

[7] Ralph H. Sprague, "A Framework for the Development of Decision Support Systems," *MIS Quarterly* (June 1980). This article describes decision support systems in detail and relates them to the more familiar, but perhaps less easily defined, management information systems.

[8] Additional information on Porter's theories of competition can be found in the references at the end of this chapter. For a complete discussion see Michael E. Porter, *Competitive Advantage* (New York: Free Press, 1985).

[9] Wiseman's book discusses five strategic thrusts. It is my contention that time constitutes a sixth thrust. Wiseman, *Strategic Information Systems*, 142.

[10] "Speed to Market," *Forbes*, May 28, 1990, 350. This special report to management by the Yankee Group presents a brief description of the strategic advantages derived from the time thrust. It contains six examples of firms that use or provide time advantages. Also see Roy Merrills, "How Northern Telecom Competes on Time," *Harvard Business Review* (July-August, 1989): 108.

[11] John Killkirk, "Firms Learn That Quick Development Means Big Profits," *USA Today*, November 22, 1989, p. 10B. Chrysler can create a 3-D computer model of a car body in about 10 seconds. It takes three to four years and costs Chrysler about $1 billion to bring a car to market; computer modeling can reduce this time by up to one year.

[12] Peter Keen, "Vision and Revision," *CIO* (January-February 1989): 9.

[13] Israel Borovits and Seev Neumann, "Airline Management and Information System at Arkia Israeli Airlines," *MIS Quarterly* (March 1988): 127. Using super minis and micros, this small airline built a high-function reservation system for only about $250,000!

[14] Dennis J. H. Kraft, Tae H. Oum, and Michael W. Tretheway, "Airline Seat Management," *Logistics and Transportation Review* (June 1986): 115. American also analyzes yield management data on no-shows to overbook optimally. In 1988, American denied boarding to fewer passengers than any other of the nation's 12 largest airlines, according to an American Airlines newsletter.

[15] Ester Dyson, "Secret Formula," *Forbes*, October 5, 1987, 242. This article discusses the vital role played by American's people in using the system to solve problems and improve customer service.

[16] *Investor's Daily*, September 26, 1989, 9.

[17] Alan J. Ryan, "Delta Lands Another CRS Partnership," *Computerworld*, February 12, 1990, 19.

[18] *Investor's Daily*, September 26, 1989, 9.

[19] *Think*, No. 3, 1989, 38.

[20] *Rocky Mountain News*, September 22, 1989, 66.

[21] Brenton R. Schendler, "How to Break the Software Logjam," *Fortune,* September 25, 1989, 100.

[22] "Baxter to Take Charge Totaling $566 Million," *Wall Street Journal*, April 5, 1990, A4: ". . . the company has been widely criticized for the slowness with which expected savings from its $3.7 billion acquisition of American Hospital Supply Corp. in late 1985 have materialized."

[23] Personal communication.

[24] "The Premier 100," *Computerworld*, September 12, 1988.

[25] Federal Express more than tripled its revenue from 1980 to 1984, more than doubled it from 1984 to 1987, and is expected to more than double it again from 1987 to 1990. According to *Value Line*, March 30, 1990, Federal Express's revenue rose from $415 million in 1980 to a projected $6.92 billion in 1990.

[26] W. Frank Cobbin, Jr., et al, "Establishing Telemarketing Leadership Through Information Management: Creative Concepts at AT&T American Transtech," *MIS Quarterly* (September 1989): 361. This is an extremely interesting example of a strategic telecommunication system.

[27] Michael Porter, "Technology and Competitive Advantage," *Journal of Business Strategy* (Winter 1985): 60.

[28] Lad Kuzela, "Efficiency—Just In Time," *Industry Week*, May 2, 1988, 2.

[29] "How Managers Can Succeed Through Speed," *Fortune*, February 13, 1989, 54.

[30] Nick Rackoff, Charles Wiseman, and Walter A. Ullrich, "Information Systems for Competitive Advantage: Implementation of a Planning Process," *MIS Quarterly* (December 1985): 285. The authors describe a process employed at GTE to search for strategic IT opportunities.

[31] James C. Emery, "Misconceptions about Strategic Information Systems," *MIS Quarterly* (June 1990): vii.

[32] Charles Wiseman, and Ian C. MacMillan, "Creating Competitive Weapons from Information Systems," *Journal of Business Strategy* (Fall 1984): 42.

REFERENCES AND READINGS

"American Hospital Supply Corporation." Harvard Business School Case 9-186-005, 1986.

Anthony, R. N. *Planning and Control Systems: A Framework for Analysis.* Cambridge, MA: Harvard Business School, 1965.

Information Week. Special issue: Strategic Systems. May 26, 1986.

Keen, Peter G. W. *Competing in Time: Using Telecommunications for Competitive Advantage.* Cambridge, MA: Ballinger Publishing Co., 1988.

Keen, Peter G. W. "Vision and Revision." *CIO Magazine*, January-February, 1988, 9.

McFarlan, F. Warren "Information Technology Changes the Way You Compete." *Harvard Business Review* (May-June 1984): 98.

Parsons, Gregory L. "Information Technology: A New Competitive Weapon." *Sloan Management Review* (Fall 1983): 3.

Porter, Michael E. "How Competitive Forces Shape Strategy." *Harvard Business Review* (March-April 1979): 137.

Porter, Michael E. *Competitive Advantage.* New York: Free Press, 1985.

Porter, Michael E., and Victor E. Miller. "How Information Gives You Competitive Advantage." *Harvard Business Review* (July-August 1985): 149.

Rackoff, Nick, Charles Wiseman, and Walter Ullrich. "Information Systems for Competitive Advantage: Implementation of a Planning Process." *MIS Quarterly* (December 1985): 285.

Stalk, George, Jr. "Time—The Next Source of Competitive Advantage." *Harvard Business Review* (July-August 1988): 41.

Wiseman, Charles. *Strategic Information Systems*, Homewood, IL: Irwin, 1988.

3

Developing an Information Technology Strategy

A Business Vignette
Introduction
Considerations in Strategy Development
Strategies and Plans
Types of Strategy
  Stand-Alone Strategies
  Business and Functional Strategies
Requirements of a Strategy Statement
Strategy Document Outline
Strategy Development Process
  The Strategic Time Horizon
  Steps in Strategy Development
Strategy as a Guide To Action
The Strategy Maintenance Process
The IT Strategy Statement
  The Business Aspects
  Technical Issues
  Organizational Concerns
  Financial Matters
  Personnel Considerations
Summary
Review Questions
Discussion Questions
Assignments
References and Readings

A Business Vignette

Information Technology Achievements
Fall Behind Expectations

At a recent user-group conference, the president of a major consulting firm division noted differences between what the chief executive officer of a firm is coming to expect from his or her IT function and what that function delivers.[1] Among the expectations that CEOs deem reasonable are these:

1. Making use of information for business strategies

2. Anticipating and dealing with change

3. Producing revenue and bottom-line results

4. Behaving like any other business function in decision-making and following procedures

While these expectations may seem reasonable, they may not be met automatically, because the IT function often works without line management involvement and because IT does have its own business identity or infrastructure.

For example, important innovations seem to come from the line managers who use IT, not from within IT itself. Applications having strategic importance for the firm are probably discovered most easily by managers responsible for line operations. IT managers must learn to reach out to their line clients to encourage creative thinking and innovation.

But line managers are not likely to uphold their part of the culture unless IT managers teach them what IT can do and encourage them to think of new ways information can be used to the strategic advantage of the line managers. In the end, the best results will come from having the line and the IT functions working closely together.

Working together is a theme at Texaco. Writing about J. Les Hodges, recently retired IS chief at Texaco, John Rymer says, "Management fads in Information Services change as quickly as the technology. Despite the turmoil of recent years, however, I/S has continued to progress toward more useful information systems. [Hodges] based his management approach on two simple themes: Teach I/S pros about the oil business, and help general managers understand how to put technology to work."[2]

However, working together is not enough; an information infrastructure is also required. IT needs a core of systems, computers, transmission devices and modes, applications, access devices, and general procedures for managing information. Unless the IT function in a company begins to design such infrastructure elements now, the company will lack them when they will be most needed—three to five years down the line, when competition is tighter than ever and the strategic needs for IT most urgent. The development of an information infrastructure begins with strategic considerations. It requires detailed planning and careful implementation.

Executives expect IT to suggest ways in which technology can be used to implement strategy and to maintain flexibility and responsiveness to changing business conditions. IT must conform to the corporate culture and make positive contributions to business objectives. The executives' expectations are entirely reasonable.

INTRODUCTION

The process of searching for an area in which a strategic system may be developed is an important endeavor for the firm, requiring considerable thought and research. However, strategic thinking covering all areas of importance to the firm is mandatory. If the firm is going to exploit information technology or other strategic opportunities for the benefit of the organization, a strategic vision must be applied systematically by the firm's senior executives in order to set organizational direction. The strategy provides focus and coordination to the activity of the firm. A well-developed strategy ensures consistency of direction among the firm's units and reduces uncertainty in shorter-range decision making.[3]

If IT managers are to be successful in most firms today, consistent strategic planning must be a part of their routine business system. Strategy development promotes actions leading to sound planning, encourages organizational learning, and provides a process to ensure goal congruence. Elements of successful strategic planning include sound preparation and implementation, line manager involvement, correct definition of business units, action steps outlined in detail, and controls integrated into the plan.[4] These objectives are a critical success factor for IT managers. Strategic planning takes organization structure into account. Deciding whether to centralize or decentralize, for example, can be handled by good planning.

This chapter develops the processes and techniques essential to successful information technology strategy development and strategic planning. It discusses types of strategies, their contents, and the organization of a strategy statement. The text develops the steps required to maintain and update a strategy document. It describes how the process of strategy development applies to the information technology function in particular, stressing the relationship of strategy statements to the ongoing activities of the firm and the IT organization. Strategy is a guide to action, but changes in the environment or other business considerations mandate a strategy maintenance process.

CONSIDERATIONS IN STRATEGY DEVELOPMENT

In the process of strategy development, thoughtful IT managers will perceive potential business opportunities and possible threats or pitfalls. To the extent possible, successful IT managers will attempt to redirect their efforts to maximize their advantages. The resulting proposed course of action may not capitalize on all opportunities available nor avoid every difficulty; but if optimally conceived, it will present a balanced approach to these conflicting influences. Since situations rarely remain static for a sustained period of time, thoughtful managers must reevaluate their situations periodically. And they must reassess and adjust their course of action in light of changing conditions.

The complete strategy expression must consider the elements discussed above, among others. Table 3.1 displays the ingredients of a strategy statement. The strategy statement sets forth objectives to be achieved and includes processes for its maintenance and use.

TABLE 3.1   Elements of a Strategy

1.   Mission statement
2.   Environmental assessment
3.   Statement of objectives
4.   Expression of strategy
5.   Maintenance processes
6.   Performance assessment

The first and most important step in formulating a strategy is to state the organization's mission and purpose. Developing a mission statement for the IT organization may be a challenging task, particularly on the first attempt. The mission statement should describe the organization's business and indicate who its customers are. The customers' needs and the organization's capabilities and resources must be considered. The IT mission should be stated in terms of meeting the needs of the customer and in terms of services the IT organization can provide. It is essential to define the IS markets within the firm itself.[5] In other words, the IT mission statement should be both reactive to customers and proactive toward customers. For example, the mission must include performing the transaction processing for the firm; but it must also include introducing new information technology for important new applications.

The process of visualizing and understanding the opportunities and threats is called "analysis of the environment." Environmental analysis (or environmental scanning) attempts to take into account the important trends impacting or likely to impact the organization. These trends may be political, economic, legal, technological, or organizational and the strategists attempt to understand them in order to position their organizations optimally. The information resulting from environmental scanning permits strategists to accomplish two tasks: Develop the objectives they intend to achieve, and formulate the course of action, or strategy expression, to be used as a guide in the attainment of the objectives. The strategy will rely on assumptions where facts are unavailable, and it will account for the nature and degree of the risks involved in attaining the objectives. The strategy must include some degree of flexibility through the inclusion of options or alternatives. The statement of strategy includes the ingredients that lead to further planning and decision making.

Strategy maintenance is the process of reviewing the environment and reassessing the course of action in light of changing events. The extent to which strategy maintenance proceeds is a direct result of the volatility experienced in the environment. Businesses in relatively stable environments may experience long-lived strategies needing only infrequent maintenance activity. However, most firms in today's economic and political environment find frequent strategy maintenance the norm. That will almost certainly be the situation for today's IT manager.

The goals, the objectives, and the strategy statement form a useful document with which to assess the efficiency and effectiveness of the organization. Thus, after the fact, a strategy statement can also be useful in assessing organizational performance and evaluating the degree to which the organization achieved its goals and objectives.

Figure 3.1 portrays the relationships of these activities to each other and to strategic and operating plans. The planning activity will be discussed in the next chapter.

The basic elements essential to a complete strategy and planning process are environmental assessment and nature of the business, goals and objectives to be achieved, strategy statement or expression, maintenance activity, and performance assessment. The development of the strategic and operational plans flows from the strategy activity. Figure 3.1 suggests the relationships of these items to each other and to the maintenance and assessment activities.

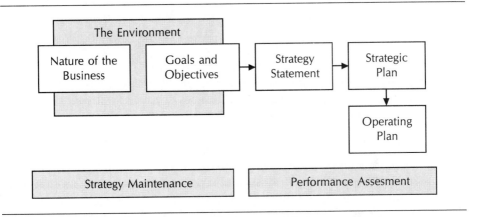

FIGURE 3.1   **The Relationship of Strategy and Planning Elements**

A statement of strategy can be useful for an individual or for a group. As we consider strategies for groups, we need to be concerned with developing formats and procedures to improve communications among members of the group. Good communication increases the group's ability to develop and use strategies effectively. Solutions to the issues of information architecture, IS's role and contribution, and the use of IS to integrate across functional lines hinge on the communication of strategic thrusts necessary to achieve strategy congruence within the firm. Succeeding sections of this chapter will deal with these aspects of strategy.

STRATEGIES AND PLANS

The strategic plan for a firm combines the strategies for its business areas and the strategies for all of its functions. The combined detail supports the overall business strategic objectives. The strategic plan calls for detailed action for the next year or two. The detailed actions comprise the short-range or tactical plan. Therefore, business and functional strategies guide and coordinate the activities of the various functional and organizational groups toward agreed-upon objectives. This guidance and coordination is expressed in statements of strategy, strategic plans, and tactical and operational plans.

A strategy forms the basis for a plan, but it is not itself a plan. The strategy spells out optimum actions to achieve general objectives, but it does not provide detail sufficient to carry out the actions. The addition of detail to the strategy sufficient for implementation by the organization creates a plan from a strategy. This is true whether the strategy encompasses a two-year or a twenty-year period. Therefore, the task of planning is substantially different from the act of strategy development. One deals with stating optimal actions to attain goals and objectives in the face of uncertainty; the other reduces the optimal actions to implementation tasks.

The processes of strategy development and strategic planning are essential to strategic management. Businesses have advanced their thinking about strategy from predicting the future to trying to create the future. Several phases of this development can be recognized. Phase 1 consists of basic financial planning, which evolves to forecast-based planning in the second phase. Phase 3 consists of externally oriented planning characterized by increased response to competitive pressures and evaluation of strategic alternatives. These phases form the basis for phase 4, strategic management.[6] Strategic management means reorganizing and redeploying resources to attain competitive advantage. It also means implementing creative and flexible long- and short-range planning systems supporting resource deployment actions. Information technology managers play an extremely vital role in this process, because the technology they manage has considerable potential for initiating future changes.

> The plan for the management of opportunity deals somewhat with the present but more with the future because in the time required to fulfill any of its plans the business will have moved into that future. The first overview specification is for an attempt to project the evolution of this future, but management is not limited to this projection. Drucker speaks of "creating the future," meaning that management has a very real power to design the future and to make it come true.[7]

The fundamentals of planning for information technology will be developed in greater detail in Chapter 4.

TYPES OF STRATEGY

Strategy expressions can be developed as stand-alone statements or as business or functional statements. The complete strategy statement for a function within the firm is the collection of stand-alone statements assembled with the functional statement. For instance, in the IT function, stand-alone strategies for improving office systems and for upgrading the mainframe computer form a part of the much larger strategy for the IT organization.

The statement of strategy for an IT function must contain the basic ingredients shown in Table 3.2. Although these elements are specifically applicable to the IT function, most of them apply to any function in the firm.

TABLE 3.2 Basic Elements of an IT Functional Strategy

Support for business goals

Technical support

Organizational considerations

Budget and financial matters

Personnel considerations

A statement of strategy for a firm includes an overall long-range strategy to achieve objectives established by the firm and more detailed strategies for each business and functional area. These detailed strategies are in support of the firm's strategy. Thus, functional strategies will be developed for sales, marketing, manufacturing, IT, and other functions. It is usual for the information technology function to have a functional strategy that is assembled with strategies from other areas for the firm.

Not all firms have adopted the strategic process described above. The IT organization must work within the norms of the larger organization and it must account for the corporate culture. If the senior managers of the firm are committed to strategic development, middle and lower level managers will see the process as having validity. They will be willing to invest their time in these activities. If the planning process adopted by the firm is forecast-based, then it will be difficult, perhaps futile, for the IT organization to shape the future in this firm.

Figure 3.2 illustrates the relationship of functional strategies to the business strategy for the firm. In a large firm there will be many more strategies than depicted in this diagram.

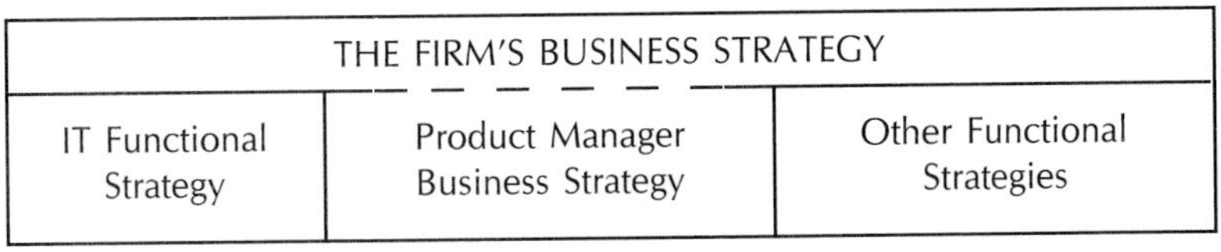

THE FIRM'S BUSINESS STRATEGY		
IT Functional Strategy	Product Manager Business Strategy	Other Functional Strategies

FIGURE 3.2 The Firm's Business Strategy

The purpose of functional strategies is to support the firm's business goals and objectives. This is accomplished through development of functional goals and objectives that are congruent with those of the firm. Effective strategies provide internal consistency between IT goals and the firm's goals.[8] The alignment of IT goals with the firm's goals, or goal congruence, is a critical success factor for the IT organization. The management processes developed in this chapter are aimed at achieving this alignment.

For example, product managers in a firm are responsible for producing revenue, and their business strategy is part of the business strategy for the firm. Each function in the firm develops a functional strategy. The collection of all the functional strategies and business strategies describes the complete strategy for the firm. This detailed document outlines the strategic goals and objectives for the firm and shows how all the units within the firm contribute to the attainment of these objectives. This is goal congruence.

In a large and complex business, the strategy is also complex and probably confidential. The strategy contains sensitive information relating to competition which must be carefully guarded in the firm. Strategy documents are produced and used by people in the top echelons of the business who are responsible for guiding and directing the firm in the long term. In some firms, a special department is in charge of the strategy development process. The personnel in this department assist the senior executives in utilizing the strategy. Strategy and plan congruence are critical success factors for executives in well-managed organizations.

Stand-Alone Strategies

Occasionally, it is desirable or mandatory for a functional organization to develop a specific strategy for dealing with a unique opportunity or threat. Generally, the subject of this kind of strategy is an embryonic question of key potential significance to the function which has not been previously considered in detail. In many cases, these specific or stand-alone strategies are developed outside the normal planning cycle in response to competitive or industry developments. In this sense, stand-alone strategies can be considered *ad hoc* actions to deal with currently emerging opportunities or threats. Once this strategy has been accepted by the firm, it will be incorporated into the strategic plans of all the organizational units that it affects.

An example of a situation that might give rise to a stand-alone strategy for the IT function would be the announcement of a new technology product by a vendor. This new product might enable the function to further some organizational objective in the short term. In this case, a strategy would be developed to capitalize on the new opportunity. If accepted by the firm, this would be followed by the appropriate planning.

Stand-alone strategies generally cease to exist as the subjects they deal with mature and become more specifically recognized in the strategic plans of the firm or its constituent parts.

Business and Functional Strategies

Business and functional strategies form the backbone of the strategy development process for the firm. Each type of strategy responds to different kinds of opportunities or objectives. Business strategies have revenue and profit objectives for the firm. Functional strategies have goals that support the firm's business strategies. Unless the IT organization is revenue producing, and some now are, its strategies will be only the functional type. That is, the IT function's strategy is in support of the firm's business objectives. If the IT function produces revenue for the firm, its strategy will also be part of the firm's business strategy.

For example, a business strategy for an IT organization that hopes to produce revenue could explore all aspects of exploiting an application in its portfolio. By making the program available to customers through employment of new telecommunications technology, revenue and profit can be realized for the firm. A business strategy details how the organization hopes to accomplish these goals.

A functional strategy, in contrast, could address the introduction of new robotic systems in support of the manufacturing plant's cost and quality objectives. This item would appear in the technical support section of the functional strategy shown in Table 3.2.

Business strategies for the firm coordinate the functions of a business to the business objectives. Functional strategies, on the other hand, coordinate activities within a function or between functions. As an example, a functional strategy within the IT organization may be designed to support and augment the firm's business strategy for reducing product development cycles and increasing revenue in the near term. This functional strategy may contain many other supporting elements as well.

The relationship of business strategies, functional strategies, and stand-alone strategies is shown in Figure 3.3. This illustration, emphasizing the IT organization, includes some revenue-producing activities. These are contained in the business portion of the strategy statement.

Well-managed IT organizations have an active functional strategy. Since the practical limitations on the allocation of functional resources must be allowed to influence business strategies, the process of developing and maintaining strategies and their associated plans inevitably becomes an iterative cycle bounded by resources of all kinds.

ASSEMBLAGE OF STRATEGIES	
IT Business Strategy	Other Functional or Business Strategies (May include stand-alone strategies for these functional units)
IT Functional Strategy	
Stand-alone Strategy A	
Stand-alone Strategy B	

FIGURE 3.3   Assemblage of Strategies in a Firm

REQUIREMENTS OF A STRATEGY STATEMENT

A statement of strategy is primarily a vehicle for focusing management attention on strategic aspects of the firm's business. It is also a means of communication to those who must review and approve the strategy and to those who use it as guidance for their actions. Additionally, the document must be available to those responsible for initiating adjustments to it. These adjustments take into account more current inputs from the environment or business. The document must also be available to those who will use it to measure and evaluate the performance of the firm or its functions.

A statement of goals and objectives alone is insufficient to meet these needs. Information must be added regarding the environment, the basis on which the goals and objectives were selected, the assumptions on which they depend, the risks present, and the options or flexibilities that are available and reasonable. The firm's long-range plans also need this kind of information, since both strategies and plans are snapshots of management's vision that direct the actions of the organization.[9]

STRATEGY DOCUMENT OUTLINE

In order to present the required information in a coherent manner, an outline containing the main points of a strategy is displayed in Table 3.3. The subheadings in the strategy do not need to be addressed separately in sequence, but the strategy must be presented in a manner that leaves no doubt about whether a statement is an assumption, a course of action, a risk, an alternative, and so on. The strategy must be a logical, coherent entity with the relevance of its parts clearly indicated.

TABLE 3.3　Strategy Outline

1. Nature of the business
2. The environment
3. Goals and objectives
4. Strategy ingredients

In the section on the nature of the business, questions such as "What is the business we are concerned with?" and "What are the boundaries of this business?" must be answered. In many instances the answers to these questions are obvious, and this section will require little development. However, the firm or organization must have a clear vision of its mission before the strategy-development process can proceed. To fail to understand the nature of the business in which the firm is engaged can be fatal.[10] Thus, the questions posed by this section are important and not trivial.

In this section, it is important to describe the IT business carefully. What purpose does the IT organization serve in the larger organization, and what type of organization is it? The roles of the IT organization can vary considerably. For instance, at Boeing, IT provides a capability to the main Boeing business but operates as a service bureau for outside customers and derives revenue. The IT group may provide major services for an insurance company and be the conduit through which all the firm's transactions flow. Ten percent of the firm's revenue may be spent on this vital function. A construction company may use data processing for design or project planning purposes and spend less than 1 percent of revenue on IT activities. The IT organization must understand what business it is in and what its role is within the firm.

The environment section states what is known and assumed about the relevant and significant factors and trends surrounding the firm. IT managers must understand those factors impacting or influencing the firm's current or future behavior. As a test of relevance, those factors should be included that have the ability to influence significantly the attainment of current goals and objectives. Another test is: "Can these factors force a change in our goals?" if later readings on the environment show changes in the analysis of those factors.

Key environmental assumptions should be reviewed at the conclusion of the environmental section, because their credibility and consistency with each other need to be understood without ambiguity. Furthermore, tracking these assumptions will enable all levels of management to adjust and update them as part of maintaining the strategy.

What is the IT environment like? What are the current capabilities and what can they be in the future? IT should know the state of the application portfolio and what new technologies can be applied to the firm's business. Is the IT organization relatively mature and disciplined or is it underdeveloped, with relatively inexperienced people and an immature management system? What is the environment today, and what can it be in the future?

The goals and objectives section must state, in concrete terms, what the ultimate objectives of the strategy are. What IT capabilities are we trying to achieve, and what long-term objectives are we going to set for the organization? For instance, is our goal to enhance the application portfolio with several new strategic systems, or are we trying to link our business units with a new and modern telecommunications system? Perhaps organizational issues such as decentralization need to be addressed as goals and objectives.

A methodology for deriving the IT goals and objectives from the firm's strategy is given by W. R. King. He begins the process by identifying the corporate strategy set and transforming the objectives into MIS objectives. For instance, if the corporate strategy calls for reductions in internal investments, then perhaps the IT organization could concentrate on inventory reductions through improved inventory management systems. Alternative means for reducing internal investments are developed and presented to management for decision making. King's methodology is complementary to that presented in this text.[11]

An important test of a statement of strategy is whether the objectives are attainable and desirable when viewed in context with the goals and objectives of the larger organization and when the environment is considered. The search for IT opportunities should be restricted to those which fall within the mission of the organization. Opportunities must serve the corporate purpose and must be aligned with corporate objectives. A second test is the degree to which the objectives can be used later to measure and evaluate unit performance.

Goal-setting benefits from some additional important considerations. Goals should be established within the resources of the firm and should have a reasonable chance of attainment. Managers must achieve a balance between easily attained goals and challenging goals with respect to resources of all kinds. Goals should be explicitly stated and should be quantified whenever possible. Subsequent planning, measurement, and control are facilitated by goal clarity. It is preferable to have a small rather than a large number of goals. A small number of goals reduces the chances of ambiguity and conflicts and permits managers to direct resources more effectively. Goals should have deadlines.[12]

The ingredients of an expression of strategy include a course of action and its accompanying auxiliary and supporting factors. A summary of these items is presented in Table 3.4.

TABLE 3.4   Strategy Ingredients

1. Course of action
2. Assumptions
3. Risks
4. Options
5. Dependencies
6. Resource requirements
7. Financial projections
8. Alternatives

The strategy statement must describe the course of action to be followed in attempting to achieve the objectives for the strategy. What steps will the organization take to achieve its goals and objectives? The steps should:

1. lead to realizing the objectives
2. be consistent with other long-range interests of the firm
3. be preferred over alternative possible strategies

For example, if a goal of the IT organization is to improve the capabilities of its people, will it accomplish this through hiring, training, retraining, termination, or possibly some combination of these actions? The course of action spells out the steps to attain these goals.

What are the major assumptions on which the strategy is based? Assumptions that are inherent in or that exercise significant influence over the strategy are technical capabilities, functional support activities, and potential competitive reactions. The test of the assumptions is their credibility. The maintenance of the strategy requires the tracking of these assumptions. Continuing with the example, one assumption might be that the current employees are trainable; that is, they possess the necessary background knowledge so that further training is possible.

Risk will always be present, and in fact should be one of the major parts of the IT functional strategy. Questions such as "What is the nature of the risks in the strategy?" and "What is their potential impact?" should be answered. In the example above, a risk of a strategy for hiring skilled employees might be that very few are available to the firm. In the event that some aspects of risk become significant, IT managers must review the optional courses of action to determine whether they offer reasonable insurance.

Since no single course of action has 100 percent probability of success, it may be appropriate to improve the probability in some cases by building options or alternatives into the strategy. These options should take into account specific risks, assumptions, or dependencies that unduly depress the probability of success. IT managers should look for options that are available within this strategy. They should ascertain how long the options are valid and upon what considerations a selection should be made. Also, they should consider whether any of the options add some cost or expense and if so, how much is added. In the personnel example above, options have been identified. They include hiring, training, retraining, and termination.

In most firms, it is likely that any single strategy will depend heavily on other related strategies. For instance, one technical strategy may depend on another technical strategy to produce a capability or a process for use by the former. "What are the key dependencies of this strategy?" "What is their nature?" and "How and in what ways are they significant to the strategy?" are some questions IT managers must answer.

The strategy must identify resources required to carry out the actions, and it must present financial projections of revenue, cost, expense, profit, and capital required to implement the strategy. What resources are required to carry out this strategy? Are any unique resources required? Will hardware, software, or people resources be available in the quantities and on the schedule required? These questions need answers in this part of the strategy statement.

Alternatives that were rejected in the selection of the strategy, and the reasons for rejecting them, should be retained for future reference.

The steps outlined above represent sound preparation and will lead to sound implementation if the planning process is successful. Action steps have been identified and details necessary for control can be developed.

The Strategic Time Horizon

The development of strategy statements within a firm usually follows a schedule dictated by the firm's corporate planning director, who is responsible for synchronizing strategy development events. This activity is usually an annual affair interspersed with planning, control, and measurement events.

A representative schedule of events depicting the strategy development process is shown in Figure 3.4. Measured from the beginning of the current year, the strategic years extend from the beginning of Year 3 to the end of the strategic horizon, as shown in part A of Figure 3.4. The strategic years may cover a period from 3 years to 10 years or more, depending on the business or industry. In this example, the strategic activity occurs during the early months of Year 1 and considers the period of the strategic years.

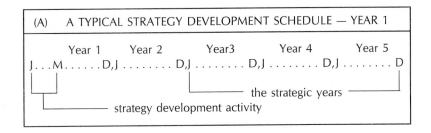

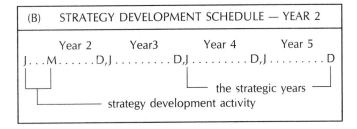

FIGURE 3.4   A Schedule for Strategy Development

The pattern is repeated annually (as shown in part B of the figure). Each succeeding strategic activity removes from consideration the first year of the preceding period and adds one year in the future. The period of Year 1 and Year 2 is covered in the tactical or operating plan for the firm. This activity is discussed in the next chapter.

The schedule suggested above is most convenient for firms using the calendar year as the fiscal year for planning and reporting purposes. Organizations such as universities would adapt this process to the academic year, starting in September and ending in August. Other organizations, such as governmental agencies, may adopt this schedule but adjust for a fiscal year starting in July and ending in June.

Steps in Strategy Development

Well-developed strategy statements take time. The strategist must take time to understand the environment and the area of opportunity or concern. The written statements should be concise and sharply focused. The statement of the opportunity or threat should highlight its relevance to the firm's future. If it is believed that new hardware is required for the data center, for example, the statement should clearly indicate why this belief is reasonable. What substantiating evidence is there, or what trend information supports this belief? What evidence supports this concern?

The IT strategist should develop, in reasonable depth, a broad understanding of the future environment that influences the area to be studied. This environmental analysis might include an estimate of future computing costs, anticipated future computing loads, technical advances expected in telecommunications or processing capability, special future business conditions, and other factors that are relevant. Objectives are set and modified, if necessary, during the iterative strategy development process. Stand-alone strategies may need to identify several possible objectives. The selection of one occurs after testing the options for credibility and attainability. Alternative strategies should span the future environment. They must be clearly expressed so that readers of the statement can gain a thorough understanding of them.

In order to exercise a reasonable choice among the alternative strategies, selection criteria and the process for using them must be established. The criteria should measure the basic, long-range effects on the firm or on the IT function. These effects may be measured in terms of profit and revenue, investment resources required, degree of risk, technological capability exploited, competitive reactions, and other factors important in each individual case.

The best strategy should be selected by using the criteria to measure the effect that could be expected from each option. The selected strategy should then be developed in greater detail. Stand-alone strategies and functional strategies and their supporting data, reasoning, and other forms of evidence must be submitted to the appropriate executives for review and approval. The review and approval process enables and ensures congruence in planning. The process forces a review of the alignment of IT and corporate goals and aids in keeping executives of all functions informed on the important strategic directions. The issue of strategy integration across functions will arise naturally, and executive or organizational learning will be fostered by the process.

Upon approval by senior executives, these strategy statements will be incorporated into the strategic and tactical plans of the firm. The process of incorporation may require minor changes because this process may be iterative as well.

To be successful, the process of strategy development expressed above requires considerable effort and thought on the part of the firm's executives. This thoughtful activity engages the best minds in the organization in focusing on the long-term health and welfare of the firm. Each function needs to be involved and the IT organization needs to be a full partner in all deliberations. The value of the resulting output is in direct proportion to the level of effort and the degree of cooperation among the senior executives. Strategy development processes and activities are vital to the firm that intends to remove IT strategic planning from its list of issues.

STRATEGY AS A GUIDE TO ACTION

An IT strategy is developed with the explicit intent of guiding management in a course of action and appraising management's accomplishments in light of this guidance and direction. Various levels of management will take direction from strategies in planning and carrying out their activities. A stand-alone strategy may require a significant amount of effort in developing and carrying out new projects, as well as reshaping old ones. Reference to the strategy can serve as a guide in keeping the overall functional effort well directed and balanced, while taking advantage of breaks and unusual opportunities as they appear.

IT managers must exert strategic control, and the strategy gives them the basis for doing this. Key dependencies and functional support activities, which have been identified in the strategies, must be tracked continuously. Significant departures from expectations should be analyzed to determine whether these variances are reason to review the strategy itself. In particular, unfavorable changes in dependencies should be scrutinized carefully. This strategic control activity preserves the vitality and the viability of the strategy between planning periods. It also helps the organization maintain its strategic focus.

Periodically, the accomplishment of management in working toward the objectives of the strategy can be reviewed and appraised. The achievability of the objectives and the degree of stretch they required must be kept in mind. A moderate degree of accomplishment against a very difficult objective must be appraised differently from a high level of accomplishment against an objective easily attained.

THE STRATEGY MAINTENANCE PROCESS

The maintenance of a strategy is an essential step in its continued usefulness. Maintenance recognizes that a strategy is constantly subject to change. Frequently, change accounts for influences and factors beyond the firm's control. A complete statement of strategy requires careful documentation of the areas where change may take place. This documentation facilitates tracking actual developments against the assumptions, dependencies, and risks in the strategy. The test here is whether the strategy has been documented in a way that clearly identifies the factors that need to be tracked.

Maintenance of the strategy is neither its implementation nor its use. Tracking will identify deviations from the strategy, which can be helpful in maintaining it as well as implementing it. It is important to realize that strategy maintenance is a vital and essential part of the total strategy process. Tracking each of the key environmental and strategic assumptions, risks, and dependencies must be a part of the strategy process. The responsibility for tracking each should be specifically assigned to someone in the organization. A diligent pattern of follow-up must be established.

Periodically, or when good judgment recognizes a significant deviation as a result of tracking, the entire strategy should be carefully reexamined and updated. The logical process for doing this is to follow the steps for developing a strategy, as previously described.

THE IT STRATEGY STATEMENT

The IT organization that follows the management process for strategy development indicated above will have met certain conditions necessary for success. The process establishes goal congruence, lays the foundation for planning, and permits the role and contribution of the IT organization to be carefully exposed. The strategy process is an important step in organizational learning, too.

The IT organization must develop and maintain a functional strategy that guides the actions of the IT function. These actions must support the goals and objectives of the firm and its constituent organizations. The IT function may also have stand-alone strategies. These strategies establish direction for projects designed to enhance the function's objectives. The IT manager is responsible for developing and maintaining these strategies and for ensuring that they are synchronized with the firm's overall strategy process.

Strategic planning is one of the top issues confronting IT managers in most firms, and it is a critical success factor for them. Success in this activity paves the way for success in many other areas, because sound and valid planning forms the base for most other management activities. Failure in this area is a prelude to failure elsewhere as well. Given the great importance of this subject to the IT management team, what issues should be considered, and how should they be explored? Table 3.5 outlines some of the important issues for IT strategies. The IT strategy statement should address these issues at a minimum; there may be other issues specific to the firm as well.

TABLE 3.5   IT Strategy Issues

Business aspects

Technical issues

Organizational concerns

Financial matters

Personnel considerations

The Business Aspects

The IT organization must maintain a keen awareness of the business goals and objectives of the firm and must develop strategies that support them. These business goals may include increased market share, improved customer service, lower production costs, or many other objectives of central importance to the corporation. As the Business Vignette indicated, CEOs expect IT to contribute to the firm's results and they expect this to happen in conformance with normal business practices. The strategy of the IT organization must be fully cognizant of and supportive of the firm's business goals.

Not-for-profit organizations, government agencies, educational institutions, and the like have goals that are perhaps different from those noted above. In all cases and in all forms of enterprise, IT must support the parent organization's goals and objectives.

IT managers, interacting with the other senior executives, are key players in the development of the functional strategy. To the extent that the IT organization is involved in attaining the firm's objectives, the IT managers must include this involvement in the function's strategy statement. They must ensure that the function's actions are in concert with the long-term goals and objectives of the firm. This involvement must be tested through an interactive review between the IT managers and their superiors. Are the strategies congruent? Are the firm's dependencies on the IT organization satisfied? If the strategies are executed as written, will the objectives be met? The answers to these and other questions will establish the validity of the strategy process and its results.

Technical Issues

The IT manager is responsible for providing leadership in attaining advantage for the firm through the use of technology. The IT strategy is one place for this initiative to occur. The strategy should reveal the practical utilization of advanced technology in support of the firm's goals and objectives. This utilization must be consistent with reasonable risks and available or attainable resources. Additionally, the IT manager must ensure the technical vitality of the organization through development and implementation of current or advanced hardware, software, or telecommunication technology.

The IT functional strategy is the vehicle for expressing the path toward improved technical health. It gives the organization the formal opportunity to establish objectives designed to maintain and enhance the technical health of the function and the firm. CEOs expect leadership from IT in discovering important technological applications. They expect IT to make efficient use of the firm's resources and to work within the norms of the organization.

Organizational Concerns

There are several reasons why organization considerations are of high importance to the IT strategy. First, the introduction of information technology tends to have organizational consequences beyond the IT organization itself. These issues are difficult to envision and, in many cases, they are hard to resolve. Not all changes are looked upon favorably by everyone. And many changes, while important to senior executives, are resisted by nearly everyone else for a variety of reasons. The firm needs training and education in the subtleties associated with technology introduction. The IT manager needs to take the lead in satisfying these training and educational needs.

Second, the role of the IT organization and its contribution to the firm must not be taken for granted in the development of the IT strategy. Not everyone appreciates the IT role, and many managers in the firm do not fully understand the contribution, or potential contribution, of the IT function. Wise IT managers will take specific actions to remedy these deficiencies. The IT functional strategy will reflect these actions. Much more will be said about this in subsequent chapters.

Financial Matters

When the IT strategy is translated into strategic and tactical plans, it must satisfy the financial ground rules of the firm. In many cases, financial constraints bound the range of opportunities for the IT organization as they do for most other functions. These resource constraints force iteration in the process of developing the business strategy for the firm. These constraints also cause successive revisions in the functional strategy for the IT organization. Wise IT managers provide guidance to their organizations on these matters in order to conserve energy and optimize results during the strategy development process.

Personnel Considerations

No functional strategy is complete without action plans that relate to the management task of recruiting, training, and retaining a base of skilled people. These personnel considerations are intimately related to technical issues, because strong technical people develop solid technical strategies and advanced technical strategies attract strong people. The IT functional strategy must develop these thrusts in coordination with one another.

IT managers must demonstrate productivity improvements in the IT organization. They have the tools to assist the entire organization in improving productivity. In many ways, information technology is the productivity engine for the firm. Skilled people and advanced technical resources provide some necessary conditions for making major productivity improvements.

SUMMARY

This chapter described the activities of strategy development from the perspective of information technology managers and their organizations. The responsibilities of IT managers were described in detail, and the processes were presented as they applied to the IT organization. These responsibilities include ensuring active participation of line managers throughout the firm. The IT strategy represents a collaborative effort. IT managers and line managers in all functions of the firm must share equally in these responsibilities, because the IT strategy is designed to further the business interests of the entire organization.

If an organization is to achieve alignment of IT and corporate goals, the managers must believe that the IT strategy is a strategy for the entire firm. The goal alignment issue is not present in firms where this belief is shared by members of the senior management team. It follows that senior managers throughout the firm have considerable joint responsibility for a sound IT strategy.

Sound IT strategies are of several types. Stand-alone strategies describe a course of action that optimizes the organization on one objective. These strategies disappear when the objective is accomplished. Business strategies provide guidance at the top of the firm for the major goals and objectives essential for the firm's long-term success. They are broad, fundamental, and comprehensive in nature. They require support from subordinate organizations within the firm. This support is described in the functional strategies.

The development of strategy statements is a fundamental responsibility of management, and success in this endeavor requires significant amounts of time and energy. Coordination between the IT organization and the other functions is essential. Because of the pervasive nature of information technology, the strategy process is more difficult and more important for the IT organization than for its peer organizations. In many instances, the IT strategy unifies and integrates other functional thrusts. This makes the IT manager's role in the strategy process of the firm especially crucial. Senior managers in all functions touched by information technology share responsibility for the validity of the IT strategy. They must be active participants in its development.

> Strategic management is a continuous, iterative process aimed at keeping an organization as a whole, appropriately matched to its environment. The process itself involves performing an environmental analysis, establishing organizational direction, formulating organizational strategy, implementing that strategy, and exerting strategic control. In addition, international operations and social responsibility may profoundly affect the organizational strategic management process. It is important that the major business functions within an organization—operations, finance, and marketing—be integrated with the strategic management process.[13]

Review Questions

1. What are the major ingredients of a complete statement of strategy?

2. Besides providing direction to the organization, what other purposes are served by a strategy?

3. How do stand-alone strategies differ from other types of strategies?

4. Under what conditions can an IT organization produce both business and functional strategies?

5. For what purposes is a functional strategy developed, and how can it be used?

6. Toward what issues is the firm's business strategy pointed? What are the elements embodied in a firm's business strategy?

7. What is the relationship of functional strategies to the business strategy of the firm?

8. What is the relationship between strategies and plans?

9. Describe the relationship of a strategic plan to business strategies, functional strategies, and stand-alone strategies.

10. Where do tactical plans fit into this relationship?

11. What are the main elements of a strategy document?

12. What role does the environment statement play in the expression of strategy?

13. What are the connections between assumptions and risks? Why is it useful to have these items separated in the strategy statement?

14. The iterative process described in this chapter is useful in resolving dependencies, among other things. Why is this so?

15. What main issues should be addressed in the IT functional strategy?

16. The strategy development process, as described, permits executives to focus on goal incongruences. When and how does this happen?

17. Why do senior managers across the firm's various functions share responsibility for the IT strategy?

18. What items should be included in the IT organization's functional strategy? What other items might well be included, and why?

Discussion Questions

1. Give some examples of IT dependencies likely to be found in the strategy that cannot be resolved through the iterative process. How can the effects of dependencies external to the firm be mitigated?

2. Compare and contrast the actions of strategy maintenance and performance appraisal.

3. Managing expectations is a key IT issue. Discuss how the strategy development process assists in this task.

4. Identify the critical success factors facing the IT manager that can be addressed through the strategy development process outlined in this chapter. Establish your own opinion regarding the vital nature of strategic development as a consequence.

5. Discuss how the strategy development process can reduce the information infrastructure stumbling block referred to in the Business Vignette.

6. An IT organization wants to develop a telecommunications capability to prepare for new services now being considered by the firm. What types of strategies will be useful in this connection, and how will they be used?

7. What are the consequences, in terms of strategy development, of the pervasive nature of information technology?

8. Describe the main strategic thrusts of the IT functional strategy for Baxter Health Care as revealed in Chapter 2. If you were to speculate on other thrusts, what might they be?

9. What are the characteristics of a high-quality IT strategy statement? Discuss these characteristics in terms of the process and the result.

10. What are the necessary conditions that must exist in the firm for the IT organization to achieve success in strategy development?

Assignments

1. Assume you are the manager of your university's information system organization, responsible for student records and the automated registration system. Develop an outline of a stand-alone strategy to incorporate the use of personal computers scattered across the campus into the system.

2. You are considering the purchase of a new, sophisticated personal computer. The computer is for use in your school activities and will be used in your job after graduation. Develop the elements of a strategy to acquire this computer.

ENDNOTES

[1] *MIS Week*, January 4, 1988, 1.

[2] John Rymer, "Texaco's Hodges: A Breed Apart," *Computer Decisions*, March 1989, 36.

[3] Henry Mentzberg, "The Strategy Concept II: Another Look at Why Organizations Need Strategies," *California Management Review* (Fall 1987): 25. According to Mentzberg, a good strategy also helps define the organization and gives it meaning and purpose.

[4] Daniel H. Gray, "Uses and Misuses of Strategic Planning," *Harvard Business Review* (January-February 1986): 89.

[5] John C. Henderson and John G. Sifonis, "The Value of Strategic IS Planning: Understanding, Consistency, Validity, and IS Markets," *MIS Quarterly* (June 1988): 187.

[6] Frederick W. Gluck, Stephen P. Kaufman, and A. Steven Walleck, "Strategic Management for Competitive Advantage," *Harvard Business Review* (July-August 1980): 154. Strategic management is largely concerned with creating competitive advantage and shaping the future, claim the authors.

[7] George C. Sawyer, *Designing Strategy* (New York: John Wiley & Sons, 1986), 169.

[8] Henderson and Sifonis, "Value of Strategic IS Planning," 187.

[9] For a slightly different approach to planning see B. Bowman, Gordon Davis, and James Wetherbe, "Modeling for MIS," *Datamation*, July 1981, 155. The three-stage model for MIS planning consists of strategic planning, organizational information requirements analysis, and resource allocation planning. The last two steps in this model are implemented in an annual planning cycle.

[10] Theodore Levitt, "Marketing Myopia," *Harvard Business Review* (September-October 1975): 26. This classic article describes the causes and results of failures. Using the marketing viewpoint, one can observe that the growth of end-user computing and the decentralization of information processing in some firms is partly the result of the centralized IT organization's product orientation, rather than customer orientation.

[11] W. R. King, "Strategic Planning for Management Information Systems," *MIS Quarterly* (March 1978): 22.

[12] Richard T. Hise and Stephen W. McDaniel, *Cases in Marketing Strategy* (Columbus, OH: Charles E. Merrill Publishing Co., 1984), 130.

[13] Samuel C. Certo and J. Paul Peter, *Strategic Management* (New York: Random House, 1988), 23.

Glueck, W. *Business Policy, Strategy Formulation and Management Action.* New York: McGraw-Hill, 1976.

Lucas, Henry C., Jr., and Jon A. Turner. "A Corporate Strategy for the Control of Information Processing." *Sloan Management Review* (Spring 1982): 25.

Marrus, Stephanie K. *Building the Strategic Plan.* New York: John Wiley & Sons, 1984.

Miles, R. E., and C. Snow. *Organizational Strategy: Structure and Process.* New York: McGraw-Hill, 1979.

Millar, Howard W. "Developing Information Technology Strategies." *Journal of Systems Management* (September 1988): 28.

Ohmar, Kenechi. "Getting Back to Strategy." *Harvard Business Review* (November-December 1988): 149.

Pearson, Andrall E. "Tough-Minded Ways to Get Innovation." *Harvard Business Review* (May-June 1988): 99.

4

Information Technology Planning

A Business Vignette

Introduction

The Planning Horizon

Strategic Planning

Tactical Plans

Operational Plans and Controls
  Planning Schedules

A Planning Model for IT Management
  Applications Considerations
  Production Operations
  Resource Planning
  People Plans
  Financial Plans
  Administrative Actions
  Technology Planning

Other Approaches to Planning
  Stages of Growth
  Critical Success Factors
  Business Systems Planning

The Integrated Approach

Management Feedback Mechanisms

Summary

Review Questions

Discussion Questions

Assignments

References and Readings

McGraw-Hill Taps One Unified System For Many Markets[1]

One hundred years after founder James H. McGraw bought his first magazine, McGraw-Hill, Inc. is leveraging its investments in information systems to capture even more business in publishing and communications markets.[2]

But the company recently decided that diversity, while good for business, is bad for systems strategy. Richard H. Shriver, head of McGraw-Hill information systems, was elected senior vice president of technology and asked to take command of a more centralized organization. "It was extremely hard to provide a corporate view of technology or corporate overview and coordination in the previous organization," Shriver says. "There were too many companies, too much autonomy; today there's commitment to work on a centralized basis."

The systems reorganization is part of a corporate restructuring that combined a half-dozen market-focused companies into three main ones. "All three companies have great opportunities and challenges in the marketplace, all totally different. My personal objective," says Shriver, from McGraw-Hill's midtown-Manhattan headquarters, "is to leverage the company's technology skills across organizational boundaries."

The companies are: McGraw-Hill Publishing Co. (book publishing, *Business Week* and TV stations); McGraw-Hill Financial Services Co. (Standard & Poor's Corp. and Data Resources); and McGraw-Hill Information Services (which includes construction, computer and communications, legal, aerospace and defense, health care, and energy and process industries groups).

One of the most powerful ways McGraw-Hill's information systems capabilities will be used is to provide a pipeline through which all the company's real-time and on-line products can flow directly to internal corporate networks. "We are closer to accomplishing that today than we have ever been before," Shriver says. "The consolidation will allow us to focus on the remaining problems." By providing integrated systems between divisions, the systems group will be able to meet customers' requirements for one access point for the company's services.

McGraw-Hill is not the only information services company that sees this market opportunity, but it hopes to get there first. "We're working on that opportunity in ways I can't talk about," Shriver says. "But they will be reflected in the projects we will be producing in the future." The reorganization will also allow Shriver to assess the technology resources of the company, providing insight into better ways to leverage the skills at hand.

Some autonomy remains, however. Product development groups still make their own decisions, but capital investments over a certain amount require Shriver's sign-off. Shriver also started and will chair regular technology review meetings with the technology chiefs of the three companies.

Among the company's accomplishments to date, Shriver rates Standard & Poor's Ticker III, the real-time financial data service, as the most successful. This tribute is particularly due to the Ticker's performance on Oct. 19, 1987, when 604 million shares were traded. "We think we were the only pricing service that made it through without serious mishap," Shriver says. "We're in the business of supplying stocks and options to traders; a second can mean the difference between profit and loss."

Standard & Poor's and McGraw-Hill use networks to keep Ticker and Marketscope, the real-time financial information service for brokers, up and running. There is redundant communication to the data center, much of it based on microwave.

Apart from integrating systems, McGraw-Hill is also trying to deliver data, no matter what equipment the customer uses. "The customer doesn't want 11 terminals on his desk to get prices," Shriver says.

Standard & Poor's financial products are not the only ones being put through the technology Cuisinart by McGraw-Hill product developers. A bond information service, bits and pieces of which were variously available—such as the Standard & Poor's Blue List, traditionally printed on blue paper—is now available electronically.

McGraw-Hill makes one of the largest investments in information systems for a company its size, with a systems budget that is about 7% of annual revenue.[3]

INTRODUCTION

The preceding chapters concentrated on strategy development and the importance of strategic thinking for IT managers and senior executives in their firms. A model for developing several types of strategies useful to the IT organization, to the firm, and to its constituent functions was presented. These strategies and strategic statements express the preferred long-term direction for the organization. Based on a thoughtful analysis of the environment, they account for risks, dependencies, resources, and technical requirements and capabilities. These strategic statements are valuable for the firm and its management because they lay the foundation for much succeeding activity.[4] The strategies serve as communication and coordination vehicles for executives and the firm's managers.

Strategic statements set the broad course of action for the firm and its functions. The strategy outlines the goals and objectives the organization intends to achieve and provides a general statement of how the firm intends to achieve them. It does not provide details. Strategy statements establish the destination and contain a general direction for reaching it. The activity of developing the detailed roadmap is planning. The specific details for reaching the goals and accomplishing the objectives are found in various plans. Formal strategy and planning documents are the vehicles for recording and communicating the firm's intended direction.

This chapter describes the elements of planning processes for the IT organization. Planning begins with developing strategic plans from strategy statements and continues through tactical or intermediate-range planning to the development of operational plans and controls. This process forms a vital part of the management system for the successful IT organization. Thus successful IT planning is a combination of four related phases: (a) agreement on the future: business vision and technology opportunity; (b) development of the business ideas for information technology application; (c) business and information technology planning for applications and architectures; and (d) successful execution of the business and IT plans.[5]

There have been several approaches to IT planning in the literature, and there is a large body of knowledge related to planning in general. This chapter explores some of the more prominent planning methods and relates them to each other and to the process. It also identifies some organizational relationships useful in IT planning and explains tools helpful to IT planners.

An organized framework and methodology upon which the IT planning activity can take place is the end product of this chapter. The methodology addresses the need for an effective planning and control mechanism to account for IT functional activities from the operational through the strategic time frame.

Planning is a critical success factor for IT managers and their organizations. It must be accomplished in an outstanding manner.

THE PLANNING HORIZON

The types of plans discussed in this chapter can be classified as strategic plans, tactical plans, or operational plans. They are differentiated primarily by the time over which they are valid. The planning horizon is the length of time considered in the plan. A typical short-range planning horizon is 30 days; the long-range planning horizon for most firms is on the order of years.

Strategic plans usually cover the period from two years to five years or more into the future. They represent the long-term implementation of strategic thinking. Strategic plans contain specific details for achieving the organization's mission and for meeting its long-range goals and objectives. Strategic plans form the basis for short-range plans. Of course, all planning activity must lead to support for the strategic direction of the organization.

Tactical plans commonly address the period from three months or so to two years in the future. Tactical plans describe actions that must be taken to begin the implementation of strategic plans. Operational plans consider the near term: from now to three months from now. Tactical plans bridge the period from the operational horizon to the strategic horizon.

Just as managers need stand-alone strategies to cover special business or functional situations, they also need plans to accompany these strategies. In some cases, it may be necessary for these plans to remain separate from the three types of plans discussed above. Usually, however, plans associated with stand-alone strategies are folded into the plan or plans covering the period of the stand-alone strategy. At some high level in the firm, however, the resource portions of these separate plans are merged.

The extended period over which the operational, tactical, and strategic plans are effective is called the extended planning horizon. A diagrammatic depiction of a five-year extended planning horizon is presented in Figure 4.1. This extended planning horizon embraces five complete calendar years. It shows the time relationship among the plan types.

Operational planning is broad-based and contains a great amount of detail, because it relates all the important items happening in the firm during the very near term. Operational plans need to be regenerated frequently in response to changing business conditions in the near term. They need to bridge the gap from the present to the tactical time period. These plans contain significant amounts of detailed information. They are assimilated and implemented at fairly low levels in the organization. They give structure to the everyday operation of the firm for first-line managers and their teams. They are the basis for taking short-term actions, and they are the reference points against which short-term results are measured. Short-term variances, requiring correction, result from the comparisons between plans and actual performance results.

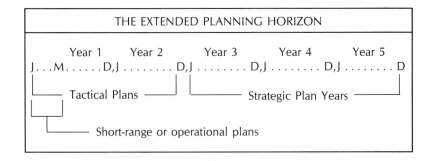

FIGURE 4.1   **The Extended Planning Horizon**

Tactical or intermediate-term plans are less detailed and are effective during a longer period than that of operational plans. Typically, these plans cover the current year plus the next year and provide overriding direction to operational planning. In the ideal situation, the implementation of succeeding operational plans, providing the implementation was accomplished without generating significant variances, will lead to the successful implementation of the tactical plan. Since activities seldom go as expected, the system will need to contend with variances in results from both operational and tactical plans.

Strategic plans extend from the end of the tactical period to the end of the extended planning horizon. Just as tactical plans give guidance to operational plans, strategic plans provide direction to tactical plans. The effective execution of tactical plans paves the way for accomplishing strategic objectives. Thus, implementation of the planning process is a succession of overlapping plans moving through time.

The three time frames must form a continuum over the extended planning horizon, and the three types of plans must form a unified picture of the future actions of the firm as well. The plans cover different time periods and contain differing amounts of detail. They portray a coordinated description of how the firm intends to carry out its strategies from now into the foreseeable future. The strategic plan supports the firm's strategy; the tactical plan provides shorter range support to the strategic plan. The operational plans give day-to-day or near-term meaning to the tactical plan. In the absence of any unanticipated changes in the environment during this period of time, and if the firm developed and carried out its plans flawlessly, then it would attain its strategic objectives as stated. Since all of these conditions are never true, the challenge of developing strategies and developing valid plans from them remains high.

Strategic planning is the activity of converting long-range strategies into long-range plans. The process takes the information found in the strategy statement and adds detailed actions and elements of various resources required to attain the stated goals. These resources consist of people, money, facilities, and technical capabilities, blended together and working toward the objectives. The strategy contains statements about assumptions, risks, and dependencies. Assumptions in the strategy must be converted to reality, and dependencies need to be accounted for in the plans. The risks noted in the strategy must be mitigated to the fullest extent possible by specific actions designed to contain and offset them. The process of planning must generate actions to accomplish the goals and objectives.

The strategic plan contains a detailed schedule of activities utilizing resources in the attainment of strategic goals and objectives. The strategic plan for a firm is also a financial statement of projected revenues, expenses, costs, investments, and profits. The strategic IT plan also includes financial information, but this is usually limited to costs and expenses. The plan is detailed enough so that it can be tracked over its lifetime and a comparison of plan to actual performance can be presented. The tracking takes place, however, in the near term when the strategic plan has been translated into tactical and operational plans. The strategic plan translates strategic goals and objectives into strategic actions two to five years or more in the future. It is the foundation on which the tactical plan is built.

Consider an IT group whose long-term goal is to improve market response times by deep automation of product design, development, and manufacturing. The strategy includes installing CAD/CAM systems for design and development engineers and for the plant. The systems are to be effective in 48 months. The strategy statement contains the assumptions for using CAD/CAM and outlines dependencies and risks. Resistance to new systems might be a risk. The strategic plan adds details of schedule, actions, and resources: space requirements for the equipment; training for the users; capital resources; and operating expenses. The plan states that space will be leased and that any resistance to the new systems will be overcome with training. The tactical plan will add more detail, particularly for the near term.

TACTICAL PLANS

Tactical plans generally cover the current year in detail and the following year in less detail. Management makes a commitment to implement tactical plans and they are used to measure management performance;

thus these plans are sometimes called measurement plans. Tactical plans form the basis for assessing the firm's performance as well. They are used by senior executives to measure unit performance. They also provide feedback on the degree to which the foundation for the attainment of long-range goals is being laid. Tactical plans contain more detail and have a more direct bearing on near-term activities than strategic plans. They provide guidance to very short-range activities and link the near-term actions to long-range activities.

To continue with the CAD/CAM example, many activities will be planned for the first two years. Some of these are: selecting hardware and software; installating hardware and software systems; training users and IT people; converting from previous processes to new procedures; and setting up measurement systems. Questions, such as where and from whom the space will be leased and what type of training is required to overcome resistance to new technology must be answered. These many activities must be scheduled, human and material resources must be assigned, and detailed budgets must be developed for the plans to be complete. Management will agree to the plan and will be measured on the degree of plan attainment. The tactical plan is the basis for day-to-day activity.

OPERATIONAL PLANS AND CONTROLS

Operational plans add detail to the near-term activities of the firm, giving direction on a day-to-day or week-to-week basis. They are used by first-line managers or by nonmanagerial employees as they carry out their assignments. Operational plans usually require much analysis, in addition to that required for tactical plans, and they must contain control elements as well.[6] They show managers how well daily or weekly activities are proceeding.

The CAD/CAM installation will contain near-term tasks. These include: identification of hardware and software vendors; setting vendor criteria; developing space alternatives; and performing systems analysis of current procedures. At the same time, planning for the next short-term goals will begin, as will budget and cost-tracking tasks.

Planning Schedules

In most firms, planning activity is regularly scheduled and is related to the firm's fiscal calendar. Usually the fiscal year and the calendar year are identical, and planning activity is seasonal. It is common and most practical to develop tactical plans so they can be approved just prior to the beginning of the tactical period. For most firms operating on the calendar year, this means that the tactical plan for the next two years is developed and approved in the few months prior to the beginning of the new year. The new calendar year begins with an implementation of the new tactical plan for that year and the following year.

In a model planning calendar, the development of strategies and strategic plans takes place shortly after the beginning of the new year and is completed and approved around mid-year. When completed, it covers the period from 30 months in the future to the end of the extended planning horizon, perhaps five or more years hence. Figure 4.2 illustrates the second iteration of planning in relation to current plans. The new tactical plan incorporates the first year of the previous strategic plan, and the new strategic plan adds an additional future year.

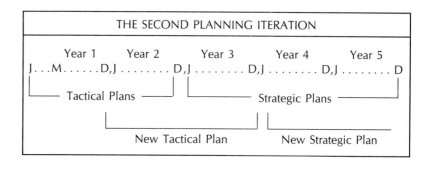

FIGURE 4.2   The Second Planning Iteration

For firms with a well-established planning cycle, planning does not start from the beginning with each iteration. As illustrated in Figure 4.2, the new tactical plan revises the second year of the previous plan and adds to it the first year of the previous strategic plan. The new strategic plan is a revision of the last years of the previous plan and adds a new year. The revisions reflect changes due to new or revised strategies, changing business conditions, or changes in the environment. Failure to achieve prior goals or objectives, or alterations related to risk management or dependency management may also lead to revisions.

The entire planning process is cyclical and repetitive, with revised longer range plans becoming more near-term with each passing cycle. Operational plans are usually not as tightly scheduled, and if the implementation of the tactical plan is proceeding smoothly, these plans may very well be of an *ad hoc* nature. The need for short-range plans depends on the circumstances and on the activities of the function and the firm.

The CAD/CAM example discussed earlier could proceed in a similar manner. Details of hardware, software, space, training, and many other items will be developed over time as succeeding plan iterations and plan implementations occur. If all goes well, the systems will be operational as planned and the firm will be reaping the benefits of shortened development and manufacturing cycles and increased responsiveness to changing market requirements. The firm will have achieved its strategic goal.

A PLANNING MODEL FOR IT MANAGEMENT

The discussion so far has related mostly to common planning elements. It is applicable to most functions in most firms. The planning horizon will vary between firms. The planning schedule could be altered considerably, depending on the industry and the desires of the firm. For example, electric utility companies usually have a long planning horizon, on the order of ten years or more, and firms in the forest products industry in the Pacific Northwest plan for 50 or more years in the future. Firms in highly volatile or highly competitive industries may benefit most from shorter range thinking. Firms on the edge of bankruptcy are interested only in survival. A year may be very long-range for them.

The information technology function within the firm must adapt to internal and external forces and constraints in searching for a management system governing its planning. The firm's planning system establishes many of the frameworks to which the IT organization must conform. The firm's planning department establishes the planning calendar and planning horizon, schedules the review procedures, and coordinates the approval processes. Usually plan formats and content outlines are also prescribed by the planning department.[7]

The IT organization must have a view of technology futures. Technology trend information is important because it permits IT to present credible technical detail in the extended planning horizon. Subsequent chapters will present methods through which IT can secure trend information. Some strategic plan elements can be specified in detail, while other items represent the manager's best judgment. Long-range planning, in particular, is a delicate blending of solid facts, intuition, and seasoned judgment. However, facts, if available, are always preferred in the planning process.

Information technology managers, in particular CIOs, have responsibilities extending far beyond the boundaries of their line organizations. Planning activity is influenced by organizational factors, and the CIO position is advantageous in this regard. One factor that aids in planning is the perceived importance and status of the IT manager. Other factors are the physical proximity of the IT group to senior managers; corporate culture and management style; and organizational size and complexity.[8] Their strategies and plans must reflect the actions of their own organizations but must also include the information processing activities of the entire firm. Their view must encompass both their staff and their line responsibilities. Their vision must include both technological and organizational factors. It must provide technical direction or at least generate practical technical alternatives. And it must consider organizational maturity and sophistication. The intelligent IT executive will not attempt to transform the firm from the contagion stage to maturity in one planning cycle.

Senior IT executives must include the entire organization's environment in their planning activity. Organizations undergo change, frequently because of information technology. IT planning activity must anticipate and respond to this change. McGraw-Hill combined market-focused companies into three main entities to utilize technology skills across those organizations. McGraw-Hill performed extensive planning to reorganize on a centralized basis.

Since there are constraints, ambiguities, facts, and uncertainties, what planning procedures might best serve the information technology function? What kinds of activities need to be planned? What tools and information will assist in the task of IT planning? A complete IT plan answers those questions; it is a detailed and thoughtful document describing the actions that will implement strategy over the planning period. It consists of the ingredients outlined in Table 4.1.

TABLE 4.1   IT Management Planning Model

Applications considerations

Production operations

Resource plans

Administrative actions

Technology planning

Applications Considerations

The applications portfolio is the complete set of applications programs through which the firm conducts its automated business functions. As we will learn later, the management of the applications assets is a very difficult task and the planning of this activity must be conducted skillfully. The plan for dealing with the application portfolio must include the selection of projects to be implemented during the plan period and the scheduling, control, and evaluation of these projects during implementation. The resources required for development, enhancement, maintenance, and implementation are also important.

Project selection describes which programs will receive maintenance or enhancement activity and what will be accomplished with this activity. Managers must know the schedule of the activity and the amount of resources it will require. Applications that are not currently in the portfolio will be defined and a plan for acquiring these programs is also required.[9]

The identification of strategic systems to augment the portfolio is difficult. Programs may be acquired through purchase or they may be developed locally or by subcontractors. In all cases, the plan describes what will happen, when it will happen, how it will take place, and how much it will cost. The plan must justify the action, must describe why the action is being taken, and must identify who is responsible for taking the action.

Managers must know the answers to "who, what, when, where, why, and how" for application planning activity. If they have considered these fundamental items in their plans for the portfolio, they will have the basis for controlling implementation of the plan and for monitoring deviations from it. Effective plans must contain control items for completeness.

There are some major difficulties in obtaining planning direction for the applications portfolio.[10] These difficulties arise from competition for resources among work items. The resolution of these issues is a fundamental IT management task. A management system designed to resolve these difficulties is presented in Chapter 7. The results of the process described there will comprise the elements for the applications portfolio plan.

Production Operations

Production operations consists of processes for running the firm's applications. Data is collected or developed; the applications are executed using current and perhaps historical data; and the results are stored and/or distributed as planned. The processes may be continuous on-line systems; periodic, scheduled systems; or, more likely, a combination of both. In many cases, the production operation is highly dependent on advanced telecommunications systems. The strategic systems we studied in Chapter 2 are of this nature. In either case, the results generated by the application are used by many members of the firm as they carry out their responsibilities. The results may also be distributed to customers or suppliers. In most cases, the firm is dependent on these outputs for successful and efficient operation.

The IT management system must plan production operations to achieve satisfaction for all users. (Users inside and outside the firm have expectations of the service provided by the IT organization, and IT must have a plan to satisfy customer expectations.) In production operations, the planning of service levels is very important.[11] The essential elements or disciplines for meeting service expectations consist of problem management, change management, recovery management, capacity planning, and network planning and management. Problem, change, and recovery management are elements for ensuring the attainment of service levels through operational management processes. Capacity planning and network planning ensure service levels through resource planning. Table 4.2 lists the planning elements of production operations.

TABLE 4.2   Production Operations Planning Elements

Service-level planning

Problem management

Change management

Recovery management

Capacity planning

Network planning

Managing production operations is vital for IT managers. Production operations is a critical success factor for them, and it must be handled properly in both the short and the long term. The basic considerations involving the elements of the production operations plan will be discussed in detail in Part IV of this text. This discussion includes the details of service-level planning and the operational disciplines of problem, change, and recovery management. Service levels are important elements of the IT plan. Problem, change, and recovery management are processes to ensure that managers attain committed service levels.

Service-level attainment depends on adequate resources with which to process the applications. The IT organization must possess sufficient computing capacity to handle the application workload satisfactorily. The physical resources include computer-processing units and all the peripheral equipment such as auxiliary storage and input/output devices. In a distributed environment where end-user computing is well developed, individual workstations and associated equipment are important components of the system's capacity. Chapter 13 discusses performance analysis and capacity planning for the computing systems serving the applications.

Network planning and management are the subjects of Chapter 14. Vision, policy, and architecture are the necessary elements for senior management involvement in the planning process.[12] Design and implementation of the network follows the planning. As with the computer components of the system, performance analysis and capacity planning are required to meet service levels. The disciplines of problem, change, and recovery management apply to networks as well. Network capacity planning is an important element of the IT plan.

Resource Planning

The next element of the complete IT plan is the resource section. The resource items of the IT plan consist of money, equipment, people, and space. Technology is also a resource, and its planning will be discussed as a separate element. Table 4.3 indicates the elements commonly found in the resource section of the plan.

TABLE 4.3 Resource Elements

Equipment plans

Space plans

People plans

Financial plans

Administrative actions

The IT plan describes the critical dependency on available resources throughout the planning horizon. A summary of the hardware and equipment required to operate the information processing activities of the firm is found in the resource section of the plan. Included in this section are the space requirements for housing the equipment and the special facilities for running the equipment. If the firm's production operation is growing, the space requirement can be substantial. The special facilities consist of heating, ventilation, air conditioning and electrical power, and perhaps raised floors for mainframes. In large mainframe installations the use of uninterruptible power sources to ensure reliability of service is also a significant cost item.

The plan must also include equipment and facilities for networks and telecommunications systems. The financial obligations of the firm consist of purchase and lease costs and expenses for services provided by third parties. The IT plan integrates the equipment and use charges with software and applications costs for the complete telecommunications system financial plan.

People Plans

The next essential ingredient of the IT resource plan is the plan for people. By any measure, this is the most crucial element of the plan. People enable the firm and the IT organization to meet their goals and objectives. All the equipment, space, and money in the country will be to no avail without skilled people. People management is a necessary condition for success, though not sufficient. The management of people resources in an effective manner is a critical success factor for IT managers. It is essential, therefore, that this portion of the plan is particularly well conceived and developed.

IT managers and administrators can provide much of the people-related information. Staffing, training, and retraining plans go hand-in-hand with the people management activities of performance planning and assessment. Managers spend a great deal of time on these matters, and their actions and anticipated actions must be reflected in the plans for the organization.

The personnel plan includes requirements for people by skill level and their deployment through the extended plan horizon. It must address attrition, hiring, training, and retraining. Sources of new individuals are identified in the IT plan, and recruiting actions are formulated. For the people currently in the organization, individual people-development plans are proposed by the managers. Plans for people are also maintained by the IT manager for skill groups such as systems analysts, programmers, support specialists, telecommunication experts, and managers. The staffing plan must be consistent with the plans to accomplish technical and organizational objectives. Working with this detail, IT planners can calculate the cost of the people resource by summing their salaries, benefits, training costs, and so on for the financial plan.

Financial Plans

The summation of the equipment, space, people, and miscellaneous costs constitutes the financial plan for the IT organization. The costs and expenses are segregated and each is identified by elements using the code of accounts for the firm. The timing of all expenditures reflects the rate at which resources will be acquired or expended. This timing must be consistent with work activities and completion of events. When the plan is approved, the budget for the planning years is constructed by funding the various accounts to the level requested in the plan. In almost all cases, the planning process is iterative in nature and the final budgeted amount is determined through intense discussion and negotiation. The early portion of the plan, typically the first year, represents a commitment. Management is measured by comparing actual performance to the planned or expected performance. This measurement is part of the performance appraisal for IT managers.

Administrative Actions

Before, during, and after the planning cycle there are many administrative actions that assist in the planning process and the implementation of the approved plan. Before planning begins, IT administration can develop planning assumptions and provide planning ground rules. For example, the statement that the number of programmers and analysts will remain constant throughout the plan period may be a plan ground rule. For planning purposes, the firm might state that it expects to increase revenue by 10 percent per year. This is an example of a planning assumption. The list of ground rules and assumptions is established so that the resulting plan remains within the bounds of reason. Without a set of reasonable ground rules and assumptions, the plan process might iterate forever.

During the planning period, IT managers must coordinate the developing plan with the many organizations that are affected by it. This happens in a variety of ways, but the administrative function must ensure that communication is complete and unambiguous. Plan-review meetings should be held with managers from all the functional areas in the firm. For example, the plan must be reviewed with marketing managers to ensure that marketing requirements for information technology support are properly planned. Such communication meetings help ensure plan congruence and assist in organizational learning.[13] The role and contribution of IT are clarified among the users through this process. A well-coordinated plan review process mitigates several of the concerns we studied in Chapter 1.

Many firms establish steering committees consisting of representatives from the functional areas in the firm. The purpose of such committees is to provide guidance to the IT planning and investment process. Effective steering committees both increase the mutual understanding of users and IT personnel and tend to increase executive involvement and understanding of IT matters.[14] Steering committees, consisting of senior executives from all parts of the business, provide valuable assistance in ensuring plan synchronization between IT and the rest of the firm.

Finally, managers and administrators are responsible for implementing the plan and for tracking activity within the organization to actions specified in the plan. Control of the organization is a fundamental responsibility of management. IT managers must install measurement and tracking mechanisms designed to ensure that the organization's activities are on target. The control functions must address the plan elements individually.

For example, actions taken with the application portfolio are routinely subject to project management control and monitoring. Control of the arrival and departure of equipment and facilities and the related space is essential. Production services must be measured and controlled to the service-level agreements negotiated with the users. The consumption and utilization of financial resources are usually tracked by the financial organization of the firm. Actual expenditures are compared with planned expenditures; variances in the accounts require explanation to the firm's controller and others. Schedules and budgets are the principle control elements in IT plans.

Technology Planning

The IT organization has a continuing responsibility to monitor advances in technology and to keep the firm informed of progress in the field. Senior professionals throughout the IT organization must maintain knowledge of the state of the art in their disciplines. Programming management must track programming technology. System support personnel and telecommunications experts must maintain awareness of current technology and must improve their expertise. These experts provide technology assessments to the user organizations in the firm through the senior IT executive. Table 4.4 lists some of the areas of technology that should be evaluated during the planning process.

Introduction of new technology leads to some predictable problems. Costs are usually underestimated and benefits frequently overestimated.[15] It is especially useful for the IT manager to fund a small advanced technology group whose purpose is to make assessments and provide models for utilization of the emerging technology within the firm. The ad tech group conducts research on emerging technologies and evaluates new concepts. The group establishes concept feasibility, which begins the development and introduction of the new concepts. Feasible concepts are transferred to development and user departments for implementation and production. These activities assist in technology introduction and provide useful insights for planning purposes.

TABLE 4.4   Technology Areas

New processor developments

Advances in storage devices

Telecommunications hardware

Operating systems

Communications software

Programming tools

Vendor application software

Systems management tools

OTHER APPROACHES TO PLANNING

The first chapter presented two ideas fundamental to the management of information technology and particularly relevant to IT planning. These concepts are Nolan's stages of growth and Rockart's critical success factors. Both are important in the IT planning process. They add value to the processes outlined above and serve to test the credibility of the results.

Stages of Growth

Nolan's theories, either the four stages or the six stages, provide predictive insight to organizations within the firm and to the firm itself. To use the stage theory, management must make an assessment of the posture of the organization or the firm at the present time. Then plans can be based on an extrapolation of the observed trends. For example, if the firm is in the control stage, it would be wise to anticipate and plan for the integration phase followed by the data administration stage. If one function is in the data administration phase and others are in the integration phase, it is advisable to understand the reasons for this disparity. Perhaps the firm should concentrate on the laggards and bring them up to the level of the more advanced function. For competitive reasons, it may be useful to understand where other firms in the industry are positioned relative to stages of growth.

Before any action is taken by IT managers, they must communicate their assessments to the functional heads involved. There may be very good reasons for conditions being as they are. In any case, the decisions must be made by all interested parties, including the steering committee, not just the IT organization. The analyses of the stage theory and its use in planning should be integrated into the more formal process outlined above.

Critical Success Factors

Critical success factors are excellent tools to use in conjunction with the formal planning process. They lend structure to planning and improve planning by focusing on important managerial issues.[16] They form an outstanding audit on the results of the process and ensure that necessary conditions for success are contained within the plan.

For instance, consider critical success factors 1a, 1d, and 3b from Chapter 1. If the IT plan is out of synch with the firm's plan, the resulting implementation will not totally support the firm's objectives. This is unsatisfactory and must be corrected. If the planning process develops unrealistic expectations or does not ensure realism in long-term expectations, the IT organization will be in trouble in later years. A third example from the production side of the organization is worth considering. If the IT plan does not contain enough resources to deliver service on time and within planned costs, the IT organization will experience difficulties in the short term. The results of the planning process can be evaluated by reviewing the plan against the critical success factors for the IT organization and the firm. This evaluation can also help level expectations within the firm.

The first chapter presented critical success factors founded on issues prominent in the industry and on other factors of importance to the firm. Critical success factors are organized into long-range items, intermediate-range items, short-range items, and business issues. This organization is designed to coincide with the planning time frames discussed in this chapter. The IT plan should ensure that all necessary conditions for success are part of the plan.

It is extremely important to know the critical success factors of the senior IT executive's superior. It is even more important to know the critical success factors for the firm. If the plan prepared by the IT organization fails to account for these factors, the IT organization will not be meeting the necessary conditions for success.

Business Systems Planning

Another popular and well-known approach to IT planning is the methodology of business systems planning, or BSP, developed by IBM.[17] BSP concentrates on the data resources of the firm and strives to develop an information architecture supporting a coordinated view of the firm's major system's data needs. The BSP process identifies the key activities of the firm and the systems and data that support these activities. The data is arranged in classes and an architecture is developed to relate the data classes to the firm's activities and to its information systems. In essence, BSP strives to model the business of the firm through its information resources. The planning emphasis changes from the applications in the portfolio to the supporting databases.

The BSP process is very detailed and time consuming. It requires a bottom-up effort and tacitly assumes that a data architecture can be developed in one step. The process works best in centralized environments where data, applications, and structure for processing are adjacent to one another. The current trends toward decentralization, broad deployment of information technology, and increasing importance and value of the technology greatly complicates the task of IT planning. A variety of planning methods must be considered.

THE INTEGRATED APPROACH

This chapter has presented the important types of plans, plan ingredients for the IT organization, and some important planning approaches currently in use. But other planning tools have been discussed in the literature.[18] All of these approaches have advantages under certain circumstances, and each has limitations and disadvantages. But most methods are weak when information technology is important to the firm, when the technology is widely dispersed, and when the organization is decentralized.

Cornelius Sullivan studied 37 major American companies to understand the effectiveness of their planning systems as related to factors indigenous to the firm.[19] Two factors were found to correlate with planning effectiveness: infusion, or the degree to which information technology has penetrated the operation of the firm, and diffusion, the extent to which information technology is disseminated throughout the firm. Firms that considered themselves to be effective IT planners were tabulated in a matrix by the type of planning system in use. Sullivan's conclusion are very revealing. They are presented in matrix form in Figure 4.3.

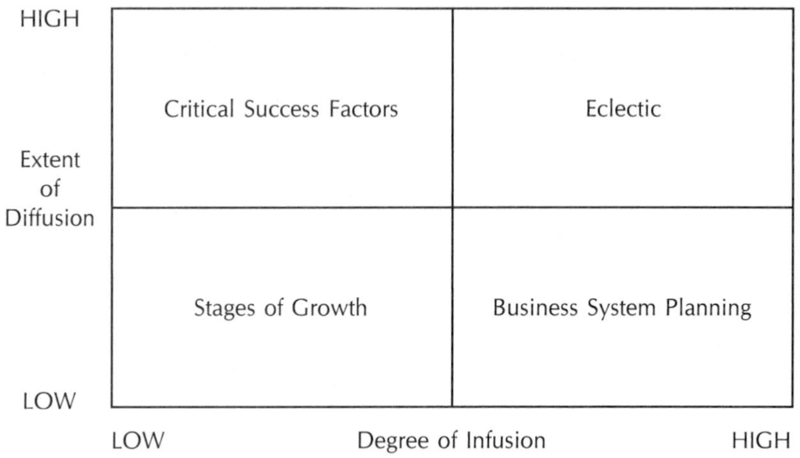

HIGH

| | Critical Success Factors | Eclectic |
| | | |

Extent
of
Diffusion

| | Stages of Growth | Business System Planning |

LOW

LOW   Degree of Infusion   HIGH

FIGURE 4.3   IT Planning Environments

Sullivan's research reveals that the stages-of-growth methodology is effective in firms in which the technology does not have a high impact and is concentrated. Most firms today will not find stages of growth to be effective as the primary planning mechanism. Critical success factors were found to be more important to firms in which the technology impact is moderate and the technology is dispersed. For many firms, the CSF methodology is a valuable adjunct to the formal planning mechanisms discussed earlier. BSP is more effective for firms in which information technology is centralized but is very important to the company. When it was developed, it was useful for centralized firms like IBM. It is much less important today.

Many of today's firms are in the upper right quadrant of Figure 4.3 or moving toward that position. For these firms, the planning process is more complex because the IT operations within the firm are very complex. Stages-of-growth concepts will be of limited value; critical success factors will continue to have importance; and concentration on information architecture will be valuable. BSP as a planning mechanism will have limited utility. What planning adjuncts will best serve the firm in the Eclectic quadrant? How can IT executives perform effective planning in the face of the fast-paced transformations increasingly common among firms today?

Firms in the Eclectic quadrant, or moving toward it, face some predictable planning issues. Major changes are taking place in application portfolio acquisition with the tremendous increases in purchased applications, joint development activity and alliances, and end-user application development. Organizations are being redesigned around information systems and information technology managers and advocates are architects of the changing environment. Dramatic adoption of telecommunications technology and increasing reliance on networked systems greatly complicates life for the IT manager. Many of these trends give rise to the issues studied earlier, such as organizational learning and information architecture.

Successful IT managers must understand these trends and must plan for them in the face of considerable uncertainty. Understanding the management issues associated with the portfolio and managing the portfolio effectively are critical. Managing end-user computing and telecommunications management are vital. And establishing relationships with executives and user organizations to facilitate change efficiently is mandatory. These topics are all addressed in subsequent chapters. Planning for these environments will, in part, be eclectic.

MANAGEMENT FEEDBACK MECHANISMS

Throughout this chapter the need for control processes or management feedback mechanisms was mentioned. Control processes consist of knowing what, when, where, why, how, and who for all essential activities of the organization. If managers know these answers, they are operating under control. If they don't, then they are operating out of control. An organization operating out of control causes serious difficulties for the managers and for the firm. Control is vital for success, especially in IT organizations, not only for the organization itself, but for the entire firm. The basis of control is found in strategy and planning activity.

The IT plan should provide data to gauge the answers to these questions. Control processes must be designed to compare the actual performance of the organization to the expected performance as detailed in the plan. The notion of performance assessment in connection with the strategy development process was discussed in Chapter 3. Plans describing how the strategy is to be implemented are used to gauge and assess actual performance. In other words, plans are the basis for assessing progress toward achieving strategic goals and objectives.

Figure 4.4 shows how these ideas fit together. Strategic processes lead to tactical processes which, in turn, are expressed in operational actions. Measurements lead to control assessments. These control assessments focus attention on tactical processes which, in turn, bring focus to strategic processes. Thus the process is iterative, and strategies and plans are validated or altered with each planning cycle.

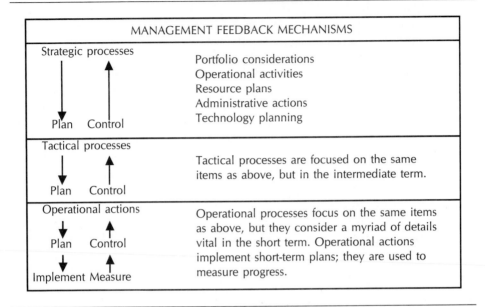

MANAGEMENT FEEDBACK MECHANISMS	
Strategic processes Plan Control	Portfolio considerations Operational activities Resource plans Administrative actions Technology planning
Tactical processes Plan Control	Tactical processes are focused on the same items as above, but in the intermediate term.
Operational actions Plan Control Implement Measure	Operational processes focus on the same items as above, but they consider a myriad of details vital in the short term. Operational actions implement short-term plans; they are used to measure progress.

FIGURE 4.4   **Management Feedback Mechanisms**

SUMMARY

The task of planning for the IT organization is complex and is best accomplished through systematic processes. The concepts of planning rely on separating the extended planning horizon into three components: strategic, or long-term; tactical, or intermediate-term; and operational, or short-term. Planning was discussed as an annual event. At each planning cycle another year is added to the strategic period and each plan is moved up one year. This systematic process is synchronized with the plan calendar for the entire firm.

The IT organization's strategic plans are developed from IT strategies. They give the strategic statements detail and specific implementation actions. Strategic plans usually cover the period from two years in the future to the end of the extended planning horizon. Tactical plans detail the actions to be taken for the next two years. They are derived from previous strategic plans and give implementation direction to previous strategic goals and objectives. Operational plans are the near-term implementation of tactical planning.

This chapter outlined a planning model for the IT organization. The model included planning for the applications portfolio and for production operations. It also included resource planning, plans for people, financial plans, and administrative actions. Technology planning is the final ingredient of the model. The model provides a framework for the complete information technology plan. As an adjunct to the model, the text suggested that the ideas of stages of growth and critical success factors should be reviewed as a test or audit of the completed plan. Business systems planning concepts are useful in planning an information architecture. The integration of these planning concepts forms an integrated eclectic approach to IT planning.

Plans are only as good as their implementation and control mechanisms. The chapter concludes with a framework that relates strategic and tactical processes and operational actions with implementation and measurement activities. Control is a fundamental management responsibility. It is especially important to IT managers. Measurements are also very important because they form the foundation of performance appraisal. Information technology planning is a critical success factor for IT managers.

Review Questions

1. What are the differences between the strategy and the strategic plan?

2. What is meant by the extended planning horizon? Describe the planning horizons discussed in this chapter.

3. What is the relationship of strategic plans to tactical plans? How are operational plans related to tactical plans?

4. For the strategic plan, what resources must be blended together and planned?

5. What financial items are included in the strategic plan for a corporation?

6. What purposes are served by tactical plans?

7. Who are the main users of operational plans? How does the firm's controller use operational and tactical plans?

8. What is meant by the planning calendar and the planning cycle?

9. What are the elements of the IT organization's planning model?

10. What considerations surround the planning for the applications portfolio?

11. What are the questions that need to be addressed for proper control?

12. What are the elements of production operations planning?

13. Which elements ensure the attainment of service levels through management processes and which through resource planning?

14. What elements comprise the resource portion of the IT plan?

15. How is the financial portion of the IT plan constructed?

16. What is meant by plan ground rules and assumptions?

17. What role does technology play in planning for the IT organization?

18. What is the relationship of stages of growth and critical success factors to the planning process described in this chapter?

19. When does the business systems planning methodology become important in IT planning?

20. What is meant by eclectic planning, and for what firms is it important?

21. What is the connection between operational plans and management control?

Discussion Questions

1. Discuss McGraw-Hill's strategy and organizational realignment in light of its information systems capabilities.

2. Chapter 1 presented a list of critical success factors for the IT manager. Which ones are related to IT planning?

3. Discuss the continuum of time frames inherent in IT planning and describe the levels of detail in the plans throughout these time frames using a mainframe installation as an example.

4. Discuss the planning cycle and the planning horizon for a university operating on the academic year starting in September and ending in August.

5. Name several planning assumptions and ground rules that might apply to the university in Question 4 above.

6. What is the role of the firm's planning department in the plan development process?

7. What additions might be included in the planning model for an IT organization that provides services to customers outside the firm?

8. Discuss the relationship between the list of critical success factors and planning for production operations.

9. Equipment plans are normally complicated by the need to include auxiliary items such as space, air conditioning, supplies and maintenance, and the like. Discuss how this might be accomplished.

10. What is the role of IT managers in the process of people planning? Discuss why this activity should take place on a continuing basis, not just at plan time.

11. Why is technology assessment vital to the IT organization? Describe how you think this assessment can be brought into the planning process.

12. Discuss some ways in which you think senior IT management should exercise control.

Assignments

1. Study BSP or one of the other planning techniques mentioned in this chapter. What are its advantages and disadvantages? How would it fit with the process we discussed in this chapter?

2. The firm for which you work would like to install a larger and more powerful mainframe to replace its present mainframe. What are the kinds of items that must be planned to accomplish this objective?

ENDNOTES

[1] Copyright 1988 by CW Publishing Inc., Framingham, MA 01701—Reprinted from *Computerworld*.

[2] McGraw-Hill earned $172.5 million on sales of $1.9 billion in 1990. It employs 15,250 according to *Value Line*, March 8, 1991.

[3] For an interesting continuation of McGraw-Hill's progress, see Johnnie L. Roberts, "Short Circuit," *Wall Street Journal*, February 6, 1990, 1.

[4] There has been much critism of strategic planning but most executives value it. See Daniel H. Gray, "Uses and Misuses of Strategic Planning, *Harvard Business Review* (January-February 1986): 89. This article describes some actions that overcome the difficulties of strategic IT planning.

[5] Marilyn Parker and Robert J. Benson, "Enterprisewide Information Management: State-of-the-Art Strategic Planning," *Journal of Information Systems Management* (Summer 1989): 14.

[6] Albert R. Lederer and Vijay Sethi, "The Implementation of Strategic Information Systems Planning Methodologies," *MIS Quarterly* (September 1988): 445.

[7] As in the previous chapter, this discussion assumes that the firm has some kind of formal planning process. The absence of a formal planning process increases the risk of plan incongruence and may lead to planning failures.

[8] James I. Cash, Jr., et al., *Corporate Information Systems Management: Text and Cases* (Homewood, IL: Irwin, 1988), 633-635.

[9] Nick Rackoff, Charles Wiseman, and Walter A. Ullrich, "Information Systems for Competitive Advantage: Implementation of a Planning Process," *MIS Quarterly* (December 1985): 285. Also see Albert L. Lederer and Vijay Sethi, "The Implementation of Strategic Information Systems Planning Methodologies," *MIS Quarterly* (September 1988): 445.

[10] Martin Buss, "How To Rank Computer Projects," *Harvard Business Review* (January-February 1983): 118. This article discusses the issue of who should make the decisions on applying resources to the portfolio. Buss endorses the formal planning process as the most effective method for prioritizing the firm's application projects.

[11] Edward A. Van Schaik, *A Management System for the Information Business: Organizational Analysis* (Englewood Cliffs, NJ: Prentice-Hall, Inc., 1985). In particular see pp. 155-225. Also see John P. Singleton, Ephraim R. McLean, and Edward N. Altman, "Measuring Information Systems Performance: Experience with the Management by Results System at Security Pacific Bank," *MIS Quarterly* (June 1988): 325.

[12] Peter G. W. Keen, "Business Innovation Through Telecommunications," *Competing in Time: Using Telecommunications for Competitive Advantage* (New York: Ballinger Publishing Co., 1988), 5.

[13] See also John C. Henderson and John G. Sifonis, "The Value of Strategic IS Planning: Understanding, Consistency, Validity, and IS Markets," *MIS Quarterly* (June 1988): 187.

[14] D. H. Drury, "An Evaluation of Data Processing Steering Committees," *MIS Quarterly* (December 1984): 257. Drury found that approximately half of the firms polled had a committee that, on average, had existed for five years. The typical committee contained six members and was most effective when it held regular meetings and when agenda items originated outside the IT department.

[15] B. Gold, "Charting a Course to Superior Technology Evaluation," *Sloan Management Review* (Fall 1989): 19.

[16] Andrew C. Boynton and Robert W. Zmud, "An Assessment of Critical Success Factors," *Sloan Management Review* (Summer 1984): 17.

[17] *Business Systems Planning: Information Systems Planning Guide* (White Plains, NY: IBM, 1984).

[18] Lederer and Sethi, "The Implementation of Strategic Information Systems Planning Methodologies," *MIS Quarterly*, September, 1988, 445. See also Barbara C. McNurlin and Ralph H. Sprague, Jr., *Information Systems Management in Practice*, 2nd ed. (Englewood Cliffs, NJ: Prentice Hall, 1989), Ch. 4.

[19] Cornelius H. Sullivan, "Systems Planning in the Information Age," *Sloan Management Review* (Winter 1985): 4.

REFERENCES AND READINGS

Anthony, R. N. *Planning and Control Systems: A Framework for Analysis*. Cambridge, MA: Harvard Business School, 1965.

Hefferon, George J. "Taking the Mystery out of IS Strategic Planning." *Information Executive*, Vol. 1 (Fall 1988): 18.

Kemerer, Chris F., and Glenn L. Sosa. "Barriers to Successful Strategic Information Systems." *Planning Review* (September-October 1988): 20.

King, William R. "Strategic Planning for Information Systems." *MIS Quarterly* (March 1978): 27.

King, W. R., and R. W. Zmud. "Management Information Systems: Policy Planning, Strategic Planning, and Operational Planning." Cambridge, MA: *Proceedings*, Second International Conference on Information Systems, December 1983, 299.

Martin, James. *Strategic Data-Planning Methodologies*. Englewood Cliffs, NJ: Prentice Hall, 1982.

McFarlan, F. Warren, James L. McKenney, and Philip Pyburn. "The Information Archipelago—Plotting a Course." *Harvard Business Review* (January-February 1983): 145.

Parker, Marilyn M., H. Edgar Trainor, and Robert J. Benson. *Information Strategy and Economics*. Englewood Cliffs, NJ: Prentice-Hall, 1989.

Rockart, J. "Chief Executives Define Their Own Data Needs." *Harvard Business Review* (March-April 1979): 81.

Sullivan, Cornelius H. "The Changing Approach to Systems Planning." *Journal of Information Systems Management* (Summer 1988): 8.

Sullivan, Cornelius H. "Systems Planning in the Information Age." *Sloan Management Review* (Winter 1985): 3.

Van Schaik, Edward A. *A Management System for the Information Business: Organizational Analysis*. Englewood Cliffs, NJ: Prentice-Hall, 1985.

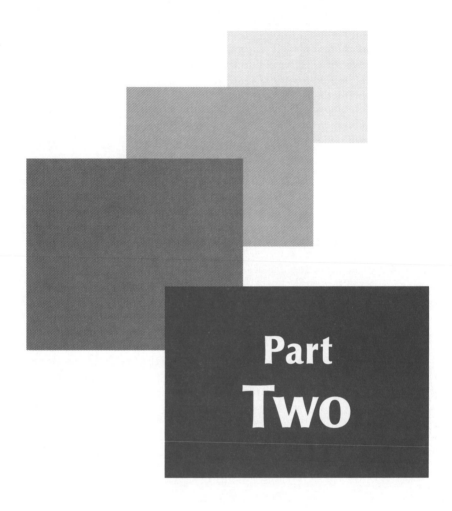

Part
Two

5 Hardware and Software Trends

6 Advances in Telecommunication Systems

Information Technology Trends

Advances in computing hardware, operating system software, and storage technology provide numerous opportunities for business managers. Simultaneous rapid advances in telecommunications systems technology allow managers to distribute data and data processing capability, integrate business systems, and leverage their returns on investments. Extensive deployment of personal workstations changes the work of employees and managers and permanently alters the firm's conduct of its affairs. Part Two chapters include Hardware and Software Trends and Advances in Telecommunications Systems.

Successful information technology managers are highly cognizant of technology trends and use trend information to prepare their people and organizations for the future.

5 Hardware and Software Trends

A Business Vignette
Introduction
The Semiconductor Industry
Semiconductor Technology Trends
Advances in Recording Technology
   Magnetic Recording
   Alternative Technologies
From Microcomputers to Mainframes
The Microcomputer Revolution
What's Happening with Supercomputers?
Trends in Systems Architecture
Current Trends in Programming
Operating Systems Considerations
   The Evolution of Operating Systems
   Contemporary Operating Systems
Communications Technology
Technology Trends
   The Meaning for Management
   The Meaning for Organizations
Summary
Review Questions
Discussion Questions
Assignments
References and Readings
Appendix

A Business Vignette[1]

Who's on Second?

Does a company that invents a popular computer chip have a duty to let others in on its good thing? Intel doesn't think so. Intel Corp.'s newest microprocessor, the 80386, could well turn out to be the biggest-selling logic chip in the history of the semiconductor industry. The 80386 is the smarts behind IBM's most powerful new personal computer, the PS/2 Model 80, Compaq's newest offerings, and a coming slew of new 32-bit machines.

This year the Santa Clara, California-based chipmaker should see revenues of over $1.1 billion from the sale of this single chip and its related peripheral chips—or nearly half its estimated 1988 semiconductor revenues of $2.3 billion.[2] More to the point, Intel's operating profit from this product family could hit $400 million, or almost three-fourths of profits from all semiconductors.

So popular is the 386, as the 80386 is known throughout the computer industry, that Intel is straining to keep up with demand, running its factories in Livermore, California and Israel full out.[3] It plans to add 386 production at two more plants this year. Why is Intel straining? In the past, Intel has ensured adequate supplies of its popular new chips by licensing other companies to manufacture its designs under so-called second-source agreements. Though the agreements helped popularize Intel's designs, they also diluted the company's profits, since its royalty was modest. With the 386, which cost $100 million to develop, Intel wants to keep all the profits to itself. Only Intel is making the 386s.

Intel second-sourced the predecessor to the 386 (the 80286, the 16-bit brain used in IBM's low-end PS/2 personal computers) among others. But by second-sourcing, four other semiconductor makers—Advanced Micro Devices, Harris, Fujitsu and Siemens—Intel created price competition for itself that started to hurt when the supply of 80286s caught up with demand. Last year, according to the market research firm Dataquest, Intel's second sources supplied a good third of the market for the 80286, taking business that Intel would dearly have liked having for itself and making it that much harder for Intel to recover its development costs.

Partly to blame for Intel's pullback are the Japanese, who drove companies like AMD and Intel out of large parts of the commodity memory chip business in the Eighties. Says Jerry Sanders, AMD's voluble president, "Suddenly there was no profit from 20% to 25% of the market."

Intel will keep the market for the 386 all to itself, at least for the time being, even if that means the chips will be in temporary short supply.

But Intel is not the only semiconductor outfit having second thoughts about second-sourcing. "There is a definite trend to having a smaller number of second sources on the same product as well as more products that are sole-sourced," says F. Joseph Van Poppelen, vice president of worldwide marketing and sales at National Semiconductor.

Second-source agreements are almost as old as the semiconductor industry itself. From the chipmakers' perspective, signing up second sources was a way for a small company to develop a dominant market position for its newly invented part. To that end second-source licenses were often given out for little or no money.

But that kind of deal is losing its attractiveness for the well-established chipmakers. Customers quickly discovered they could drive down the price by playing one source against the other. Intel Chairman Gorden Moore stated in 1986: "The investment in designing some of these chips is growing astronomically, so one can't afford to second-source as broadly as in the past."

For smaller companies, conventional second-sourcing can still be a way to turn their chip into an industry standard. Sun Microsystems has been licensing its new reduced-instruction set microprocessor to outfits like Cyprus Semiconductor and Fujitsu. But Sun is in a different position than Intel is with the 80386; Sun is only one of a dozen companies with versions of such a microprocessor.

What's at stake here is not the morality of second-sourcing, which people will argue when it is convenient for them. What is at stake is Intel's chance of making a good profit on its unique product. Windows of profit opportunity don't remain open long in this industry.

INTRODUCTION

During the past 30 to 40 years, unprecedented advances in digital computing capability have been fueled by rapid technological innovations. These innovations occurred in the fields of semiconductor technology, recording technology, telecommunications capability, and, to a lesser extent, in the programming arena (see Table 5.1). Taken together, these four activities are the foundations of the digital computer revolution.

Separately, each endeavor is subject to limitations imposed by the laws of physics; human or organizational limits; system or interrelational barriers; fabrication limits; and limitations reflecting economic reality. During the last three decades practitioners in these fields have marshaled intellectual, financial, and production resources to push practical limitations of these technologies toward their theoretical limits.

TABLE 5.1   **Technical Foundations of Advances in Digital Computers**

Semiconductor technology

Recording technology

Telecommunications

Programming

They have been extraordinarily successful. Progress over the past three decades is measured in orders of magnitude (factors of ten). Technical capability is frequently reported as doubling in performance and halving in cost over a period of a few years. Size, as measured by area or volume, has declined proportionately, and size is one of the underlying reasons behind the performance increases and cost reductions. Only the ability to produce computer instructions—the task of programming—has proceeded at a more normal pace.

Additionally, and again with the exception of programming, one can confidently predict that the next five years or so will reflect a continuation of the rapid innovative pace of the past. We can be less certain about the practical utilization of these technologies within our social and business structures. Certainly the results will be dramatic and highly meaningful for most of us and for the organizations in which we operate.

> There is great opportunity in the growth of global information networks, in medium- and smaller-sized businesses, and in the needs of operating departments and individuals in larger organizations. Software and service continue to lead the industry's growth. In technology, there are no limits in sight for improvements in performance, function and value. The race is on to make computers simpler and easier to use. Emerging technologies such as image processing and artificial intelligence offer great promise for the future.[4]

This chapter will explore these trends and make careful projections concerning their future form and direction. In particular, the text will relate these trend data to the activities of our organizations and will extract meaning from them for ourselves as managers, for our employees and associates, and for our organizations. Successful managers remain alert to trend information; exceptional IT managers use leading technological indicators to prepare themselves and their firms for the future.

THE SEMICONDUCTOR INDUSTRY

The invention of the transistor in 1948 set in motion a stream of innovations and inventions that spawned the semiconductor industry.[5] The industry currently employs nearly 300,000 in the U.S. and generates U.S. revenues of about $28 billion. Advances in semiconductor technology have helped fuel the growth of the electronic data processing industry. The electronics industry is now the third largest in the U.S., behind automobiles and petroleum. The U.S. computer industry creates annual worldwide revenues of about $150 billion.[6]

Semiconductor technology is important to IT managers because it forms the foundation on which the information industry is built. Semiconductor chip technology makes today's microprocessors possible and permits system designers to pack ever more computational performance into systems of all sizes at steadily decreasing costs. Advances in chip design fuel the growth of the information handling industry and make possible the information age as we know it. IT managers must have knowledge of the trends in this industry if they are to be effective in their jobs.

There are more than 100 companies competing for a share of the $50 billion market worldwide. Most output of the semiconductor industry is a commodity product. Product development and manufacture is very costly; this commodity business is very capital-intensive. And, as we learned in the Business Vignette, the dynamics of this intensely competitive industry are complex and exciting.

Major companies around the world participate in the semiconductor business, but Japanese firms hold about 50 percent of the market. North American companies got about 10 percent of the Japanese market in 1988, the highest figure since 1984, but Japanese firms captured 20 percent of the North American market, the highest share ever.

Because the industry is capital-intensive, the semiconductor market is dominated by large companies. In 1988, the top ten companies accounted for more than 50 percent of the worldwide market; 84 percent of the market was held by the top 25 firms. Many small firms, however, have found a niche and are making a contribution to the growth of the industry. The appendix to this chapter displays the major players in the industry listed by sales volume.

Although the electronics industry is growing at exceptional rates, the growth is dependent on the ever-decreasing size of the integrated circuit itself and to continued increases in the density of devices on the carrier. Industry growth rates are directly related to circuit density increases or growth in the number of devices per unit of carrier. There are several other important trends as well. In addition to the increasing density of circuitry on the chip, the reliability of the semiconductor devices is improving with time. In addition, the cost per unit is rapidly declining. All of these trends are favorable for the ultimate consumer and for the semiconductor industry. Table 5.2 summarizes this trend information.

TABLE 5.2   Semiconductor Technology Trends

High industry growth rates

Declining device size

Increasing density of devices on chips

Rapidly declining prices

Increasing reliability

Figure 5.1 displays the growth of circuit density versus time. The density of integrated circuits on a chip as a function of time from the late 1950s to the present and projected into the first fifth of the next century is shown in this figure. Notice that the vertical scale is in powers of ten, ranging from one to 10 billion circuits per chip. The number of circuits per chip during the 30-year period from 1960 to 1990 grew by seven orders of magnitude, from about three to 30 million during this time! Projections to the year 2020 range from slightly less than 1 billion to more than 10 billion circuits per chip, depending on assumptions regarding the processes to build these superchips.

The reasons behind the thrust to shrink the size of integrated circuits are to reduce overall cost and to minimize the switching time of the devices, i.e., the time needed for the device to go from one binary state to the other. Ever decreasing circuit dimensions yield lower unit cost and higher unit performance, thus fueling continued semiconductor industry growth.

Present manufacturing technology produces circuit lines on a silicon chip that are 2 microns wide. (A micron is one ten-thousandth of a centimeter, or roughly one one-hundredth the width of a human hair.) Advanced production techniques are capable of reducing the line widths even further, in some cases to less than 1 micron. Reducing the line dimensions from 2 microns to 1 micron quadruples the number of circuits per unit area and reduces signal travel distances, thus improving chip performance. Therefore industry growth rates, which depend on increasing performance and diminishing unit costs, are inversely related to the size of the fundamental circuit.

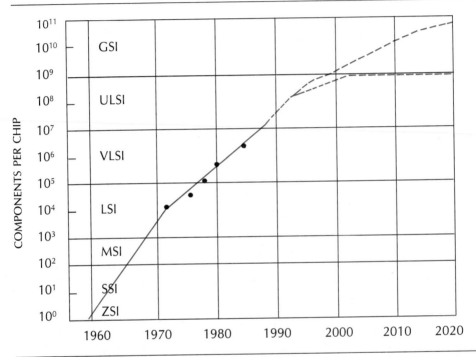

FIGURE 5.1   Integrated Circuit Chip Density vs. Time

Through the late 1980s, a period of 30 years since the commercial introduction of the single transistor chip, the number of transistors that can be fabricated on each chip has grown to more than a million—a factor of 10 to the 6th power and an average growth rate of 100 times each decade! And, while the performance increased by more than 10,000 times, the cost per chip has remained substantially unchanged. All the important factors in chip production have been made favorable. Unit size has substantially declined; unit performance has remarkably increased; and cost to the consumer has declined very dramatically.

As the dimensions of silicon devices shrink and approach their minimum practical dimensions, scientists are searching for new technologies and materials to maintain the downscaling process. One such material, gallium arsenide, is believed to possess these physical properties. Gallium arsenide will permit another factor of 100 in device size reduction, increasing switching speeds up to 200 billion times per second, and gaining more orders of magnitude reduction in cost per unit of function. Research on this technology is under way at industrial, government, and academic laboratories around the world.

There are enormous challenges to overcome to develop manufacturing tools and processes to cope with the unique problems of size reduction. Scientists in many countries are tackling these problems. In Japan, scientists have fabricated experimental one-million-bit chips that can retrieve a bit of data in 22 nanoseconds. The development of silicon circuits that can switch 75 billion times a second, a switching speed previously unattained in silicon, was made possible by reducing the size of circuit parts down to 0.1 micron.[7] This technological achievement helps close the gap between silicon and gallium arsenide. Technologists believe silicon will continue to be used for chip fabrication for the next decade. With extensions of existing technology, it will be possible to construct 256-megabit-memory chips.[8]

The invention of the transistor and its continued development over more than 50 years, has been enormously important in many ways, and its importance will grow. Chip densities will continue to increase by orders of magnitude through advances in physics, metallurgy, and manufacturing tools and processes. We can plan on smaller, faster, lower cost, and more reliable memory and logic circuitry. IT systems, however, require more than logic and memory to function. Large-capacity storage devices are a vital and important adjunct to powerful computer chips. Fortunately, scientific advances abound in the field of recording technology as well.

ADVANCES IN RECORDING TECHNOLOGY

We have witnessed rapid progress in logic and memory circuit development over the last few decades, and, fortunately for digital computing, recording technology progress has been equally dramatic. Very high speed processors have enormous appetites for instructions and data, and they rely heavily on high performance auxiliary memory for system throughput. Approximately equal progress in both fields is essential to advances in systems development.

Magnetic Recording

Magnetic recording devices allow permanent storage of massive amounts of data, measured in gigabytes, at lower cost per byte than semiconductor devices. Access to the data is in the range of 20 milliseconds and the data transfer rates can be on the order of 4 megabytes per second. For example, the IBM 3390 direct access storage device holds up to 22.7 gigabytes of data. Average seek times are 12.5 milliseconds; average rotational delay is 7.1 milliseconds. Cost per megabyte of storage is 18 percent less than that of the model it replaces.[9] Although the sophisticated rigid disk devices are not removable or portable, high performance tape devices allow virtually unlimited amounts of data to be stored and/or physically transported.

The technology is impressive. A rigid disk about 12 inches in diameter spins at more than 3600 RPM. The data is read or written by a read/write head flying over the surface on a cushion of air about 10 millionths of an inch thick. The speed of travel approaches 100 mph. Tracks of data separated by less than 1/1000th of an inch hold close to 100K bytes of data each.

Manufacturing tools and processes required to construct rigid disks in the factory rival those found in the semiconductor plants. Tolerances on surface finish approach 1 millionth of an inch, and the purity of materials in the thin-film heads is similar to that of the silicon used in chip production. The disk manufacturing plant is highly capital-intensive.

Advances in head design, physical architecture, recording materials, and production practices will ensure continued growth in capacity and performance of both rigid disks and magnetic tape devices. As with silicon chip technology, advances in magnetic recording technology will migrate from the higher performance and capacity devices to the lower performance, more widely available devices. Cost/performance improvements at the high end of the line will find expression at the low end at a later time.

Alternative Technologies

Magnetic recording will remain important for a long time, but optical devices are beginning to play an important role. The recent announcement of the NeXT personal computer included optical storage devices and omitted removable magnetic devices entirely. This implies program distribution on optical diskettes or via telecommunications.

The compact disk, known mostly for its ability to record sound to near perfection, is also a powerful medium for storing computerized data and information. One disk less than five inches in diameter can hold

600 million bytes or characters of information, equivalent to about 30 sets of encyclopedias. If this isn't enough for an application, devices resembling jukeboxes are available for accessing many CDs from a single computer. Read-only disk drives are being installed at a rate approaching 100,000 per year. They will account for $300 million in sales for computer applications in 1990.[10]

CD-ROM storage costs per megabyte are 10 percent of magnetic tape and 2 percent of floppy disks. Write-once read-only memory costs about 50 percent of the price of equivalent storage on tape.[11]

Numerous large databases are being made available on CDs for computer users. Newspaper and magazine references, abstracts of academic research, legal and medical files, and large databases of financial information are available at prices that will decline with improvements in the technology. Eventually this storehouse of information will include audio and pictorial information augmenting the printed media; this will greatly improve the process of information retrieval.

Table 5.3 illustrates some of the important characteristics of magnetic and optical storage devices. The devices are those included in the newly available NeXT personal computer. Both devices are contained in the cube, the box housing most of the electronics, which measures one foot on a side.

Optical storage devices are relatively new and their availability is limited, but magnetic rigid disk drives are available with a wide range of capabilities. North American revenues from the computer storage market are projected at approximately $40 billion annually in 1993. Approximately $28 billion is generated by rigid files. The remainder is derived from flexible files, tape devices, and optical storage devices.[12]

TABLE 5.3 **Recording Technology in the NeXT Computer**

	Magnetic	Optical
Megabytes per device	660	256
Access time in milliseconds	16.5	92
Data transfer rate in mb/sec	4.8	4.6

FROM MICROCOMPUTERS TO MAINFRAMES

The construction of computers from building blocks of logic and memory, on-line and off-line storage devices, and a variety of input and output devices spans a performance spectrum of several orders of magnitude. The spectrum is typically but somewhat arbitrarily divided into microcomputers, minicomputers, and mainframes. This division is based on central processing unit (CPU) performance measured in millions of instructions per second, or MIPS. The spectrum is typically extended on the low end to include embedded micros and on the high end to incorporate supercomputers. Embedded micros are small computing devices found in machine tools, automotive or aircraft control systems, or household appliances such as microwave ovens.

A portrayal of this spectrum and its progress over time is shown in Figure 5.2. The horizontal axis covers the period from 1960 to 1995, and the logarithmic vertical scale in millions of instructions per second or MIPS ranges from less than 0.1 to 100. Supercomputers throughout this time have ranged upward from the high performance side of the area labeled "mainframes."

Several important trends are clearly delineated in this figure. Mainframe computers, which originated in the 1950s, did not attain one million instructions per second until around 1965, but they have steadily increased in performance at the rate of about one order of magnitude every 12 years since then. Prior to the advent of minicomputers in 1966, mainframes were our only CPUs. The idea of personal computers originated shortly after 1970, and it was at about this time that small computing devices were being embedded into equipment.

Minicomputers, those CPUs intermediate in performance between mainframes and personal computers, are a vital component of the computing spectrum. They serve small and medium businesses as central processors or large businesses as departmental computers. As shown in Figure 5.2, the performance of minis has increased at least as rapidly as that of mainframes.

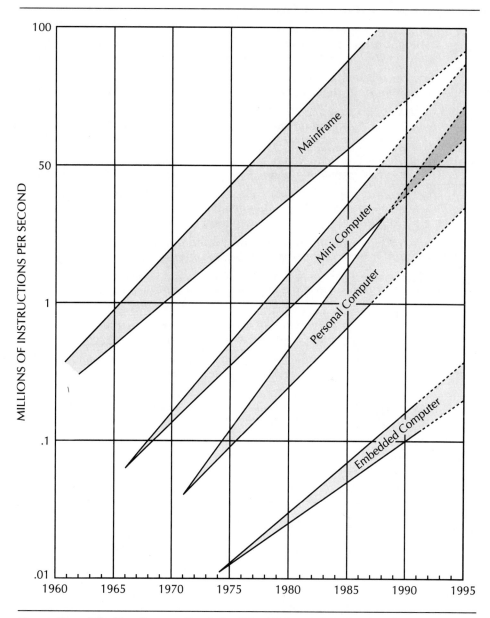

FIGURE 5.2 **The Evolution of Digital Computers**

THE MICROCOMPUTER REVOLUTION

The performance increase of personal computers has been most rapid of all. Some are available with more than 20 MIPS capacity. Personal computer performance in 1985, as measured in MIPS, equaled that of mainframes available in 1968; that is a lag of 17 years. The lag in performance between PCs and mainframes narrows to 15 years as time progresses. The growth of PCs in both performance and sheer numbers will continue unabated into the next century. This trend is highly important. It changes forever the role of computing in modern society.

The personal computer puts the power of yesterday's mainframe on the desks of millions of today's workers. And the trends of exponential growth rates are sure to continue with the power of the 80486, the RISC 6000 processor, and many other implementations. The growth in numbers of PCs rivals their growth in power. It is estimated that the total number of PCs installed in 1992 will be nearly 93 million and that nearly half of these will be part of a local network. U.S. purchases of microcomputers in 1987 were 40 percent of the $26.7 billion CPU market.[13]

Current PCs range in speed from 1 to 10 MIPS, but within 10 years desktop computers will be capable of 100 or perhaps 1000 MIPS. Gigabytes of storage will be available within the machine. The desktop computers will be connected through high-speed data links, giving users access to many more gigabytes of information. Office workers 10 years hence will have enormous computer power at their fingertips. The challenge for them, their managers, and their firms will be to use this capability to advantage. The field is ripe for inventors, innovators, and visionaries.

WHAT'S HAPPENING WITH SUPERCOMPUTERS?

Supercomputers, the highest performance systems of all, are manufactured by firms in the U.S. and abroad to serve the needs of scientific laboratories, government agencies, and large commercial customers. Leading-edge technology and architecture are employed in these machines, a precursor of what is to be found in more widely available machines years later.

Uniprocessors having capabilities exceeding 50 MIPS are commonplace today, but increasingly designers are turning to multiprocessors to achieve performance enhancements. Multiprocessors are not new, having been in existence since the 1960s, but present-day processor technology is substantially different from that of two decades ago. Multiprocessors used in commercial applications frequently consist of a front-end machine serving as a job scheduler for one or more back-end production machines.

In some applications the workload is divided by type between the processors; one processor handles communications and another performs batch production, for example. The trend of advanced technology, however, is toward multiprocessors having as many as tens of thousands of interconnected processors.

Since these large machines are typically used for scientific calculations, their performance is measured in floating point operations per second. The abbreviation for this is flops. Floating point is a term used to describe the ability of the computer to keep track of the decimal point over many orders of magnitude while doing arithmetic calculations. One million flops is one megaflop or mflop and one billion flops is one gigaflop or gflop. Supercomputers having a large number of parallel processors are capable of one trillion flops or one tflop. Since floating point instructions take longer to execute than many other types of instructions, the MIP rate of these supercomputers generally exceeds the flop rate by a factor of two or three.

Table 5.4 displays performance capability for a wide range of computers. Supercomputers head the list, but desktop computers, starting with the RS/6000 at 27.5 MIPS, have considerable capability.

TABLE 5.4   Computer Performance Comparisons

TF-1	1 terraflop
Connection Machine	24 gigaflops
GF-1	11 gigaflops
ETA-10	1.2 gigaflops
RP3	1300 MIPS
Alliant FX/8	12 megaflops
INTEL 386s	512 MIPS
CRAY Y-MP	275 megaflops
IBM 3090 600E	76 MIPS
RS/6000 320	27.5 MIPS
SUN 370	16 MIPS
386-based PC	3 MIPS

Small-parallel processing computers have come to be called mini-supercomputers. They range in size from the Alliant FX/8, which has eight processors, each capable of over 12 mflops, to increasingly more powerful machines. The ETA-10 supercomputer, for example, is an eight-processor model with over two gigabytes of memory; each processor is capable of 1.2 gflops.

Thinking Machines Corp. has recently shipped an 8192-processor Connection Machine model which can be upgraded to 65,536 processors incrementally. The processors are interconnected in a 16-dimensional hypercube that provides extremely rapid interprocessor communications. Performance of the Connection Machine is 24 gflops. Transfers of data to storage occur at a peak rate of 320 megabytes per second. Significant amounts of research are required in order to gain the full potential of these large machines.

Large experimental parallel processors have been constructed for special purposes. IBM, in cooperation with New York University and with support from the Defense Advanced Research Projects Agency, has delivered a machine called RP3, for Research Parallel Processor Prototype. This is a Unix-based computer having 512 individual processors and is rated at 1300 MIPS. The processors employ the Reduced Instruction Set Computing (RISC) architecture.

A faster, more capable machine has been produced at IBM's Yorktown Heights research facility for specialized research. The machine, named GF-11, has 576 processors. Its performance capability is 11 gigaflops. This machine will be used to perform calculations to derive the mass of a proton; these calculations are expected to take about four months. Ordinary supercomputers would require tens of years to do the same task.[14] Even larger machines are being considered.

Computer manufacturers and government agencies are discussing the development of ultra-fast computers with projected processing speeds of 1 trillion flops. Processing speeds of this magnitude can be accomplished only by interconnecting large numbers of processors, each having high instruction execution rates. Computer scientists are considering machine designs that have more than 32,000 processor nodes, each node capable of executing independent instruction streams at the rate of 100 mflops. Large amounts of main memory and high-bandwidth internal communication channels will be required by these computers. Machines of this type are intended to be used for scientific research, but they will be applied to business applications later.

Semiconductor manufacturer Intel, in its Scientific Computer division, is putting together 128 of its 80386 chips to create a new 512 MIP machine at significantly lower cost than that of current supercomputers. Several firms, using chips from domestic suppliers, are producing parallel processors for what is estimated to be a $200 million market.[15]

Supercomputers are precursors of mainframes. The technology and architecture embodied in them will be found in mainframes of the future. Advances in supercomputer technology show no signs of slowing. Mainframes are precursors of personal workstations. Personal workstations of the future will have 10 to 100 times the power of today's personal computers. The technology of computing is advancing rapidly and relentlessly.

TRENDS IN SYSTEMS ARCHITECTURE

As we have seen, performance or speed of computer systems continues to increase while unit size declines. Overall, cost per unit of function is declining significantly. Given that there are some theoretical and practical limitations that will impact these trends in the long term, what alternatives are available to prolong or extend them? What systems considerations or architectural innovations can be employed to attain continued cost/performance improvements?

These questions, among others, are the grist for the research mills at computer firms, government agencies, and public and private research labs in this country and abroad. The trends include ever faster circuitry, increasing numbers of nodes in parallel architecture, and revolutionary software and operating systems. The goal of the research is to achieve practical realization of inherently high performance through new architectural and hardware designs. Research breakthroughs must occur in programming and operating systems too.

CURRENT TRENDS IN PROGRAMMING

Parallel computer architecture raises significant challenges for programmers and language designers. In the past, compiler designers focused their attention on the efficient execution of sequential operations. Conventional languages specify that tasks be carried out one at a time and that program instructions and data be brought together for discrete sequential operations. Faster operation required increases in the speed of purely sequential activities. Parallel computers, however, demand languages designed to attack problems at the outset in a simultaneous fashion in order to capture the benefits of parallel hardware.

Because there are many sequential programs in existence, and perhaps because it was easier to work that way, first attempts at parallel programming tried to convert sequential operations into simultaneous operations. For example, programs containing loops were examined for parallel operations with some success, particularly in scientific vector or matrix calculations. Computer programs have been written to search for parallelism in what normally are sequential instructions. These programs generate code that makes use of the implicit parallelism they have discovered. Other methods based on data flows have also been implemented.

In the Connection Machine described earlier, the main memory is divided into approximately 65,000 increments, one increment for each processor. The data are distributed over the memory increments, one data element per increment. Then, as each processor carries out the same instruction in synchronized fashion, the data are transformed in all memory increments simultaneously. This parallel approach to problem solving works well on certain classes of problems. The machine has a supervisor processor that coordinates the activities of all the other processors and itself is controlled by a supervisor program.

Parallel-processing compilers are being developed for commercial applications. IBM clustered FORTRAN allows jobs to be executed across twelve processors with speed increases up to ten times faster than would be possible with a single processor.[16] This is an area of active research and significant advances are expected.

The task of designing language translators and making parallelizing compilers is formidable indeed, but the control programs needed to manage large parallel computers pose even more demanding challenges. Solutions to these challenges will permit us to capitalize on the enormous potential of parallel-hardware architectures.

OPERATING SYSTEMS CONSIDERATIONS

An operating system is a program that acts as an interface between the computer users and the computer hardware. The purpose of an operating system is to provide a friendly environment in which users may execute programs. Thus the primary goal of an operating system is to make the computer system convenient to use. A secondary goal is to permit the computer hardware to be used efficiently.[17]

The Evolution of Operating Systems

Early computers were operated by a programmer or operator who performed all the tasks needed to solve the problem at hand. The operation was called "hands on" because the user wrote the program and operated the computer devices. The system was reserved for use with a sign-up sheet. Each of the I/O devices was managed by the programmer/user through device drivers. The device driver, usually a small deck of punched cards, was included in the deck of punched cards for the application program. Setup time played an important part in this type of operation, as did the scheduling system itself. Some programs ran longer than anticipated, inconveniencing the person next in line. Other programs failed to run at all, causing the machine to remain idle until the next person arrived.

The operation of personal computers or departmental minicomputers in many organizations today resembles those early experiences. But, like the evolution of hardware, the future of operating systems for micro- and minicomputers can be predicted from the evolution of operating systems on large mainframes. What are the stages through which operating systems advanced, and what is likely for personal computers? Table 5.5 displays the steps through which operating systems have progressed.

TABLE 5.5 Operating Systems Advances

Programmer/operator

System monitors

Uniprogramming operating systems

Multiprogramming operating systems

Multiprocessing operating systems

System monitors and permanent computer operators solved many of the problems indicated earlier. The operator batched similar jobs. This reduced setup times and improved efficiency by managing the peripheral operations for the user. The system monitor contained all the device drivers, compilers, assemblers, and the job control system. To improve throughput, peripheral computers were available to perform card-to-tape and tape-to-printer tasks, therefore more efficient use of the CPU was achieved.

Further improvements in CPU efficiency can be attained by multi- programming application programs in main memory simultaneously. The operating system schedules execution of these programs according to pre- determined algorithms, which are designed to reduce delays to a mini- mum. The operating system must contain the functions of memory management, task scheduling, main memory and data protection mechan- isms, timer control, and others. Multiprogramming operating systems are extremely complex and require large amounts of time and resources to develop. These operating systems are now available for large personal computers.

Multiprocessor hardware configurations have been in existence for two decades or longer, and operating systems to manage them have evolved with the hardware. These operating systems handle all the functions mentioned above; in addition, they manage several interconnected multiprogramming CPUs. The degree of complexity in these systems qualifies them as the most complex undertaking of human beings. The complexity of operating systems supporting large parallel processors will probably increase by orders of magnitude in the future.

Contemporary Operating Systems

Contemporary operating systems contain extensive communications and telecommunications software that permit applications programs to utilize networks with ease. They also provide data-management systems for use by application programmers to handle and organize the large amounts of data associated with many modern applications. Most organizations have specialists who install and update the operating system. These skilled individuals, called systems programmers, assist applications programmers and users with the intricacies of the operating system.

Contemporary operating systems for mini- or microcomputers are also communications oriented. Small computers are frequently networked to each other and to large systems throughout the firm. It takes many years for sophisticated operating systems to evolve for large computer systems. But the pace of evolution for small computer systems is much more rapid. The technologies of large systems are migrating to small systems with increasing rapidity. Alert IT managers will observe these trends and profit from them.

The software associated with future systems will be extremely complex, and it will be substantially different from that in common usage today. Systems software designers and perhaps systems programmers must change their vision of data processing problems if they are to realize the full potential of hardware advances. Their new vision will make it possible to solve ever more sophisticated problems, problems thought intractable in the sequential world of today. The users of computing technology will benefit considerably from their vision and inventiveness.

COMMUNICATIONS TECHNOLOGY

The worldwide telecommunications industry has been capitalizing on the same technology as the computer industry, and for the same reasons. Digitization of what formerly were analog communication signals is permitting the industry to apply the enormously productive digital technology to its benefit. New devices, new media such as satellites and fiber optics, and changes in the industry structure and environment such as the court-mandated breakup of AT&T are all operating in concert. These and many more factors are bringing growth to the industry. They also benefit users by providing enhanced communications capability. Advances in telecommunications will be explored in the next chapter.

TECHNOLOGY TRENDS

The technology of information handling is advancing rapidly. IT managers must be alert to technology trends and must be prepared to take advantage of the trends for themselves and their organizations. What are the trends and what do they mean for managers? Table 5.6 summarizes computer technology trends.

TABLE 5.6   Computer Technology Trends

1.   Vast increases in computational capability
2.   Availability of huge data stores
3.   Complex, easy-to-use operating systems
4.   Extensive integrated telecommunications networks
5.   Increasing function at declining costs

The Meaning for Management

The economics of technology advances will make vastly increased computational capability available to the firm. This capability will result from larger numbers of mainframes, mini- and microcomputers with improved performance, compared with today's systems. Future managers will have orders of magnitude more computing capability than today's managers. They will be continuously challenged to achieve the full potential of the rapidly evolving capability.

The increased power of computer hardware will be accompanied by huge data stores containing all kinds of information. Much of this information will be generated within the firm, but increasing amounts will originate from public sources available to wide audiences. Access to these data stores will be rapid, easy, and inexpensive. The potential for information overload will increase.

Advanced hardware with great power and complexity will be made functional by enormously complicated operating systems. Operating systems will be designed specifically to make computing power available to a wide audience in an easy-to-use manner. These hardware and operating system advances will take place on systems of all sizes, from supercomputers and mainframes to small personal workstations. Extensive, fast, and easy-to-use telecommunications networks will enable parallel processing at the firm level.

Electronically interconnected computing devices for all the firm's employees will permit rapid communication and nearly simultaneous parallel problem solving. In some firms, employees are beginning to engage in parallel processing with local networks. The firm will be linked electronically to its customers and suppliers on a global basis. Telecommunications will shrink time and distance and will enable the efficient functioning of international alliances and global partnerships. Alternative forms of organization will flourish in this environment.

The meaning for management is significant. In 20 years, the potential available to solve business and other problems will exceed current capability by orders of magnitude. It will challenge management to capture this potential for the benefit of their firms.

These potential benefits will not come without some eventual pitfalls. Management must not only harness this growing wave of potential, but must also manage the inevitable personal and organizational problems riding it. Recalling the Drucker quotation with which we began the first chapter, we realize that one of the challenges will be to accomplish technology assimilation in the face of reductions and dislocations of managers.

Skillful general managers will face severe challenges. They must develop strategies to introduce emerging technology. They must capture the advantages of the technology for their organizations, and they must contend with a cornucopia of difficulties along the way.

The Meaning for Organizations

The first chapter emphasized the fact that the conversion to knowledge-based organizations will cause major organizational changes. One of these will be the reduction of management positions by two-thirds, according to Drucker. Some scholars thought the introduction of digital computers 30 years ago would generate widespread unemployment from the automation of thousands of routine tasks. Instead, the creation of new jobs and the upgrading of old ones offset employment declines. The possibility now exists that the anticipated job reduction will occur among managers. Ironically, these are the same people who are expected to usher in the new knowledge-based age.

Perhaps, as in the past, we are being too pessimistic and our knowledge-based society will again create new and better jobs for those middle managers predicted to be displaced. In any case, our organizational lives will not remain static for long. Managers can look forward to evolutionary, if not revolutionary, changes in organizational structure and in the content of managerial roles. Adapting to the ever-shifting currents of change via organizational and structural adjustments will challenge future managers.

Other organizational changes are under way. Firms are creating alliances and forming joint ventures to exploit technology. And they are transforming themselves in other ways too. Baxter International and IBM joined forces in the health care field; Volvo's subsidiaries throughout Europe and North America are tightly coupled to headquarters in Sweden; and Corning's structure is a radical departure from the norm of 20 years ago. These and other manifestations of information technology are discussed in subsequent portions of the text.

Rapid advances in technology are impacting organizations in another way. Alliances are forming among major corporations to speed the development of new technology by sharing financial and intellectual resources. For example, an organization called the Consortium for Superconducting Electronics is being planned.[18] AT&T Bell Laboratories, the Research Division of IBM, the Massachusetts Institute of Technology, and MIT Lincoln Laboratories are forming a consortium to investigate applications of high-temperature superconductivity. The organization will coordinate the activities of researchers in the areas of high-speed signal distribution, superconducting devices, advanced integrated circuits, and thin-film materials and technology. Consortia, alliances, and joint ventures have been formed by other companies both here and abroad to produce advanced technology for the information processing industry.

SUMMARY

The rapid technological advances experienced for more than three decades will continue during the next decade or two. These advances will require solutions to many technical problems in metallurgy, mechanics, systems architecture, programming systems architecture, and in other scientific and engineering disciplines. Information processing technology will advance rapidly, challenging our ability to utilize it fully. There will be technological solutions looking for problems. Management and enterprise-wide issues will continue to restrain our ability to capitalize fully on the rapidly emerging technology.

Achievement of these technological advances brings huge opportunities for managers in all lines of work. These opportunities can be exploited by many managers; their employers will obtain advantage from the exploitation of technology. The employment of sophisticated technology is not an optional activity. In our competitive global economy, failure to stay technologically current will be a high-risk option.

"Is the environment today riskier than it used to be? You bet it is," says John Akers, chairman and CEO of IBM. "Why? Technology is the key answer. Its pace is quickening, and any competitor who won't take the risks to be right on the curl of the wave won't survive for long."[19]

Review Questions

1. What are the technological foundations for advances in digital computers?

2. Which of these technologies has advanced most moderately and what do you think is the reason for this?

3. What are some of the characteristics of the semiconductor industry worldwide?

4. What trends do we observe about the semiconductor industry and what evidence is there that these trends will continue for the next decade or more?

5. Why are advances in logic and memory technology *and* simultaneous advances in recording technology essential for progress in systems development?

6. Why are rigid-disk manufacturing plants similar in sophistication to semiconductor plants?

7. Why are optical recording devices for use in digital computing systems increasing in popularity?

8. What range of performance in computing systems is now available to businesses?

9. What are the important trends in supercomputers?

10. What challenges do parallel processors pose to operating system developers?

11. How does the operation of modern-day PCs resemble the operation of mainframe computers 20 years ago? What are the implications for IT management?

12. What functions are performed by modern operating systems developed for mainframe computers?

13. What is the relationship of communications technology to other parts of information technology?

14. How would you summarize the dominant trends in computer technology today?

15. What meaning do technology trends have for management? What meaning do they have for organizations?

Discussion Questions

1. Using concepts from Part I, compare and contrast the strategies of Intel and Sun Microsystems.

2. Discuss the strategies that are being employed by the Japanese manufacturers in the commodity memory chip business.

3. With financial data given in this chapter and making some careful estimates of other necessary data, construct a break-even chart for Intel's 386 parts. What are the significant factors upon which break-even depends?

4. The Intel 80386 cost about $100 million to develop. It is estimated that the follow-on product, the 80486, took $300 million to develop. What risks are involved in this development work, and how might these risks be quantified?

5. What is the annual compound growth rate in unit performance per dollar for a technology that doubles in unit performance and halves in cost over a four-year period?

6. What is the comparative cost per MIP of a present day personal computer and the TF-1? What might account for the disparity? (Use approximations.)

7. Discuss why technology advances increase management risk. What actions might managers take to mitigate the risk factors?

8. What factors have encouraged the rise in alliances, consortia, and joint ventures? What are the international implications behind these movements?

9. Discuss the elements of risk and reward for corporations engaged in high-technology development.

Assignments

1. Perform research to determine the amount of money spent by U.S. corporations on research and development activities in the areas of semiconductor technology, recording technology, telecommunications technology, and programming. Compare these amounts with the R&D activity in the pharmaceutical industry. What conclusions or inferences can you draw from the data?

2. Obtain descriptive details on one of the advanced computers mentioned in this chapter and summarize them for the class. In particular, describe where this computer is expected to be used.

ENDNOTES

[1] Adapted by permission of *Forbes* Magazine, March 7, 1988. © Forbes Inc., 1988.

[2] Intel ranked 266 in sales on the *Forbes 500s* list in 1989. Revenues were more than $3.1 billion, and Intel earned $391 million. It employs 21,500. Intel's productivity of $18,200 profit per employee places it fifth out of 31 companies in computers and communications hardware. "Ranking the Forbes 500s," *Forbes*, April 30, 1990, 286.

[3] David Coursey, "IBM Shows First PC Based on Intel 486," *MIS Week*, April 17, 1989, 1. Intel has other advanced products in its line. In early 1989 Intel announced the 80486 and vendors rushed to support it. The 486 uses memory caching; it runs at approximately double the speed of the 386.

[4] John F. Akers, *IBM Corp. Annual Report*, January 31, 1989.

[5] John Bardeen, Walter Brattain, and William Shockley invented the transistor at Bell Laboratories in 1948. In 1956, they received a Nobel Prize for their invention.

[6] During the 1940s and 1950s, analog computers were undergoing development and were thought to have great promise. The invention of the transistor and the development of high-density silicon chips allowed digital computers to eclipse their analog counterparts.

[7] "IBM Scientists Create Fastest Silicon Circuits," *The Denver Post*, March 17, 1988, C5.

[8] "Silicon's Sway in Chips Expected to Continue," *The Wall Street Journal*, February 25, 1988, 29.

[9] Barbara Depompa, "IBM Unveils Fastest Mainframe Storage," *MIS Week*, November 20, 1989, 14.

[10] Gary Slutsker, "Search Me," *Forbes*, January 25, 1988, 90.

[11] Mary E. Thyfault, "Distributed Computing Users Set for CD-ROM, Study Says," *MIS Week*, January 29, 1990, 19.

[12] *MIS Week*, April 5, 1989, 1.

[13] "The New Rules of the Game," *Solutions*, special issue, 1989, 7.

[14] "The Next Computer Revolution," *Scientific American*, October, 1987, 64.

[15] "Intel: The Next Revolution," *Business Week*, September 26, 1988, 74.

[16] *Second Quarter 1989 Report to Stockholders*, IBM, 1989.

[17] James L. Peterson and Abraham Silberschatz, *Operating Systems Concepts*, 2nd ed. (Reading, MA: Addison-Wesley Publishing Company, 1985). This book is a good source of information on operating system fundamentals.

[18] *Second Quarter 1989 Report to Stockholders*, IBM, 1989.

[19] John F. Akers, "Letter from Chairman and Chief Executive Officer," *Think*, Vol. 54, no. 3 (1988).

REFERENCES AND READINGS

Brandt, Richard, Otis Port, and Robert D. Hof. "Intel: The Next Revolution." *Business Week*, September 26, 1988, 74.

Communication News, August 1987. Special report on fiber optics.

Gelernter, David. "The Metamorphosis of Information Management." *Scientific American*, August 1989, 66.

Mackentosh, Allen R. "Dr. Atanasoff's Computer." *Scientific American*, August 1988, 90.

Meindl, James D. "Chips for Advanced Computing." *Scientific American*, October 1987, 78.

Peterson, James L., and Abraham Silberschatz. *Operating Systems Concepts*, 2nd ed. Reading, MA: Addison-Wesley Publishing Company, 1985.

Wiegner, Kathleen K. "Who's On Second?" *Forbes*, March 7, 1988, 158.

The U.S. electronics industry achieved sales of approximately $268 billion in 1990 and is expected to increase sales to $580 billion in 2000 according to Standard & Poors. Computers, office equipment, and communication equipment accounted for 55 percent of sales in 1990 and are projected to increase to 63 percent of the much larger total in 2000. The remaining sales are for instruments and controls, consumer products, and components.[1]

Merchant semiconductor suppliers to the electronics industry worldwide are dominated by well-established Japanese firms. The top three suppliers are NEC Corp., Toshiba, and Hitachi Ltd., and Mitsubishi, Fujitsu, and Matsushita are sixth, seventh, and ninth respectively. Motorola, Texas Instruments, and Intel ranked fourth, fifth, and eighth in this industry. The top nine suppliers had combined sales of approximately $29 billion in 1989. Some of these firms manufacture computer products and use their semiconductor components in them. However, some major computer companies also manufacture semiconductors for their own computer products but do not sell semiconductor components to others.

In addition to component logic and memory products, some firms design, develop, manufacture, and sell microprocessors or microcontroller chips for others to use as computer building blocks. In some cases, a second-source supplier is licensed to manufacture products designed by others. The top ten suppliers of microprocessors and microcontrollers in 1989 are listed below.[2]

1. Intel Corp.
2. NEC Corp.
3. Motorola Inc.
4. Hitachi Ltd.
5. Mitsubishi
6. Texas Instruments
7. Toshiba
8. Matsushita
9. Advanced Micro Devices
10. SGS-Thomson

[1] *Standard & Poor's Industry Surveys*, May 24, 1990, Volume 158, No., 20 Sec. 1. E-16/E-31.

[2] Op Sit

6

Advances in Telecommunications Systems

A Business Vignette
Introduction
The Telecommunications Services Industry
Telecommunications Equipment Suppliers
Telecommunications Systems
    Cluster Controller
    Line Adapters
    Telecommunications Lines or Links
    Communications Controller
A More Sophisticated Telecommunications System
Service Offerings
Some Current Developments
    Integrated Services Digital Network
    The T-1 Service
Network and Communications Standards
Open Systems Interconnect
Vendor Network Architecture
    IBM Systems Network Architecture
    DEC Digital Network Architecture
    Transmission Control Protocol/Internet Protocol
Additional Types of Networks
    Local Area Networks
    Wide Area Networks
Summary
A Business Vignette
Review Questions
Discussion Questions
Assignments
References and Readings
Appendix A
Appendix B

A Business Vignette

Radical Changes Are Standing the No. 3 U.S. Industry on Its Ear. They Will Transform the Way We Work.[†]

John Sculley, chairman of Apple Computer, turns to a VCR in his gadget-crammed office and pops in a cassette. "Let me show you how the Macintosh will work in a large corporate network," he says. On the videotape, a manager flicks on his computer and sees a diary entry for the day, reminding him that one of his employees is due for a salary review. The manager types a password that links his desktop Mac to the company's mainframe computer, which handles personnel records. The Macintosh retrieves the files, runs a program that compares the employee's salary with those of his peers, and displays the results as a bar chart on the manager's screen. The manager decides how big a raise he wants to give and alerts the personnel and payroll departments electronically. Elapsed time: three minutes.

Declares Sculley: "That's a perfect example of where we think the industry is going. We're trying to create the links that will bring the desktop Macintosh into the mainstream of corporate computing."

At age 40 the computer industry—now the third largest in the U.S., after only automobiles and oil—is undergoing a mid-life crisis brought on by two developments long in the making. Desktop computers have become so powerful that they can perform many of the functions of their bigger minicomputer and even mainframe brethren. Customers have become downright adamant that manufacturers figure out how to make their systems work together.

The industry that has transformed the way most people work is about to be transformed by the way most people want to work.

As in earlier upheavals that redrew the map of the computer industry—the advent of minicomputers in the 1960s and that of personal computers a decade later—the immediate catalyst for the turmoil is new technology. Already reaching the market are powerful microcomputers that cost as little as $5,000 to $10,000. They can handle large, complex tasks once reserved for machines 20 times more expensive.

Breakthroughs in performance are opening up new markets for microcomputers. Many customers who have bought the souped-up new PCs for such

[†] Stuart Gannes, excerpted from *FORTUNE*, August 1, 1988, 43-45.

uses as financial spreadsheets and word processing, for example, are discovering that desktop machines can handle all sorts of additional computing tasks, from engineering design to newsletter publishing.

Equally impressive is the increasing versatility of communications networks. Says John Young, chief executive of Hewlett-Packard: "The central reason customers haven't been getting as much value out of their computers as they would like is simply the inability to tie them together. That barrier is about to disappear."

For most companies, the era of network computing is not yet a reality. But it is arriving with stunning speed. A strong synergy is emerging between the new desktop machines and modern computer networks. Each makes the other more useful, and thus more attractive. Computer industry analyst Mark Stahlman of the Wall Street firm of Sanford C. Bernstein estimates that in just four years, sales of high powered microcomputers—along with sales of other components of highly integrated networks—have grown to constitute 10% of the computer business. "By the early 1990s these systems will account for the majority of growth in the industry," he says.

With desktops emerging as the crown jewels of computerdom, personal computer manufacturers are hurrying to improve their machines, especially by adding networking capabilities. American businesses have bought more than 15 million personal computers. At present, fewer than 30% of them are hooked up to networks. Soon, however, the great majority of them may be. "By 1991, 90% to 95% of our machines will be used in networks," says Bill Lyons, IBM's general manager of personal systems merchandising. Indeed a much ballyhooed feature of IBM's new PS/2 series of advanced personal computers, the so-called Micro Channel, helps the machines link up with IBM minicomputers and mainframes. The company contends that very shortly personal computers will not only be connected through local area networks, but also function as terminals for large data-processing systems.

Says Sculley: "There are two major differences between the last generation of personal computers and the models available today. The first is that the new PCs are far easier to use. The second is networking, which is a big part of how these machines will operate. When you get to sophisticated networking, if anything requires more performance than your computer has, you can farm it out to other machines on the network. The personal computer will become the epicenter of the whole industry."

The preceding chapter presented trends in semiconductor, recording, telecommunications, and programming technologies. It explored trends in the development of electronic digital computers and discussed the meaning of these trends for users of the technology. This trend information gives us a perspective on the future capabilities available to managers and to their firms. Technology advances enable systems architects to construct ever more capable computing devices at steadily decreasing costs to the end user. Systems and applications programmers have created a hundred billion or more lines of code to apply this technology to problems in virtually every area of human endeavor. The pace of innovation has increased over the decades. It promises to continue into the near future at least.

The expression of technological advances is nowhere as evident as in the explosive growth of desktop computers. For instance, Apple Computer earned $438.4 million on sales of nearly $5.4 billion in 1989, from products that did not exist 20 years ago. The Business Vignette points out that the genesis for the microcomputer revolution is new technology, and microcomputers based on new innovations are changing the nature of work in industrialized society. However, the tremendous impact of personal computers depends on telecommunications and networking to a considerable extent.

The communications industry worldwide has experienced enormous growth for the same reasons that the data processing industry has. Advances in logic and memory technology, coupled with the ever-increasing digitization of analog signal transmission, are offering tremendous opportunities to the industry. Advances in transmission devices and in media and their implementation are providing increased bandwidth to a broader audience. The economics are favorable, and the industry is experiencing significant growth in demand for services.

Although these technological advances have been shaping the progress of the telecommunications industry, nothing has had a greater effect in the United States than the divestiture of AT&T. The breakup of AT&T and the Bell System, starting in 1984, about 100 years after the company was formed, ranks as one of the most significant business events in the U.S.[1] It changed forever the business climate in one of our largest and most important industries, moving the industry from monopoly to one of competition. The creation of the Regional Bell Operating Companies from the divestiture of AT&T set in motion a stream of events intended to enhance and improve our telecommunications industry. These events have had important national and international consequences. Social, regulatory, and technological factors are combining to increase the pace of change to the providers of the service and to the users of the service.

This chapter focuses on the factors shaping the industry. It will consider the business, regulatory, and technological forces behind the trends. It will also review the industry, its characteristics and makeup, and the systems and services it provides. The text will consider emerging service offerings and the importance of standards organizations. And it will demonstrate how these factors provide opportunity for managers and their organizations.

THE TELECOMMUNICATIONS SERVICES INDUSTRY

Today's telecommunications services industry is shaped primarily by the antitrust case brought against AT&T during the early 1960s. The case was tried before U.S. District Court Judge Harold H. Greene who ruled that, effective January 1, 1984, the American telecommunications monopoly had to divest its local telephone companies and must compete in the long-distance and data communications markets. The seven operating companies created at that time now serve the telecommunications needs of their respective geographical areas. AT&T and many competitors now provide long-distance service to the U.S. and beyond.

More than 200 million telephones and millions of other communications devices in the U.S. obtain local and long-distance service from AT&T and the seven regional Bell operating companies (RBOCs) and a large number of non-Bell companies. The non-Bell companies range in size from large interstate corporations to small local companies. Altogether, the annual revenue for the group is estimated to be $164 billion in 1990. A summary of the major firms in this industry, displaying their revenue estimates for 1989, is shown in Table 6.1.

TABLE 6.1 **Revenue of Selected U.S. Telecommunications Companies, 1989**

Firm	Revenue (in millions)	
AT&T	$37,250	
Ameritech	10,700	RBOC
Bell Atlantic	12,300	,,
Bell South	14,300	,,
NYNEX	13,600	,,
Pacific Telesis	9,700	,,
S.W. Bell	9,100	,,

(Continued)

Part Two: Information Technology Trends

TABLE 6.1   Continued

Firm	Revenue (in millions)	
U.S. West	10,000	,,
Centel Corp	1,150	
COMSAT	450	
GTE	21,400	
MCI	7,700	
S. New England Tel.	1,600	
United Telecom	8,300	

Of the $180 billion in revenue generated in 1990 by these and other companies not listed, $79.7 billion, or 44 percent, is derived by the seven RBOCs. AT&T garners 21 percent, while all others capture the remaining 35 percent. Since divestiture, the RBOCs have increased the total number of installed lines by 13 percent. They have grown their revenue by 23 percent and have decreased their employee base by 5.4 percent. All the firms listed above are profitable, and most of them are growing modestly.

TELECOMMUNICATIONS EQUIPMENT SUPPLIERS

The revenue for equipment suppliers in 1988 approaches $8 billion, not including equipment produced by the firms in Table 6.1. A prominent company in the business of producing telecommunications equipment is Northern Telecom. This Canadian firm is expected to generate $6.8 billion in revenue in 1990. NT spent 12 percent of sales on research and development in 1989 and is the first company to initiate a fiber-optic network product line.[2] There are also many small companies producing equipment for the telecommunications marketplace. These firms provide equipment for services ranging from small local area networks to large multinational satellite networks. They enable the use of the media and technology by companies of all kinds and sizes.

TELECOMMUNICATIONS SYSTEMS

The purpose of telecommunications systems is to connect information processing devices for the purposes of exchanging data and information.

The information exchanged can be voice, data, or video communication originating or terminating with humans or machines. Figure 6.1 displays a typical telecommunications system.

The system depicted in Figure 6.1 may be found within a firm supporting office systems, for example, or may be used in an order-entry application. Many types of applications utilize large mainframes to coordinate information from many terminals, or support many terminals accessing databases on a mainframe. Some of the strategic systems we studied earlier utilize a telecommunications network of this type. What are the main parts of this system and what are its characteristics?

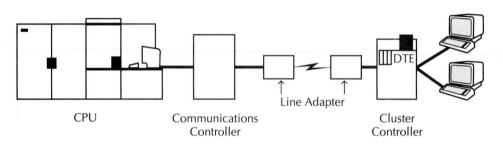

FIGURE 6.1   A Typical Telecommunications System

Telecommunications systems are characterized by numerous input and output devices serving multiple simultaneous users. The users expect real-time responses from the system; therefore, the network must be dynamically loaded. Some or all of the components of the system may be owned by the firm using the network; but, in most cases, the firm engages a third party to provide the transmission facility. These facilities may consist of long-haul fiber networks, microwave links, or other facilities. These capabilities are obtained from the firms in the telecommunications service industry.

The equipment and services available to firms today allow the integration of voice, data, and image communication. This integration improves the efficiency and the cost-effectiveness of the services; and, as in the case of digital computing technology, causes changes in work patterns and alters organizational structures. These integration opportunities allow companies to leverage their networks and to use their networks for competitive advantage.

Information can be transmitted over telecommunications media in either analog or digital format. In analog or voice format, information is represented by transmission of a signal that varies in frequency and amplitude. In the North American telephone system a standard voice channel accommodates a range of frequencies up to 4000 cycles per second. Frequencies above 4000 cps are attenuated. Another way to characterize the range is this: The bandwidth of the system is 4K Hz, i.e., communication is restricted to frequencies below 4000 cycles per second. The telephone equipment at the transmitting end of the circuit converts vibrations received in the microphone into electrical signals. At the receiving end of the circuit it converts the electrical signals into mechanical vibrations in the receiver.

In digital transmission, information is represented by a bit stream in much the same manner as it is in the computer. Computer signals or data are digital. They must be converted into analog format prior to transmission over the mostly analog circuitry of the local telephone network. The process for accomplishing this conversion is called modulation, and the reverse process is called demodulation. The process of modulation and demodulation, or conversion of digital signals to analog and back to digital again, becomes unnecessary when the telephone system is totally digital. Totally digital networks are replacing analog systems rapidly. AT&T's network, originally analog for voice transmission, is nearly 90 percent digital at this writing.

Cluster Controller

As depicted in Figure 6.1, the cluster controller is an important hardware device that interacts with the telecommunications line and manages communication devices attached to it, typically terminals used to initiate or receive messages. The purpose of the cluster controller is to handle inbound and outbound traffic from the devices, collating and decollating the messages, and providing control functions. It scans the devices for activity and monitors the line coming to it. In some instances, it buffers simultaneous messages to or from the input/output terminals. This equipment is frequently referred to as data terminal equipment.

Line Adapters

Line adapters perform the function of matching the digital signal of the data terminal equipment to the signal characteristics of the telecommunications line. Most low-speed data communication occurs over lines designed for voice communication, which operate in analog rather than digital mode. The line adapter modulates the analog signal with information originating in digital format and demodulates or reverses the process at the receiving end of the link. The line adapter is frequently called a *modem*, a term derived from the functions of *mo*dulation and *dem*odulation.

Modems have surged in popularity along with personal computers. Personal computers connected through the phone lines via modems communicate with each other, with computers of all types, and with other networks. This type of operation will increase as the trend toward distributed data processing accelerates. Modems for these purposes operate at speeds from 300 bits per second up to 2400 bps. Modems for much higher operations are available in a wide variety of capabilities and at prices related to their performance potential.[3]

Telecommunications Lines or Links

The medium used to transmit the signals between the line adapters or modems is called the line or link. The important characteristics of the link are its length, its directional capability, its ability to share signals from more than one source, and the physical material from which it is constructed. These factors all affect the ultimate capability of the link. They also play a role in the pricing algorithm for the link.

Although there is no theoretical limit to the length of a telecommunications link, there are some practical limitations. Electrical signals travel approximately 300,000,000 meters per second, or one mile in 6.3 microseconds, or about one foot per nanosecond. The signal suffers a time delay, albeit very slight in most cases, as a result of the distance traveled. The signal strength is attenuated or diminished by line losses along the path. Hence, the power required for signal transmission is a function of the line length. There are some other small effects as well. These limitations or considerations apply to each segment of the link.

For physically connected lines or links, the media can consist of copper wire, coaxial cable, or optical fibers. Two insulated wires wrapped around each other are called a twisted pair. Twisted pairs are common in local telephone circuitry. A coaxial cable consists of a metallic sleeve through which runs an insulated circuit wire. The braided metallic sleeve protects the conductor from physical damage and isolates it from electrical noise. The bandwidth or signal-carrying capacity of the coaxial cable is much greater than that of the twisted pair. Coaxial cable is used for many applications including TV. Cable TV is named for the communication medium it uses. Optical fibers transmit signals in the form of light waves or pulses. They can have very great bandwidth.

Links can consist of microwave transmissions that use the atmosphere as the medium, or satellite transmitters and receivers in which the atmosphere and space are the media. Each medium has different transmission characteristics and capabilities, and each can be used efficiently in the telecommunications system.

Communications Controller

The communications controller is a complex, specialized unit that controls telecommunications links and is the termination point for the lines. It provides the means for the mainframe computer to access the remainder of the telecommunications system. The communications controller is a very advanced programmable digital computer designed specifically to perform the task of interfacing between many line adapters and the main CPU. It contains large amounts of logic and memory circuitry, and it can be reconfigured to accommodate changing network designs.

Sometimes called the communications front end, the communications controller manages high-speed inbound and outbound traffic and performs numerous control functions. It scans the lines for messages or traffic and monitors and coordinates the computer communications activity. The controller has high-speed buffer memory that is used to store temporary message queues. The controller also contains logic circuitry designed for error detection and some form of error correction. A great deal of logic is required to perform line control. This device is the heart of the computer-based telecommunications system.

Telecommunications applications run on the CPU, supported by the hardware discussed above and by the operating system, which provides communications and system-management functions. These functions contained in the hardware and software are relatively transparent to the applications systems designer and the system user. Systems programmers or telecommunications specialists support the network hardware and software for the firm's applications programmers.

A MORE SOPHISTICATED TELECOMMUNICATIONS SYSTEM

The telecommunications system described above is useful in applications such as reservation systems or order entry. Many large office automation systems are configured in this manner as well. Many large international applications make use of satellites as part of a network. A typical telecommunications system utilizing a variety of links is depicted in Figure 6.2.

The telecommunications network shown in Fgure 6.2 uses most of the types of links discussed above. The messages originate at terminals connected by copper wires to a microwave relay station, which sends them to a satellite transmission system. The messages are transmitted to a satellite 22,000 miles above the earth's surface and beamed back to a satellite receiving station several thousand miles away. The messages continue via fiber-optic cables to a switching station, where they enter the copper system again. The network of copper wires routes the messages to the receiving terminal. This type of system is typical of those used for transacting international business.

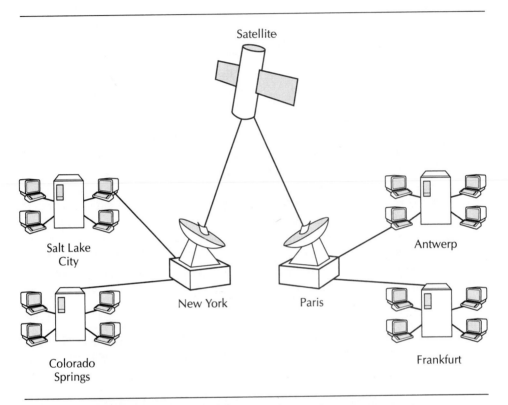

FIGURE 6.2   Components of a Telecommunications System

SERVICE OFFERINGS

Firms that sell telecommunications services to private individuals, businesses, and industry are called common carriers. They provide a wide variety of service offerings, ranging from the dial phone network through broadband satellite transmission. The bandwidth for local voice grade service is limited to about 4K Hertz; satellite video communication requires a bandwidth from 1M Hertz to 6M Hertz. The telecommunications industry provides many services between these extremes.

Transmission distances range from local and interoffice up to 30 miles or so, to intercontinental. Some multinational firms have telecommunications systems connecting their offices around the globe. The bandwidth in these systems may be available over these distances continuously or on demand.

The most familiar form of on-demand service is the telephone dial network. The line or link is permanently installed and attached to switches, the telephone itself, and other equipment. The phone is available for use on demand for local or long-distance communication. The cost of the local transmission media and of the trunk lines between central switching stations is recovered via a fixed monthly rate regulated by the state public utility commission. Business users pay a higher rate than residential customers. The long-distance rate is a function of length of connection time, time of day and day of week, and distance. Since deregulation of the U.S. phone service, firms compete for the long-distance business through pricing and service tactics involving these three parameters.

Continuous service is also available from common carriers via leased line or link. The common carrier provides the link to the lessee for a fixed fee, usually on a monthly basis. For this charge, the common carrier provides a transmission path according to stated specifications including bandwidth, noise levels, and modulation characteristics. Lessees, through the use of appropriate line termination equipment, apply the line to their particular application. Fully utilized leased lines are at lower cost to the lessee on a unit-of-transmission basis than on-demand service.

Common carriers may fulfill their obligations to the lessee through the use of several types of media. A long-distance phone call, for example, may start its journey over copper wire, move through coaxial cable or fiber-optic cable, move interstate via microwave, and conclude its journey over copper wire. The providers of service may vary the media from time to time to optimize their business or to improve the quality of service to their customers. The customer usually is indifferent to the choice of media, providing the technical specifications that were agreed to are being met.

Some leased lines today are used for the transmission of digital information. Some of this information is digital to begin with and some, perhaps most, is analog information that has been converted to digital format for transmission purposes. Most local lines within the U.S. phone network system today send and receive analog information, but the remainder of the system is digital. Within the analog portion of the network, digital signals must be converted to analog format prior to transmission using line adapters or modems.

Digitized information can be transported over a range of media through the common carrier network. The transmission requirements dictate the type of equipment, the media to be employed, and the resulting cost.

SOME CURRENT DEVELOPMENTS

Integrated Services Digital Network

Integrated services digital network (ISDN) technology integrates digitized voice and data of all types for transmission over the copper circuitry and other links currently installed in the phone system. ISDN eliminates the analog portion of today's networks and makes the network digital from end to end. It offers enhanced communications services and high-speed data transmission over one network element.[4] These services are now being offered to a few commercial customers, but they are proposed to be available to all subscribers who desire them.

ISDN is now part of a project within the International Telegraphic Union for standardization of parameters and interfaces for networks. Standardization will allow the network to handle a variety of mixed digital services. Implementation of the project is proceeding in the U.S. and abroad. All of the RBOCs have run experiments with the standards. The outcomes of the trials and experiments have intensified ISDN activity in the U.S.

There are two classes of service offered by ISDN: the basic rate and the primary rate. A depiction of these services is shown in Figure 6.3. The basic rate, called 2B + D, consists of two 64K bps B channels and one 16K bps D channel. The B channels can transmit either digitized voice or data; the D channel is used for signal control and routing information. An additional 48K bps is also available for other purposes.

The primary rate service, named 23B + D, consists of 23 64K bps B channels and one 64K bps D channel. The primary rate supports 1.544M bps bidirectional transmission. The B channels can be used for a mixture of voice and/or data communications needs. This type of service is useful for interconnecting data processing facilities within the firm, for connecting sites within the firm for voice communication, or for integrating voice, facsimile, and data transmission.

There are several advantages to ISDN technology. Today's traditional telephone and network system is inefficient because it contains a mixture of analog and digital technology. Since the 1960s, the local phone companies have been upgrading their switches to digital systems, but the circuitry to the individual phones remains analog in nature. ISDN is totally digital. ISDN equipment will be able to communicate with the central switch during routine calls using the D channel; this permits performance improvements and paves the way for new and different functions.

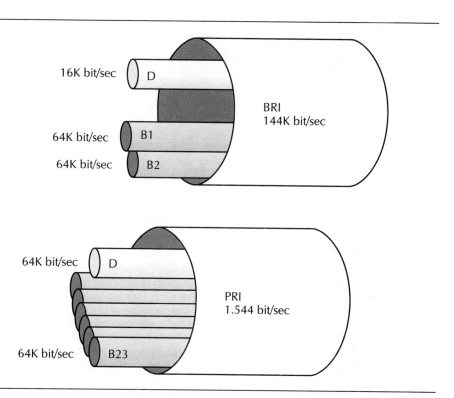

FIGURE 6.3   ISDN Basic and Primary Rate Interfaces

ISDN is not without problems. Equipment manufacturers have not totally standardized their products, and incompatibilities are causing problems for some implementations. Some customers' equipment that is analog in design must be changed or accommodated during the conversion. Additionally, the incompatibilities between the ISDN subscribers and the analog customers will need to be addressed. These difficulties will retard the rate of adoption of ISDN.

Since adoption of ISDN in the U.S. must proceed in the face of a huge investment in copper and analog circuitry, and the relatively effective phone system must remain operable during the transition, there are obvious financial barriers to a speedy implementation. Some countries with a less well-developed phone system or with a system in need of replacement may be motivated to proceed with ISDN implementation on an accelerated schedule. Parts of Europe will adopt ISDN earlier than the U.S. does.

The rise of ISDN promises to provide increased business volumes for the semiconductor chip manufacturers who design and fabricate ISDN chips. The market for these chips was about $17 million in 1988 and is projected to grow to more than $650 million by 1992. Depending on the ultimate penetration into the marketplace of ISDN technology, and including the switching systems and other equipment required for ISDN, some are forecasting the overall market driven by ISDN to reach $50 billion toward the end of the next decade. Whether these forecasts are accurate or not is the subject of intense debate within the industry.

The T-1 Service

In response to its new competitive environment, AT&T brought out a service called T-1 in September, 1983. (T-1 had been implemented internally for many years within the phone system to connect exchanges, but had not been made available commercially prior to 1983.) A T-1 link can be constructed of copper wires enhanced to handle digital signals at rates exceeding 1.5M bps. A T-1 link may also be implemented from other media. The link can handle up to 24 simultaneous phone conversations at approximately four times the cost of an ordinary line. If a firm can utilize the line for more than four phone conversations simultaneously, it can reduce costs significantly. More than 1000 firms, among them American Airlines and Merrill Lynch, have been able to capitalize on this technology, and it is expected that many more will do so in the future.[5]

T-1 services now being offered have been extended to include very much higher bandwidths than originally proposed. Table 6.2 displays the current structure of multiple service in North America. In addition, there is widespread interest in fractional T-1 service. This service permits customers to obtain a portion of a T-1 link for their use.

The conventional format for utilizing the basic T-1 capability divides the line into 24 channels. Using time-division multiplexing and pulse-code modulation, each channel can carry one voice transmission. Thus up to 24 simultaneous conversations can be transmitted over a basic T-1 link. Considerable increases in communication bandwidth can be obtained through the use of more capable T-type links with appropriate increases in lease charges. (Many firms cannot use the bandwidth of a T-1 line; they are purchasing fractional T-1 service from others.)

TABLE 6.2   T-1 Services

Digital Facilities	Transmission Rate	Number of T-1 Equivalents
T-1	1.544M bps	1
T-1C	3.125M bps	2
T-2	6.312M bps	4
T-3	44.746M bps	28
T-4	274.176M bps	168

Two steps are required to digitize an analog signal. The first step is sampling the analog signal, called pulse amplitude modulation. The second step quantizes the sample, called pulse code modulation. (A more detailed description of pulse amplitude modulation and pulse code modulation is presented in Appendix A of this chapter.) Twenty-four individual analog signals are sampled and quantized in turn once each 125 microseconds as shown in Figures 6.4 and 6.5. These quantized signals are assembled into one frame for transmission over the T-1 link. Each frame contains 193 bits of information for a total transmission rate of 1.544M bps. The construction of the frame is illustrated in Figure 6.5.

The process for constructing the frames by alternately interleaving digital signals from the 24 sources is called multiplexing. Devices for performing this operation are called multiplexers. Multiplexers for T-type links are electronic devices, frequently programmable, which assemble signals from a number of low-speed sources. For example, the signals from 24 voice-grade phone lines are assembled and interleaved for transmission over a single high-speed T-1 link. An additional multiplexer is required at the remote end of the T-1 line to unfold or decollate the signals into their 24 constituent parts. In addition, some administrative data is transmitted for synchronization purposes, message routing requirements, and the like.

Telecommunications equipment suppliers build and market multiplexers for handling one or many T-1 lines. These devices manage the traffic, detect and isolate defective lines, and provide redundancy, thereby improving reliability. Multiplexers are becoming increasingly sophisticated with the continued growth in functional capability of logic and memory circuitry such as that employed in computers. This sophistication improves throughput because it makes voice compression possible. It also increases reliability through error detection, correction, and redundancy techniques, and it increases customer satisfaction. These functional improvements are increasing demand for telecommunications services. Nearly one-half of the locations owned by U.S. *Fortune 1000* companies now employ T-1 technology.[6]

1 The signal is first shaped so it occupies a discrete set of values.

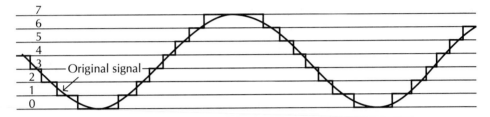

2 It is then sampled at regular intervals and the resulting signal coded for transmission.

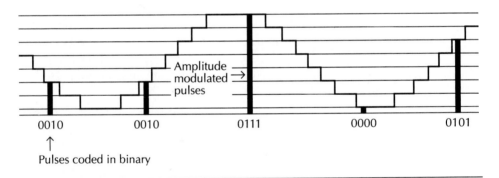

FIGURE 6.4   Pulse Amplitude Modulation

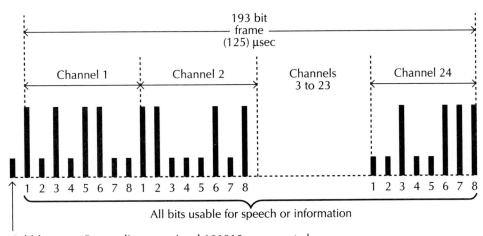

FIGURE 6.5   **The T-1 Frame**

NETWORK AND COMMUNICATIONS STANDARDS

Several important standards bodies and numerous standards govern the details of networking and communications. The International Standards Organization (ISO), operating through committees, establishes basic and far-reaching standards on a global level. National standards bodies such as the American National Standards Institute (ANSI) set certain standards for the U.S. In addition, equipment manufacturers and vendors propagate standards through the design, development, and sale of communications equipment. (These standards are not necessarily the same as those established elsewhere.)

Some widely known and widely supported standards include the ISO seven-layer Open Systems Interconnect standard; IBM's Systems Network Architecture; DEC's Digital Network Architecture; and the Defense Department's ARPAnet protocol TCP/IP (transmission control protocol/internet protocol). Others include the EIA (Electronics Industry Association) cable-pin configuration and use standards, the ANSI communication control characters standard, and the IEEE (Institute of Electrical and Electronic Engineers) standards for Ethernet and for other products and services. Some major corporations such as Boeing and General Motors have established *de facto* standards involving certain applications.

An important standard in today's environment is the International Standards Organization's Open System Interconnect Standard (OSI).[7] The purpose of the OSI standard is to create an open or nonproprietary computer communications discipline. The standard or model describes a network architecture capable of interconnecting different types of computers, networks, and communication technologies into an open or integrated network. The architecture describes how the network operates, the form it takes, and the protocols it uses. The architecture includes hardware, software, and data-link standards. OSI describes an approach to seven layers or levels of communication.

The seven layers of the ISO/OSI architecture are shown in Table 6.3. Figure 6.6 illustrates the seven layers of the OSI model and shows how one application communicates with another, using the conventions of the model. A more detailed description of the seven-layer architecture of OSI is presented in Appendix B of this chapter.

TABLE 6.3 Seven-Layer Architecture

Application

Presentation

Session

Transport

Network

Data link

Physical

OSI is important because it provides structure and modularity for a very large problem. Each layer specifies or provides services and each has at least one protocol or definition of how networks establish communication, exchange messages, and transmit data. The layered protocols reduce complexity, allow for peer-to-peer communication from one node to another, and permit changes to be made to one layer without affecting another.

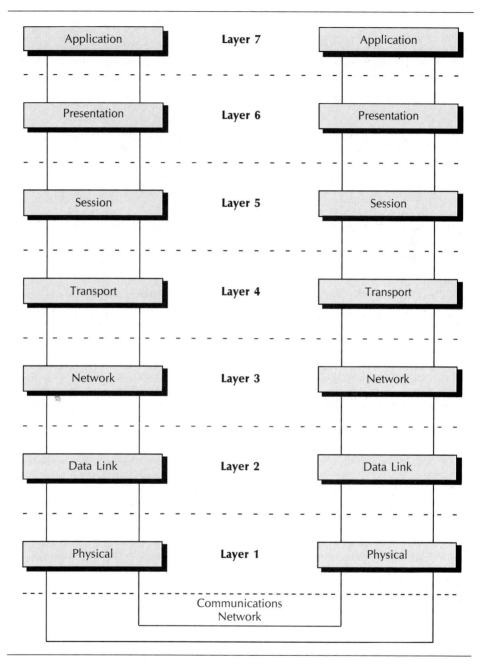

FIGURE 6.6   The Open System Interconnect Model

VENDOR NETWORK ARCHITECTURES

Several widely used network architectures do not correspond completely with OSI. The U.S. Department of Defense sponsored TCP/IP; Digital Equipment Company and IBM initiated two others. These architectures are important because of their popularity and widespread adoption. They form standards for many organizations and applications.

IBM Systems Network Architecture

IBM introduced Systems Network Architecture (SNA) as a strategic product offering in 1973 and has secured a large user base among major companies in the U.S. SNA is an IBM proprietary product. It is considered a *de facto* standard for IBM customers worldwide. SNA is a hierarchical, layered model supporting distributed mainframes, token-ring local area networks, and peer-to-peer communication. It is useful for application products such as CICS (Customer Information Control System) and IMS (Information Management System, a database management and retrieval program). The layered architecture is similar to that of the OSI model.

SNA is driven from a mainframe that manages the telecommunications process through a program called VTAM (Virtual Telecommunications Access Method). VTAM controls communications between terminals and the CPU through the communications controller, or front-end processor, which hosts a network control program. The cluster controller manages the individual terminals at the remote end of the link. In the SNA network, the mainframe supervises all events within the network.

DEC Digital Network Architecture

DNA (Digital Network Architecture) is Digital Equipment Corporation's proprietary network standard. Like SNA, it is a layered architecture roughly comparable to ISO/OSI. DNA supports distributed, interconnected DEC minicomputers and terminals attached to minis. DNA is popular for local area networks within a building or group of buildings. It uses the Ethernet vendor standard, which is similar to the IEEE 802.5 standard. DNA provides rapid data access and high data transfer rates, up to 10 megabits/second.

Digital Equipment Corp. stated its intention of modifying its network in the next few years. It wants to comply with the international standard protocols in order to further communications between brands of computers. DEC intends to use the lowest four layers of the OSI network model in its DNA architecture. This decision is seen by many as a strong endorsement for the OSI architecture.

Transmission Control Protocol/Internet Protocol

TCP/IP is a set of protocols developed by the Defense Department for its ARPAnet network; therefore, it is vendor-independent. Many vendors support TCP/IP. UNIX and Ethernet environments use TCP/IP and IBM, DEC, and Apple Computer support TCP/IP products. TCP/IP is widely used as an approach toward network interconnectivity; some vendors, including IBM, have TCP/IP software packages. This architecture is also of the layered variety.

Figure 6.7 develops a comparison between the ISO/OSI models and the three prominent vendor models described above.

ISO/ OSI	IBM/ SNA	DEC/ DNA	DOD/ TCP/IP
7. Application	End user (not part of SNA)	User	Process
		Network mgt	(File transfer protocol)
6. Presentation	Presentation services	Network applications	
5. Session	Data flow &	Session control	Host-to-host
	xmision ctl		
4. Transport	Path control	End communication	TCP
3. Network		Routing	Internet prot.
2. Data link control	Data link control	Data link	Network access (IEEE 802.3 or Ethernet)
1. Physical	Physical (not part of SNA)	Physical link	

FIGURE 6.7   ISO/OSI Model *vs.* SNA, DNA, and TCP/IP

There are three chief differences between the architectures described in Figure 6.7. The functions in an OSI layer may differ from the function in the equivalent layer in the *de facto* architecture; the interpretation of the function may not be identical; and some particular functions may be found outside of the layered structure. There are no significant performance differences.[8]

To attain closer conformance to network standards, executives of several computer and communications companies established the Corporation for Open Systems (COS) in 1986. Members of the COS include Apollo Computer, Apple Computer, AT&T, Control Data Corp., Digital Equipment Corp., General Electric, IBM, the National Bureau of Standards, Sun Microsystems, UNISYS, and Wang Laboratories. COS plans to test products for conformance and interoperability to OSI and ISDN standards. This effort, not yet fully implemented, will result in certification for products that pass the COS tests.

ADDITIONAL TYPES OF NETWORKS

Networks are frequently classified by the distance over which they operate or by the types of service they provide. Thus we have local area networks (LANs) and wide area networks (WANs), in addition to large national or international networks. We also have the telephone network, cable television networks, and satellite networks for television or other purposes. There are means for connecting these networks together, when required.

Local Area Networks

Instead of switched networks, architectures have been devised for local networks. The topologies of these networks are based on ring, star, branch, or other configurations. Users can exchange information or access information in databases using wide-bandwidth local networks. Each terminal on the network has a unique identification or address to facilitate information exchange. The network system contains some form of concentrator and some software to permit efficient and effective use of the network by terminal operators. These local networks are owned by each firm and, therefore, are private networks.

LANs usually serve microcomputers located within several thousand feet of each other. Each microcomputer contains a hardware device called a controller board to connect the micro to the communication lines. The controller board serves as the line adapter and also coordinates the internal processing of the micro with message processing. Frequently LANs have more expensive devices connected to them, such as large disk files, color graphic plotters, or laser printers. One microcomputer on the network, known as a file server or a print server, will act as the manager of the I/O devices. The LAN permits sharing of the peripheral devices by all users of the network and improves service to the user at modest costs.

Local area networks are very popular. A recent survey of individuals responsible for more than 500 corporate and institutional microcomputer installations across the U.S. revealed that 81 percent of all sites either had LANs installed or intended to do so. Ninety percent of sites with 250 or more PCs had or intended to have LAN networks. At the end of 1987, for all sites surveyed, 30 percent had no LANs, 24 percent had 1 Lan, 13 percent had 2 LANs, and 33 percent had 3 or more LANs installed. Of this same group, at that time only 19 percent projected having no LANs installed within 12 months.[9] Thus, local area networks are rapidly becoming common.

Figure 6.8 illustrates LAN network configurations consisting of branch, ring, and star architectures. Protocols to facilitate the use of these configurations are implemented in hardware and software. The purpose of protocols is to organize the communication among and between the micros and to solve any contention problem that such communication may generate. The medium for LAN communication is generally twisted pair or coaxial cable. When high-speed communication is required, fiber optics can be used.

Communications specialists in the systems support department of the IT organization understand the technical advantages and disadvantages of the different configurations and media. They develop network requirements and conduct network systems analysis for units needing LAN installations. In addition to a thorough understanding of user requirements, the installation of LANs requires a thoughtful analysis of the technical trade-offs and a clear understanding of the cost considerations.

Wide Area Networks

Wide area networks differ from local area networks in the speed of the links, the link media, and the architecture. The wide area network or WAN is built so that it can be expanded easily to incorporate larger geographic areas. A WAN is owned by a firm and therefore is a private network. Local area networks can be connected to each other or to WANs through gateways. The gateway is a microcomputer used to match the protocol of the LAN it serves with the protocol of the network with which it communicates. The gateway is subject to communication standards that make the interconnections possible for various LAN protocols. The objectives of the standards also intend to make the operation of the network transparent to the user.

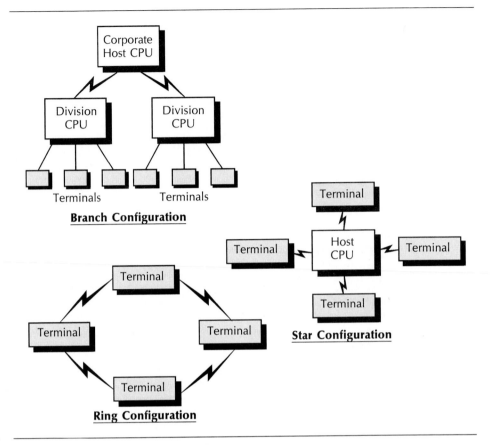

FIGURE 6.8   LAN Configurations

Nearly every mid-sized or large business has its own communication switch—an extension to the local telephone system called a Private Branch Exchange or PBX. The PBX can switch both voice and data in most cases, but data exchange through the PBX tends to be slower than it is with most LANs unless the PBX is designed for full data switching. The PBX is useful for applications such as E-mail or other office communication in which the volume is low and subsecond response time is not mandatory.

IT managers must manage the process for specifying and installing local area networks. They must ensure coordination between application developers and system users. IT managers must plan the network applications and must develop the plans from the strategy statements. Networks are critical to strategic applications and to productivity improvements for the entire firm. The IT organization and its management team must play a strong leadership role to ensure success with network technology for the firm. The importance of networks and network managment is illustrated by the Vignette describing Volvo's global network and the role it plays in Volvo's international operation.

SUMMARY

Telecommunications is one of the important ingredients of advances attainable with information technology. Semiconductor technology and programming are used extensively in the development and use of telecommunications equipment. The telecommunications service industry and equipment suppliers are providing increasing function at declining costs. Their businesses are growing at modest rates. Telecommunications hardware and software are being integrated with data processing equipment and voice, image, and data communications are being integrated on a large scale.

This chapter discussed telecommunications systems useful for major applications such as reservation systems or order-entry applications. The telecommunication services needed for these applications were presented. Special note was made of emerging technologies such as ISDN and the commercial availability of T-1 capabilities. Firms striving to attain advantage are employing these technologies to reduce costs or to provide improved service for their customers.

Standards are necessary for the efficient and effective use of networks. In particular, the ISO/OSI model is useful because it specifies a standard for the communication of information from one application to another. It details the communication through seven layers, starting at the physical layer and continuing to the application itself. Utilization of the OSI model is intended to make the interconnection of equipment from different manufacturers easier. As we learned, there are many standards for network applications. Some vendor-defined standards are very popular. However, they are not totally compatible with the standards issued by standards-setting bodies here and abroad.

Hardware and software that serve customer needs are available for communication between offices with LANs or to link LANS together forming larger networks. Most business organizations in the U.S. find these capabilities attractive and have adopted them. On a larger scale, telecommunications systems serve national or international needs. Because of this, they are subject to standards and policies established by international organizations and national governments. Multinational firms and firms engaged in international commerce are rapidly implementing these technologies.

A Business Vignette

Volvo's global network is a major factor in the Swedish automaker's success in international markets.[10]

Volvo's Net Gains

Data communications is the lifeblood of Swedish auto manufacturer Volvo. The company relies on foreign sales for 85% of its Skr84 billion ($12.5 billion) revenues. Consequently, its business, more than many of its rivals, depends on an effective means of keeping in touch with its far-flung operations. "We differ from other manufacturers in that we don't have a large home market. So it's very important to us that we have good communications," says Karl H. Hubinette, president of Volvo Data, Volvo's corporate network.

The network is based on IBM systems and SNA, and it is one of the most sophisticated IBM setups outside the U.S. Volvo has been working with SNA since 1978. It now operates a backbone network that links 20,000 terminals and some 25 mainframes located at more than 10 data centers. Over 500 leased lines connect the boxes. The bulk of Volvo's IBM-compatible terminals—13,000 of them—are in Sweden.

In addition to the IBM systems, Volvo boasts another 200 processors and 3,000 workstations with Digital Equipment Corp. badges on them. These systems are connected to the SNA network either by protocol convertors or via dial-up lines to central systems in Sweden. They are mainly involved with engineering applications. Operational costs alone for the network amount to some $5 million for data and $10 million for voice.

Hubinette is quick to sketch out the business advantages of data communications. "The network helps us primarily in two areas: product development and marketing," he says. As far as product development is concerned, Volvo wants to cut by 25% the lead time between the design of a product and its appearance in showrooms.

Hubinette says data communications will play an important role in achieving this by smoothing the transfer of data between designers in Gothenburg (Volvo's headquarters) and production staff, either in Sweden or in one of Volvo's overseas plants. When a component needs modification, the files can be updated at headquarters and transmitted to production systems. This occurs with less risk of the errors and delays associated with traditional blueprints and machine tapes. The best results will come when Volvo's suppliers can also send information electronically.

The other parts of Volvo's business—marketing, distribution, and sales—are equally important. They account for half the cost of producing a vehicle. Volvo's corporate network is already used by the company's importers and dealers to check out the availability of spare parts.

Volvo set up a network board with 10 general managers drawn from the main subsidiaries of the company. The board's role is to settle strategy for the network. Often the board has to decide whether a newcomer to the group should be allowed to use the network. That was no problem with the recently acquired heavy truck division of General Motors with its 200 North American dealers. Since it was wholly-owned, the risks were minimal. But the decision was tougher with an associated truck company, Clark Michigan, in which Volvo eventually decided that the business benefits outweighed the risks to proprietary data on the network and linked up with the U.S. firm.

One of the big spurs to network growth was the arrival of Memo, an electronic mail system developed by Volvo Data in the early days of mailboxes. The system, unique when it was developed in the late '70s, "sneaked out of Volvo Data," recalls Roland Linderoth, Volvo Data's director of product development. "We had two bright guys in technical support who had the idea and designed a simple tool. There was no high-level management decision involved." Memo is now used by some 300,000 people and is sold as a commercial product. But it was senior management who got to try it first. "You can only introduce something like this from the top down," maintains Linderoth.

Keeping the network up is often a matter of good relations with the local PTTs.[11] A manager has been assigned to this public relations task in each country where Volvo operates. "I would very much like to see an open, international network with quality and reasonable prices," says Linderoth. "I don't have the impression that unit costs are going down. Hardware is cheaper, but you need more and more of it, and software is getting more expensive, too. I'd prefer a public network to our private one. But today it is better to have leased lines because you have total control and can help your PTT to find faults on them." According to Linderoth, however, Volvo has been discussing quality with PTTs since 1979, "and we still have not finished."

"In the end, it's quality not quantity that counts," says Linderoth: "Data communications is nothing to the end user unless you maximize the availability of the right information at the right time."

Review Questions

1. Describe Volvo's network architecture. Who governs the use of the Volvo network?

2. What important management issues surface in the Volvo vignette?

3. Describe the external and environmental forces that are shaping the telecommunications industry.

4. What are the major elements of telecommunications systems?

5. Describe the function of a line adapter. Why are line adapters becoming popular today?

6. What forms of media are found in telecommunications links?

7. Geosynchronous earth satellites are 38,500 kilometers above the earth. What is the time delay for round-trip communication?

8. What are the important characteristics of a communications link?

9. What services are provided by a common carrier?

10. What is T-1 service?

11. What is a time-division multiplexer and how does it work?

12. What are the advantages and disadvantages of ISDN?

13. What are the principal standards bodies discussed in the text?

14. What are the main features of the OSI model?

15. What are the principal vendor architectures and why are they important?

16. What is the purpose of the Corporation for Open Systems? How does it intend to accomplish this purpose?

17. How are terminals linked together in a LAN? How are LANs linked together to form a WAN?

Discussion Questions

1. Using concepts developed earlier, describe Volvo's network in terms of its strategic role.

2. How are the international considerations of Volvo's network managed?

3. Elaborate on the advantages of having Volvo's suppliers send information electronically over the network.

4. Describe the elements that could be used to develop a network system linking two facilities, one in the U.S. and one in Europe, having the capacity to handle 23 19.2k bps data communication transmissions and 23 phone conversations simultaneously.

5. In what ways is the telecommunications industry dependent on the semiconductor industry for advances in telecommunications technology?

6. Discuss the importance of the OSI architecture.

7. Vendor network architectures differ from each other and from the ISO standard in some important ways. What are some of the consequences of this for IT management?

8. Discuss the role of standards in telecommunications. Why is it so difficult for organizations to set standards for the industry?

9. Discuss the reasons for the widespread use of vendor standards in the telecommunications industry.

10. Making some assumptions regarding the amount of data per airline reservation and knowing that the Sabre system processes about 500,000 reservations per day, calculate the average instantaneous data rate at the communications controller.

Assignments

1. Using library references, review Judge Greene's modified final judgment and outline the major significant effects it should have on American business.

2. Develop a business and financial summary of one of the RBOCs from information sources in the library reference section.

3. Obtain a description of the standards that have been developed for LANs and gateways. Summarize their main features on one or two pages.

4. Write a brief description of the X.25 protocol and one of its applications in a network. What is the main advantage of packet switching?

ENDNOTES

[1] The Bell Telephone Company was formed in 1877, Western Electric was acquired in 1882, and American Telephone and Telegraph was formed from these companies in 1885.

[2] Compiled from *The Value Line*, April 20, 1990.

[3] John H. Humphrey and Gary S. Smock, "High-Speed Modems," *BYTE*, June 1988, 102. This article describes the standards associated with modems and presents many of the technical details of these devices. There is a good description of the electronic fundamentals involved in modulating and demodulating signals. An evaluation of products from six companies is also provided.

[4] James C. Brancheau and J. D. Naumann, "A Manager's Guide to Integrated Services Digital Network," *Data Base*, Spring 1987.

[5] Flemming Meeks, "I Can Get It for You Wholesale," *Forbes*, February 6, 1989, 120.

[6] J. B. Miles, "Stratcom Releases Low-End T-1 Multiplexer," *PC Week*, February 6, 1989, C9.

[7] Eric Hindin and John Helliwell, "OSI's 7 Levels of Standardization Ease Information Exchange," *PC Week*, March 12, 1987, C31.

[8] Kornel Terplan, "Performance Impacts of Network Architectures," *Journal of Capacity Management* (September 1984): 226.

[9] Laura Cooper McGovern, "Local Area Networks Are Now the Rule, While MultiLAN Use Continues To Grow," *PC Week*, February 6, 1989, 126.

[10] Reprinted from *DATAMATION*, October 1, 1987. © 1987 by Cahners Publishing Company.

[11] Telecommunications is regulated at the national level in most countries and, in some, the government owns and operates the postal service and the telephone and telegraph service. In France and Switzerland this government agency is known as *Poste Telegraphe et Telephone* or PTT. In Germany it's the *Deutsche Bundespost*. The Office of Telecommunications regulates service in the United Kingdom. Japan has a Ministry of Posts and Telecommunications.

REFERENCES AND READING

Black, Hyless D. *Data Communications and Distributed Networks*. Englewood Cliffs, NJ: Prentice Hall, 1987.

Blumberg, Donald F. "Looking Ahead: Network Planning for the 1990s." *Data Communications*, February 1988, 185.

Duncanson, Jay, and Joe Chew. "The Ultimate Link?" *Byte*, July 1988, 278.

Feldman, Robert, *et al.* "The Divestiture of AT&T Reaches Five-Year Mark." *MIS Week*, January 2, 1989, 13.

Gulick, Dale. "What ISDN Brings to the Office of the Future." *Data Communications*, August 1988, 151.

Keen, Peter G. W. *Competing in Time: Using Telecommunications for Competitive Advantage*. Cambridge MA: Ballinger Publishing Company, 1988.

Lew, H. Kim, and Cyndi Jung. "Getting There from Here: Mapping from TCP/IP to OSI." *Data Communications*, August 1988, 161.

Rochester, Jack B. "There's a Rosy Future in EDI." *CIO Magazine*, January-February 1988, 20.

Stallings, William. *ISDN: An Introduction*. New York: Macmillan Publishing Co., 1989.

Stamper, David A. *Business Data Communications*, 2nd ed. Redwood City, CA: Benjamin/Cummings Publishing Company, Inc., 1989.

Terplan, Kornel. *Communications Network Management*. Englewood Cliffs, NJ: Prentice-Hall, Inc., 1987.

APPENDIX A

Two steps are required to digitize an analog signal: sampling the analog signal (pulse amplitude modulation) and quantizing the sample (pulse code modulation). The amplitude of the analog signal is sampled at discrete intervals in time. Then these heights are used to represent the original signal. This sampling and encoding process is portrayed in Figures 6.4 and 6.5. Nyquist's sampling theorem states that the original signal can be reconstructed from the samples if the sampling frequency is at least twice the frequency of the analog signal.[1] The sampling frequency in the telephone network is 8000 times per second, and this is what limits analog frequencies to 4K Hertz or less. Figure 6.4 illustrates the pulse amplitude modulation process.

The amplitude modulated pulses are then converted into a series of bits that define the height of the sampled waveform. One conversion method assigns an 8-bit byte of data to represent the amplitude of the sample. A digital transmission rate of 64,000 bps, or 64K bps, represents 8000 pulse amplitude samples per second, converted into one-byte digital signals.

Twenty-four individual analog signals are sampled and quantized in turn once each 125 microseconds. These quantized signals are assembled into one frame for transmission over the link. Each frame contains 193 bits of information for a total transmission rate of 1.544M bps for the T-1 link. The construction of the frame is illustrated in Figure 6.5.

There are several advanced variations of the digitization process which increase data transmission rates and improve signal reliability.

APPENDIX B

A brief explanation of the functions in each of the OSI layers is presented in Table 6.4. The description starts with the physical layer and proceeds to the final or application layer.

Layer 1 establishes the physical and electrical connection between the network and the device or devices. It specifies physical connections, number of wires, and signal voltage levels. Layer 2 packages data for transmission and unbundles it on receipt. This layer uses protocols or rules to maintain understanding within the communication system. These rules facilitate synchronization, package information into frames, establish sequence and transmission control, and detect errors.

---

[1] Another important part of information transmission theory came from Claude Shannon, a pioneer in information theory, who related transmission speed to error rates and signal power. For a brief, fascinating sketch of Claude Shannon see John Horgan, "Profile: Claude E. Shannon," *Scientific American*, January 1990, 22.

TABLE 6.4   ISO/OSI Model

Layer	Function
7. Application	Enables communication between applications by utilizing services supplied by lower levels, e.g., electronic mail, file transfer.
6. Presentation	Represents information to communicating applications, preserving meaning while resolving syntax differences. Is the data interpreter for applications.
5. Session	Provides the means for cooperating presentation entities to organize and synchronize their dialogues and manage their data exchanges. Handles error recovery for abnormal session terminations.
4. Transport	Provides transparent transfer of data between session entities. Provides logical connection between network nodes. Includes flow control, prioritization, and message assembly.
3. Network	Provides for the transparent transfer for all data submitted by the transport layer. Connects network address with physical address.
2. Data link	Provides means to establish, maintain, and release data link connections. Provides services to the network layer. Detects and possibly corrects error conditions from the physical layer.
1. Physical	Defines mechanical and electrical interface to the media for the transmission of the raw bit stream.

Layer 3 transfers and routes data through a network and matches logical addresses with physical addresses. This layer also selects links appropriate for the traffic, performs load leveling, and manages priorities. Layer 4 provides transparent transmission and controls transmission quality. It provides users with the means to establish, maintain, and release transport connections required for two-way communication.

Layer 5 establishes and controls the dialogue between presentation entities. The act of establishing a session is complicated by the need to consider authorization and agree on addresses and communication conventions. Layer 5 provides the means for performing these tasks. Layer 6 translates message formats for the network and the application. It deals with security and encryption, maintains security algorithms, and allows the users to ensure secure communication. Layer 7 receives messages from the sending application and communicates them to the receiving applications.

Part

Three

7 The Existing Portfolio Resource

8 Managing Application Development

9 Alternatives to Traditional Development

10 Successful End-User Computing

The Application Portfolio Resource

The application portfolio and related databases comprise large and critical resources for the firm. These resources are growing in value and form a base from which the firm can gain strategic competitive advantage. The firm must continue to add value to these assets and must mitigate the effects of technical and business obsolescence. IT professionals and skilled users participate jointly in this activity. This Part addresses the management issues associated with the application portfolio and data resources. Chapters include The Existing Portfolio Resource, Managing Application Development, Alternatives to Traditional Development, and Successful End-User Computing.

Managing application and data resources is a challenging task. Successful IT managers view this activity as a cornerstone of their mission.

7

The Existing
Portfolio Resource

A Business Vignette

Introduction

The Application Resource
  Depreciation and Obsolescence
  Maintenance and Enhancement

The Data Resource

Applications as Depreciating Assets

Spending More May Not Be the Answer
  The Programming Backlog
  The Need To Prioritize

Enhancement and Maintenance Considerations
  The High Costs of Enhancement
  Trends in Resource Application

Spending Wisely

A Process for Portfolio Management
  Satisfaction Analysis
  Strategic and Operational Factors
  Costs and Benefits
  Some Additional Important Factors

Managing the Data Resource

Why The System Is Valuable

Summary

Review Questions

Discussion Questions

Assignments

References and Readings

A Business Vignette

The information management industry has created another industry: rescues. Specialized shops under the direction of corporate consultants and IM gurus have made millions in the last few years—$30 million in 1987-88 for one accounting firm, according to a story in *Business Week*.[1] These shops catch and tame runaways, which are systems-development projects that get out of control.

Businesses in the United States have many tales to tell of the horrors that can occur when large-system development is not planned well; when conflicts arise among developers and users; and when planners do not take into account the downstream problems that can and do arise. Among these problems:

1. Systems must connect with other systems, especially in businesses that have large databases that may be accessed by many applications and users. In the early days of system development, the developers could get their arms around individual systems and they built stand-alones with a degree of success. Now, however, different skills and techniques are required for developing systems that can talk to other systems.

2. Developers need business, so they tend to oversell what they can deliver. The people who do the grunt work of writing the applications programs may be overworked, understaffed, and unmotivated, but their executives will promise unrealistic performance, including low cost, short development time, high reliability, and ease of use. What happens when the programmers can't or won't deliver? In companies where reliability of data is especially important—health care firms, for instance—lost or incorrect data can mean not only loss of subscribers but also hardships and danger to those subscribers and loss of their business to the firm.

3. The more complex the application, the longer development will take and the more it will cost. These factors leverage each other; over a period of several years, the basic cost of development services will continue to grow. Therefore, a project that takes four years instead of two to complete will be billing far higher developer rates as the calendar leaves flip over. Delays mean more cost. For example, *Business Week* cites an Allstate Insurance project begun in 1982 with an $8 million projected cost and completion in 1987. By 1988 the cost had risen to $100 and completion was pushed to 1993.[2]

How to solve these problems? First, according to *Business Week*,[3] the decision makers—CEO, CIO, or any others responsible for paying the bills—must get involved and stay involved in the process from the beginning.

Second, the project must be reviewed every six to twelve months. At each review, not only progress reports but all budget items and all performance criteria must be reviewed against the requirements placed on the system.

Third, if outside contractors are used, some way must be found to guarantee the high standards and abilities of the people who will do the design and the programming for the system. To do this may require the hiring firm to specify qualified individuals by name in the contract.

Fourth, the people who will use the system must be part of the development effort. In effect, a constant series of user trials is needed; it will not suffice to wait until the end of a project to bring the users on board. A surprising number of development projects have failed to involve users; in those cases, transaction problems, culture problems, and lack of general acceptance have resulted. Knowledgeable users can alert the developers to possible conflicts in transactions; they can provide information about how data flows and is used in the organization; and they can assist general buy-in of the new system by serving as champions for it in the trial and introduction phases.

INTRODUCTION

Part Two discussed systems used for the delivery of information. It concentrated on advances in semiconductor and recording technology, operating systems, and hardware and operating system architecture. The strong, dominant trend of rapidly declining unit costs is the basis for the rapid infusion of new technology into nearly every niche of our society.

Continuing in support of hardware advances is the inexorable growth of software, such as operating systems, telecommunications programs, and utility programs. Application programs developed within the firm are growing too. They are being supplemented by a proliferation of commercially developed programs. These application programs are typically business utilities, such as payroll, general ledger, billing, and inventory control. They are designed for applications across a broad variety of user organizations, and they support some of the fundamental activities of the firm. Recently many of these application programs have been made available for implementation on personal computers. This software and the hardware it supports comprise the delivery system base for the firm.

The remaining ingredients of the information system for the firm are the large databases used in concert with the application portfolio. This vital information, on which the firm thrives, is typically unique to the firm. These program and data resources represent an enormous investment acquired over a long period of time. The applications and the data are of critical value to the firm both operationally and strategically.

This chapter discusses the technical and management opportunities and difficulties inherent in the portfolio and related databases. It provides management processes and methodologies for handling the portfolio difficulties in a disciplined manner.

THE APPLICATION RESOURCE

Depreciation and Obsolescence

The application portfolio and related databases are growing in magnitude and importance over time.[4] The firm must continue to make investments and apply resources to these assets to prevent depreciation and obsolescence. Depreciation results from the accumulation of functional inadequacies due to gradually changing business conditions. For instance, the work-in-process inventory system designed for a labor-intensive process will lose effectiveness as labor-intensive processes are replaced by robotics. Obsolescence results from the introduction of changes to the firm that reduce the appropriateness or value of current applications. For example, the introduction of labor accounting terminals on a plant floor may render the previous labor accounting system largely obsolete.

Program and data assets represent a foundation from which the firm desires to obtain strategic competitive advantage in the long term. These application and data assets are vitally important to the firm.

Maintenance and Enhancement

In contrast to those for hardware technology, costs associated with maintaining, enhancing, and improving the portfolio are mostly people related. They are rising over time. New tools to improve the productivity of individuals engaged in the enhancement process are being developed. But for the bulk of the applications written in third-generation languages, no major cost breakthroughs are anticipated. Additionally, and again in contrast to hardware, there is no easy, low-cost way to move from an older generation of application software to a new and modern generation.

The worth of the firm's information system is gauged by the total system function performed for the firm. To a considerable extent, the functions provided by the application programs represent the yardstick by which the system's value is measured.

THE DATA RESOURCE

The databases that accompany the application programs are essential assets to the firm. Over time, they grow in size and importance to the firm. These databases are the result of continued accumulation of information, and they represent an investment of resources. The firm's databases form the foundation on which future advances in applications technology can proceed. These data systems provide the lifeblood which sustains the day-to-day operation of the firm. These data resources and their inherent database management systems, coupled with the hardware on which they reside, are a vital and critical adjunct to the application portfolio.

The data resource may also be incurring loss of value as the applications themselves depreciate. The database management system, which may be satisfactory for the current applications, may constrain modernization of the portfolio and make enhancements much more difficult. In addition, large databases are probably involved in numerous interactions among and between applications. This coupling through the databases greatly complicates the revitalization or enhancement of the application programs themselves.

In some firms, the database resource became dispersed over wide distances as the firm expanded geographically. In most firms, local dispersion of the data accompanied the introduction of individual workstations. The information architecture for the firm is an important issue, as we observed in the first chapter. In all cases, the intrinsic value of the data demands that protection mechanisms be installed and that they operate effectively. The data resource is vulnerable to loss and destruction, and it suffers from systemic forms of obsolescence.

APPLICATIONS AS DEPRECIATING ASSETS

In actual practice, applications pose a management dilemma for the firm. The application portfolio represents a resource base on which future applications can be developed. At the same time, improvements and enhancements are mandated by ever-changing business conditions. This implies depreciation or even obsolescence. Increasing amounts of time and money are required to capitalize on the stream of opportunities available to the firm through its application programs. Resources are required to avoid the negative consequences of obsolescence and depreciation. Since the continuing demand for resources to fuel the growing asset base of applications shows no signs of diminishing, most firms face difficult and important decisions regarding resource allocation.

SPENDING MORE MAY NOT BE THE ANSWER

The applications programming department in most organizations faces a large workload backlog; in many firms it ranges from one to four years in length.[5] The backlog stems from the organization's need to keep the business competitive via new applications. Also, there are requirements to keep the current application portfolio functionally modern. In addition, the process for maintaining and enhancing applications is inherently inefficient; and, for reasons discussed later, programmer productivity in this activity tends to be low. The program development organization typically struggles with the management of the backlog and, in attempting to do something for everyone, performs a less-than-satisfactory job for the organization as a whole. Well-managed organizations can overcome many of the difficulties. They can be successful in today's environment.

The Programming Backlog

The application programming backlog is defined according to the formula:

$$\text{Backlog} = \frac{\text{Work to be accomplished in person-months}}{\text{Number of persons to do the work}}$$

For instance, if the organization has two programming tasks to perform for a total of 36 person-months of work and has three programmers to do the work, the backlog is 12 months. Some assumptions are made in calculating the backlog this way. One assumption is that there is no wasted time due to skill imbalances during the development cycle. This is most likely to be true for large development organizations. Another is that none of the code is reusable, so it must be uniquely developed for each application. This is less likely to be true for large organizations.

In addition to the identified backlog, there frequently exists an unidentified or "invisible" backlog, which occurs when the identified backlog is large. An invisible backlog develops because using departments, in need of programming work, do not make their requirements known. They are reluctant to add to an already long list of outstanding work requirements. The true backlog facing application developers is the sum of the identified and the unidentified outstanding work. In many firms, the identified backlog is on the order of two to three years or more. The invisible backlog may be very large indeed, depending on the dynamics of the firm.

The Need To Prioritize

For many firms, the alternative of applying more resources to application development is not reasonable. Good programmers are difficult to obtain, they are expensive, and many firms are reluctant to make a long-term investment in programmer development. Newly acquired programmers are not immediately effective, because they need to learn the culture and business aspects of the organization. Therefore, the issue is the difficult question of prioritizing the application development resources.

Programming prioritization is important for other reasons also. The application portfolio is a strategic resource. It has been developed over a long time, and it will remain important in the future. Expenditure of application development resources demands high-level consideration. In addition, the portfolio is a source of management expectations. IT managers need a thoughtful process for managing these strategic resources and expectations over the long term.

A development plan is needed. The plan must describe the application of available resources, application programmers, and available financial support to the tasks that the enterprise wants to accomplish. Figure 7.1 shows the process in which the organization is involved and portrays several alternatives available to it.

The application portfolio in Figure 7.1 consists of existing applications owned by the firm and some required applications that do not yet exist. The portfolio analysis process described below provides an organized and businesslike approach for decision making. The first level of decision making selects among the choices of maintenance, enhancement, acquisition, or termination. Additional decisions between abandonment or replacement, as well as alternatives to acquisition, are displayed. The process depicted in Figure 7.1 and discussed below provides a rational approach toward maximizing the value of the portfolio. Upon completion of the process, resources are identified for the separate tasks of maintenance, enhancement, or acquisition in an optimal manner for the firm.

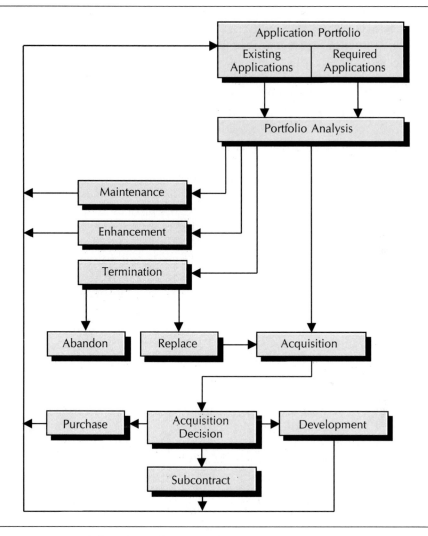

FIGURE 7.1   Portfolio Alternatives

ENHANCEMENT AND MAINTENANCE CONSIDERATIONS

It is common for most programming resources in firms to be deployed in maintenance and enhancement of existing applications.[6] Generally, fewer resources are available for new development. It is estimated that there are about 70 billion lines of COBOL code in operation in the U.S., and most of this application code is depreciating because of new and changing business conditions. This huge program asset base requires a continuous investment of resources to prevent depreciation and obsolescence. Dealing effectively with program depreciation leads to important positive consequences for the firm.

Vendors have produced several generations of hardware over the years, and each generation had significant cost and performance advantages over its predecessor. The hardware vendors have also eased the transition from one generation to the next for the application programmer. Computing systems are created to make the old programs relatively easy to migrate to the new hardware. While this upgrading of the hardware has taken place with cost and performance advantages, the applications supporting the firm have been much more difficult to upgrade. For many firms, the result is that old application programs are running on modern hardware. For example, the FBI's National Crime Information Center runs 900,000 transactions per day. The program was written in the 1960s in System/360 assembly language. The FBI's payroll and personnel system, written in COBOL, has been enhanced and maintained for more than 20 years.[7] It is relatively easy to upgrade the CPU or to add new input or output devices; but it is much more difficult to upgrade the application software.

The High Costs of Enhancement

Many firms feel locked to the past. Their applications require significant modernization, yet the resources to do the job are not available. The expenditure of money on hardware brings early returns, but the cost to correct the deficiencies in the application portfolio remains extremely high. What are the principal reasons for the situation as it exists today?

There are a number of reasons for application maintenance and enhancement requiring large investments in time and money. They are summarized in Table 7.1.

The old programs in the portfolio were built with programming techniques that are now considered unsatisfactory. The techniques that were used were not designed for ease of modification. Many old programs aren't well structured. They may contain strings of code arranged in a convoluted and unstructured manner. These old programs have had a continuous series of enhancements over the years, probably performed by many different programmers. Because of the enhancements and the enhancement process involved, the program documentation is generally poor or absent. In some cases, more time is spent trying to understand the program than fixing the programming problem.[8]

TABLE 7.1   Why Maintenance Costs on the Existing Portfolio Are High

Obsolete programming techniques were used.

Documentation is obsolete or absent.

Many programmers made modifications.

Old versions of languages were used.

Languages were mixed within the program.

Unskilled programmers made enhancements.

Architecture changes are required.

File structures need major changes.

Many of the older programs are written in earlier versions of a current programming language and may even contain embedded assembly language code. Modification to these applications is difficult to perform and is an error-prone process. Modification introduces risk into the operation of the programs. It is a low productivity activity and is not considered to be very desirable work.

In many firms today, enhancement and maintenance programming is frequently considered good training. It is therefore assigned to the least experienced programmers in the organization. Because of their inexperience, junior programmers may produce less efficient code, and the code contains more errors. Skilled, senior programmers usually prefer new development work. Attracting good people to perform maintenance is becoming increasingly difficult. In some firms, individuals are reluctant to take responsibility for program maintenance.[9]

Older programs generally require major architectural improvements. For example, the old program may operate in sequential batch-processing mode and now must be converted to on-line operation. Perhaps the original input media was punched cards; now terminal input must replace card-image input. The implementation of a new database management system may be required. It may impact data that currently exists on sequential tape files. Changes such as these impact the architecture and foundation of the applications and require massive amounts of effort. Architectural alterations may demand more resources than were invested during the initial development cycle.

In many instances, the files associated with each application are unique. They may consist of sequential tape files copied to disk. This means that a change in the application's function may dictate large and expensive changes to the database or to other programs that use the database. In total, these enhancements demand large investments. Because of the complexities involved, modifications and enhancements to the portfolio proceed at a slow pace. For the same reasons, the maintenance and enhancement process is inherently error prone and risky.

Trends in Resource Application

The conditions described above result in the continued expenditure of large sums of money, increased dissatisfaction with the program development department, and likely deterioration of application quality. Figure 7.2 portrays the deployment of funds over time in maintenance and enhancement and for new development. This picture is typical of many organizations today.

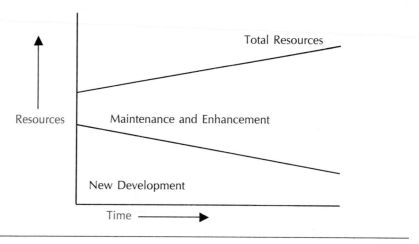

FIGURE 7.2   Typical Expenditures of Funds on the Portfolio

Figure 7.2 shows that the typical firm is increasing its total expenditures on programming because of escalating costs per programmer, additions to the programming staff, or both. However, the amount of resource devoted to new development is declining over time because of the demands for changes and alterations to current applications. Maintenance and enhancement activity consumes increasing fractions of the available resources. It may account for up to 80 percent of the total.

There are some reasons to be hopeful about finding solutions to the applications problem. However, the current situation in business and industry is far from satisfactory. Most large firms have several thousand applications in the portfolio, and even small to medium-sized companies own several hundred to a thousand or more programs. These applications may be recent additions to the portfolio, or they may be 10 or more years old. Occasionally, an application will be found that was designed for now-obsolete hardware and that requires some form of hardware simulator for operation. In some cases, the program cannot be modified because the documentation or skills are no longer available.

Full-time maintenance programmers usually support many applications. For example, one maintenance programmer may be responsible for financial applications, another for marketing applications. Several of them may support manufacturing applications, and others respond to work requests from other departments. As problems arise or when enhancements are required, the maintenance programmer moves from program to program. Since there is a large backlog of work, the programmer may take shortcuts and ignore some important aspects of the work. In the interest of pleasing the greatest number of users, maintenance programmers frequently fail to document changes completely. Many simply ignore documentation, further complicating future maintenance.

The term *maintenance* itself takes on several meanings in this situation. Some effort is devoted to repairs. These repairs consist of fixing bugs in new programs and finding and repairing bugs in enhancements to old programs. As the portfolio ages, this effort increases in magnitude. Frequently, additional effort is required to interface or bridge enhanced programs to old databases. These bridge programs add to the portfolio size and also require enhancement as other programs or databases change. Bridge programs may also introduce errors, further complicating the situation. Maintenance may include minor rewrites. In some cases, it becomes easier to rebuild portions of the application than to work with old code. Application maintenance is highly inefficient. Firms engaged in this activity are on a course depicted by Figure 7.2.

For all the reasons illustrated above, software maintenance is not viewed as highly desirable work, and programmers usually prefer other assignments. However, the prevention or deferral of obsolescence to the application portfolio is very important to the firm. Poor quality workmanship or untimely maintenance and enhancement activity can severely impact the firm's ability to conduct its affairs effectively.

In many firms the situation is not improving. Maintenance and enhancement consumes more than 50 percent of the budget for programming. And, programmers continually add to the amount of code in use by approximately 10 to 40 percent per year. Many of the people working on these applications were not involved in the original development activity, and many of the difficulties noted above continue to plague maintenance programmers today.

Many firms lack an organized and disciplined approach to this thorny problem and rely on one or more reactive methods for prioritizing the application programming team's work. Table 7.2 itemizes some common, default approaches to programming prioritization.

TABLE 7.2   Common Approaches to Prioritizing Programming Resources

"The squeaky wheel gets the grease."

React to perceived threat of failure.

Recover from embarrassing situations.

Adjust to threats from competition.

In the first instance, managers throughout the firm contend for application development resources by making personal appeals to application development managers or directly to programmers. The success of the appeals depends on the degree of persuasion employed, the relative position of the person appealing, or the degree of implied or actual threat involved. Seasoned managers are more likely to succeed than new managers. Friendship frequently plays a role. A system of score-keeping (who owes what to whom) may develop. The process is rich with emotionalism, high on anxiety, and very low in objectivity. In an attempt to calm troubled waters, IT managers frequently try to do something for everyone. This usually raises expectations, generates overcommitment, and yields unsatisfactory results. Effective IT managers must use disciplined processes successfully in this situation.

In spite of these difficulties, the squeaky wheel theory is rather popular. Most managers believe they can negotiate a better deal for their organizations than can their peer managers. Old-timers are especially reluctant to give up their favored position. They tend to resist a more rational process. In some firms, the squeaky wheel is the corporate culture; in others it is the default management system. Firms with this culture or this management system are characterized by a lack of teamwork and frequent finger pointing.

In a reactive environment, priorities are frequently readjusted to avert difficulties. When problems are recognized within the squeaky wheel environment, corrective action usually follows promptly. Unfortunately, the firm often rushes from one disaster to the next in a futile attempt to keep all systems from collapsing at once. The inevitable result is a terrible waste of resources, generally unsatisfactory performance in the long run, and the conclusion by even the most casual observers that management does not know what is going on.

Since the process depicted above is often not successful, embarrassing situations are likely to develop for the firm. Payroll processing goes astray for all employees to observe firsthand, incorrect or duplicate billings are sent to customers, or, worse yet, collectibles are not received. Occasionally they are received, but improperly recorded. These situations demand immediate action. Resources are deployed from around the firm to correct the problems post haste.

When activities within the firm appear to be on a steady course, competitors may be marshaling their IT forces for offensive action. Detection of these competitive actions by the firm elicits a prompt reaction. Again, resources are deployed to contend with these threats and to ward off their consequences, if possible. When survival of the firm is at stake, all other activities have lower priority.

These reactionary actions and out-of-control situations are not desirable, however popular they may be. Is it possible to develop a process that better serves the firm? What management tools and techniques are available to assist in prioritizing the resources? Given the limited and constrained resources available to most firms, what approach is preferred under these circumstances? Obtaining sound answers to these questions must be a high priority for business managers, because major consequences derive from the actions taken or omitted under these circumstances.

A PROCESS FOR PORTFOLIO MANAGEMENT

The following discussion presents an approach that allows the strengths of the firm's senior managers to be combined with those of the IT organization to yield a preferred course of action. The methodology focuses on business results. It presents managers with alternatives for their consideration. The prioritization process requires significant communication among and between the various players. It causes them to view the problem from the level of the senior people in the organization. This approach demands a general management perspective because it focuses on considerations fundamental to the firm. For these and other reasons, this approach is likely to achieve superior results.

Several factors are important in prioritizing the backlog. Among these are the firm's business objectives and the financial benefits derived from the applications. Frequently, the applications have important intangible benefits. Some, in fact, may be leading-edge technology and have technical importance.[10] Usually, there are interrelations among and between these factors. For example, a common situation is an application that is required to meet business objectives but does not show a positive financial return. The benefits may be great but intangible. Likewise, an investment to attain technological leadership may be valuable for competitive reasons.

The questions posed by these considerations make resource allocation decisions difficult. Users alone do not have enough information to allocate resources. A combination of IT managers and user managers may lack the view that top executives have. To resolve these issues, many firms utilize a steering committee of top executives to prioritize the application development backlog. If the steering committee does not approach prioritization in a systematic manner, but acts informally and with a political orientation, the results are likely to be mediocre. If the committee incorporates the advice and counsel of senior executives as part of the firm's formal strategy and planning process, the results are likely to be superior. This text strongly favors the planning approach.[11]

How does this happen in practice? What steps can the IT organization take during the strategy and planning process to prioritize the programming work? What is an appropriate management system to carry out these difficult and important tasks? The following analyses present an organized and disciplined methodology to resolve these issues.

Satisfaction Analysis

The first step in this disciplined approach is to perform an analysis of satisfaction on each application in the portfolio. Satisfaction ratings for each application are obtained from the using organization and from the IT organization. The satisfaction analysis focuses on attributes important to each of these organizations and quantifies what otherwise might be emotional perceptions. Typical factors used in the satisfaction analysis for the IT organization and the user organizations are shown in Table 7.3.

TABLE 7.3 Factors Used in Satisfaction Analysis

User Factors	IT Factors
Sound function	Good documentation
Easy to use	Modern language
Good user documentation	Ease of operation
Healthy cost/benefits	Trouble free
Sound architecture	Easy to operate

This list is not intended to be exhaustive, but it illustrates most of the attributes contributing to satisfaction. In some cases, there may be firm-dependent attributes that should be included.

The applications in the portfolio are scored from 0 to 10 by each organization; then they are sorted in several different ways. The scorings or ratings are used in a first attempt to attain a consensus from the organization on which applications will or will not receive additional funding. Applications for which it is agreed that no additional investment is required are removed from further consideration.

Some interesting results of the analysis to this point can be revealed by plotting the results in a diagram such as that displayed in Figure 7.3. This figure portrays the analysis conducted by the application users and by the IT organization. The applications have been evaluated with a closed list of attributes, such as shown in Table 7.3. Each application is plotted to show its score by both groups. The results for three programs are shown on Figure 7.3. For discussion purposes, the programs are identified as A, B, and C.

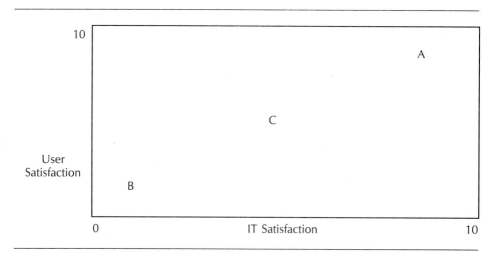

FIGURE 7.3   A Graphical View of Some Satisfaction Analysis Results

Newly developed and recently installed programs generally reside in the upper right quadrant of Figure 7.3. One such program is A. It is not uncommon to find new programs, such as program A, scoring lower than 10-10. Because of trade-offs made during development and changing business conditions during the elapsed development time, newly developed applications are not completely satisfying. Frequently, requirements become clear only after the program is implemented!

Many applications, perhaps those currently undergoing enhancement activity, will be found in the lower left quadrant of Figure 7.3. An example of this type of program is B. In many instances, the bulk of the enhancement resource will be devoted to programs having low satisfaction to both IT and the users. Whether this should be the case is not obvious, given the information available to this point.

Program C is an average program in the portfolio. It is not very satisfactory to the users or to the IT organization. Programs of this type probably make up the bulk of the applications in most firms. When all the results of this analysis are plotted, they tend to fall along the diagonal of the diagram. It is unusual for an application to be very satisfactory to the using organization and also to be very unsatisfactory to IT. The converse is also true. Generally, difficulties with applications affect both users and IT people.

This analysis alone provides insufficient information for deciding how to prioritize the backlog. One missing ingredient is the short-term versus the long-term perspective on the value of each application. The next step supplies this information.

Strategic and Operational Factors

To delineate more clearly which of these applications merits additional resources, managers must obtain insights from a different perspective. The organization must understand the strategic or operational value of the applications. The importance of the applications should be considered from both the short-range and the long-range perspective. The firm's managers must evaluate both the strategic and operational value of each application. As in the previous analysis, the applications are scored 0-10 by the user managers on their strategic importance and on their operational importance. Figure 7.4 portrays the results of the strategic versus operational analysis for selected programs.

Some examples of applications that illustrate the insights provided by this analysis are identified in the figure. The examples have been selected to show how this analysis forces differentiation between applications in the organization.

Typical programs indicated by 1 are applications such as payroll, accounts payable, or student accounting. These applications are very important operationally to the firm, but they generally provide little or no strategic advantage. Number 2 identifies programs such as the Merrill Lynch CMA, the Sabre reservation system, or a seismic data processor for an oil exploration company. These programs are important both operationally and strategically. Applications such as these are vital to the firm on a daily basis. They provide strategic competitive advantage as well.

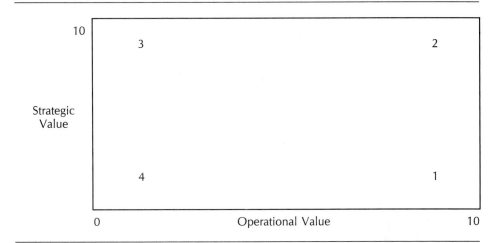

FIGURE 7.4   Strategic *vs*. Operational Value

Other programs such as the CMA in 1977, computer-aided instruction programs for elementary schools, or the next-generation vector processor are indicated by Number 3. These applications have relatively low operational importance to the firm at this time, but they have high potential strategic advantage in the long range. Programs in the Number 4 position are relatively unimportant strategically and operationally. Examples of these applications are office furniture inventory programs and one-time statistical analysis programs. These applications are needed, but they are not highly important to the mission of the firm.

Application owners do most of the evaluation. They must be sure to take into account strategic goals and objectives of the firm. IT input should be obtained during the process, especially in relation to anticipated technology improvements. This input is especially useful for emerging applications or for applications that are candidates for major enhancements. The list of programs included in this analysis should not be restricted to those currently in the portfolio. It must include new requirements presently in the backlog as well.

The scoring is again 0–10, but the plotted results will show considerable scatter as compared with the previous diagram. This is normal and desirable, because it allows for discrimination among the candidates. When this analysis is used in conjunction with the previous analysis, a clearer picture of the candidates for resource deployment becomes available. Either analysis taken by itself provides some useful information but is not sufficient for making decisions. Together they build a base for the next step.

Costs and Benefits

The additional element, which completes the prioritizing process, is an analysis of the cost and benefits of the actions proposed for each application. The analysis must include the cost of failure to do the work and the benefits of intangible results.[12] In most cases, it is not necessary to have highly refined data to reach final conclusions. Most analyses can proceed if the costs and benefits are known to within 20 percent or so. Additional refinement may be required in the event of a close decision. The time value of money must be incorporated into this analysis to ensure financial integrity. Some examples of various analysis methods can be found in texts on financial accounting.[13]

Benefits information is provided by user managers, and development and implementation costs are provided by IT. The benefits may be purely financial, or they may be largely intangible. If a program later becomes a candidate for resource expenditures, the firm's controller should "book" the benefits and the costs. That is, the expected benefits are committed by the user organization and the costs and expenses are committed by IT. Both are tracked by the controller. This process ensures a higher degree of integrity in the data than do other methods. It places commitment on the organizations where it belongs.

A tabulation of all the applications, after completion of the scoring and the cost-benefits analysis, can be revealing. Table 7.4 illustrates some of the conditions that may be found in a large portfolio of applications.

TABLE 7.4   Compilation of Program Ratings

Program Name	User Sat.	IT Sat.	Strategic Importance	Tactical Importance	Costs	Benefits
prgm 1	8	8	1	8	10K	14K
prgm 2	2	3	1	3	30K	40K
prgm 3	7	9	8	4	100K	200K
.						
.						
.						
prgm 2000	. . .					

Three examples of the analysis results have been listed. Program 1 is a recently completed application with fairly high tactical importance to the firm. Additional investment in Program 1 will yield a modest return.

Program 2 is an old program that is not very satisfactory to either the user community or to IT. Although this program is not strategically or tactically important, modest financial benefits will occur if additional expenditures are applied to it.

Program 3 is a relatively new program for which a great deal of strategic importance is anticipated. Both the user group and the IT organization are quite satisfied with this program as it currently is. Additional investments are required to achieve the relatively high future payoff anticipated for this application. In a large portfolio, there are many individual differences between applications which permit decision makers to discriminate.

Sufficient information is now available for making decisions. The decision makers are the firm's senior functional managers and their superiors. The appropriate time to conduct this activity is after the strategy has been established and just prior to the beginning of the planning cycle for the firm.

Using all the factors brought out by this analysis, the management team prioritizes the list of programs from high to low. Their judgment and experience are combined with the data to reorder the list until agreement is reached or a decision is made. This process serves to obtain consensus and commitment. It puts decision making at the management level where it belongs and reduces overcommitment. The list of work items is closed when the sum of resources required equals that available to the firm. This may mean, for example, that the first 400 applications of 2000 receive resources and the bottom 1600 remain at status quo. This process points out what will be done and what will not be done, and the firm obtains an achievable plan.

Some Additional Important Factors

Given the analysis outlined above, the firm's senior decision makers are in a good position to allocate critical resources. This analysis should be directed through the IT organization. The IT organization must obtain the users' view of the applications in the portfolio. It must also take the lead in developing the discussion preceding the decision to deploy resources. IT managers are responsible for bringing a completed analysis to senior executives.

Each senior member of the review team will obtain a clearer understanding of the business needs. If the proper environment has been established by the top individual, the needs of the firm will be well served. If this process is followed rigorously, it eliminates other less-disciplined approaches, such as the squeaking wheel method. This approach puts the decision making where it belongs, at the top of the organization. It eliminates many small *ad hoc* deals between individuals at lower levels in the organization which could reduce the firm's effectiveness. This methodology sheds light on the complex problem of resource use. It leads to a much better-informed organization, because it creats organizational learning. And it prepares the organization to use information technology to advantage.

Required resources usually far exceed available resources. As illustrated in Figure 7.1, if money is in greater supply than people, the alternative to purchase applications or to subcontract program development may appear attractive. Although each alternative raises some additional issues, many organizations prefer these choices to the option of hiring and training more analysts and programmers.

An additional alternative not shown in Figure 7.1 is to utilize a service bureau for certain of the firm's data processing needs. Information of the type displayed in Figure 7.4 will be most helpful in deciding whether to buy the service, buy the application, subcontract the development, or perform the development in-house. Routine tasks such as payroll processing can perhaps be performed best at a service bureau. Many other routine data-handling activities are service-bureau candidates also.

Strategically valuable applications or those that utilize proprietary processes or data are best completed within the firm. This local development will probably be completed by the resident application programming department. In some cases, the applications can be developed by the users themselves. In many cases, end users can perform many maintenance or enhancement changes. This activity, called end-user computing, can be very effective. It is discussed in Chapter 10.

When the prioritization process is complete, IT managers have all the necessary information to complete the applications program portion of the IT plan. The management systems for prioritizing and for IT planning are essential portions of strategic application management. Tactical plans for the firm contain cost and benefit information developed during this process. This tracking information becomes part of the performance measurement of the managers who own or control the applications. In a similar manner, the costs expected to be incurred on applications are committed and tracked. IT development managers are measured on this commitment. The firm's normal cost-accounting system is used for measurement purposes. This process incorporates strategy, strategic planning, and operational control elements. It ensures a high degree of integrity, secures commitment, and places responsibility where it belongs.

The process introduced above offers one final major advantage. The disciplined process, involving the firm's senior executives, provides a splendid opportunity to identify and fund new strategic systems. Using the framework developed in Chapter 2, executives can begin the task of identifying strategic opportunities. At this time, they have a great deal of valuable information on the portfolio and the environment is favorable. This is the proper time to search for strategic opportunities. A structured methodology for accomplishing this task has been documented by Rackoff and colleagues.[14]

The prioritization methodology presented in this chapter provides an ideal background from which to begin the search for strategic information systems opportunities. The right people are present, the timing is right, and a decision-making process has been established. New strategic opportunities can be introduced and resources allocated to their implementation. All conditions necessary for sound decision making are present.

MANAGING THE DATA RESOURCE

The data resource accompanying the applications, the database management system, and certain hardware items must form an integral part of the prioritization process. Applications enhancements, conversions, or rebuilding activities will require resource expenditures on the database in many cases. In some instances, database conversions may be important enough in their own right to be considered as separate items in the methodology. That is, a database conversion or reorganization activity should be considered for prioritization in the same way that an application is.

In many cases, database considerations and resources will bear on the sequence (and perhaps on the priority) of the actions planned for the applications. An information architecture will begin to emerge from this work, if the firm does not already have one. In the final analysis, the interplay among the applications and data must be clearly understood. Data sources within the firm must be well defined for the process to be most effective.

Application specialists in the IT organization understand the technical nature of the database system and the interactions among the applications. The architectural considerations of the portfolio need to be developed for the firm's senior managers by IT professionals. This element of executive education must be conducted thoroughly and skillfully to improve the executive decision-making process. This activity provides another opportunity for the IT executives to obtain agreement on direction and expectations among their peers and with their superiors. It's an opportunity that should not be lost.

WHY THE SYSTEM IS VALUABLE

The methodology developed in this chapter organizes the decision-making process and provides a framework for judging how best to deploy scarce resources. This methodology causes decision makers to focus on issues fundamental to the well-being of the entire organization. It discourages parochialism. The process puts decision making at the senior executive level where it belongs. It serves to integrate the IT function into the fabric of the business. Executives at many levels in the business will be educated in IT issues during this process. The results are embedded into the strategizing, planning, and control operations of the firm.

The management system for dealing with the application portfolio attacks some of the critical issues discussed in Chapter 1. It offers the opportunity to advance organizational learning and helps define IT's role and contribution. It focuses on data as a corporate resource. This process must be used in conjunction with strategy and plan development for maximum effectiveness. This management system forces congruence between the firm's strategic plan and the strategic plan for IT regarding the application resources. Finally, the process ensures realistic expectations regarding the portfolio. All senior executives know what to expect from IT; and IT knows unambiguously what it must deliver. Many critical issues faced by the IT organization and its clients are best attacked by employing these processes effectively.

SUMMARY

The application portfolio and related databases form a large and important resource for the firm. These resources are frequently accompanied by significant problems and unique opportunities. They require senior management attention, and they relate to several critical success factors for IT managers. IT managers must handle issues presented by the application portfolio skillfully.

This chapter exposed the issues surrounding the application portfolio and presented an approach for dealing with them effectively. The process requires a high degree of interaction among individuals at all levels in the organization. This management approach requires that senior executives focus on the issues with the most fundamental concerns of the firm foremost in their minds. The process causes the portfolio to be scrutinized with the strategic and tactical directions of the organization as a reference base. It represents a disciplined, businesslike approach to a significant business problem.

There are many advantages to this methodology. It puts decision making in the hands of those most responsible and most capable of making well-informed decisions about important issues. It makes IT managers jointly responsibile for the results and forces proper accountability throughout many management levels. This methodology eliminates many other less-effective methods for dealing with the priority issue. The development of unrealistic expectations regarding the applications is less likely with this process. This management system causes decision makers to consider alternatives not likely to be revealed by other, less well-disciplined methods.

Managers must have knowledge of some fundamental attributes about the applications and about the requested enhancements for them. This knowledge may not be readily available, but informed decisions cannot be made without it. If the firm is not accustomed to handling resource questions in a disciplined manner, this approach will require some executive education. This prioritization process may be useful in other situations within the firm as well. Lastly, this process yields substantial educational benefits regarding information technology for many of the participants. For all of these reasons, the IT organization and its management team will be well served by this management approach.

Review Questions

1. What are the ingredients of a firm's information system?

2. Identify the intangible assets of the information system in a firm.

3. What is the difference between depreciation and obsolescence with regard to application programs?

4. Why are the firm's databases growing in magnitude over time? Why is the database management system an important asset?

5. Distinguish between *maintenance* and *enhancement* as these terms apply to application programs.

6. The terms *depreciation* and *obsolescence* are used in connection with programs. How does this compare with depreciation and obsolescence of more tangible assets?

7. What is the definition of application backlog?

8. What is the invisible backlog, and under what conditions is it likely to be large?

9. What courses of action for a program can result from the application of the process discussed in this chapter?

10. Describe the dilemma posed by the application backlog.

11. Why does maintaining and enhancing the application portfolio consume large amounts of resources?

12. What additional complications to the enhancement process are caused by the fact that data files are frequently shared by application programs?

13. What are the essential elements in the methodology presented in this chapter?

14. Why is a plot of the satisfaction analysis (as in Figure 7.3) likely to display scattered along the diagonal? Why is it unlikely that any application will be rated 10-10?

15. How should the IT organization participate in the strategic/operational analysis?

16. Why are costs and benefits important to the process?

17. How does the process outlined in this chapter cause executives to focus on a variety of alternatives?

18. Under what conditions may the firm's databases restrict the choice of options or otherwise impact the decision-making process?

Discussion Questions

1. Discuss the causes of the computer runaways in the Business Vignette. What additional causes may be present?

2. Using data presented earlier in this text and through appropriate estimates, establish a relative value for the application programs and data resources in a typical firm. Be prepared to discuss your results.

3. Discuss the pros and cons of using application maintenance and enhancement as a training vehicle for junior programmers.

4. What additional attributes might be useful in the satisfaction analysis? Under what circumstances might additional attributes be required?

5. Using information from Chapter 2, trace the evolution of Merrill Lynch's CMA program on Figure 7.4 as you believe it occurred.

6. How does the application backlog relate to the issue of expectations discussed in Chapter 1?

7. Discuss the problems that will arise if satisfaction analysis alone is used to prioritize the backlog.

8. Discuss the issues associated with compiling the cost and benefit data for the applications. Discuss the effect on decision making of intangible costs and intangible benefits.

9. How can the time value of money be incorporated into the cost-benefits analysis?

10. Discuss how the management process described in this chapter helps accomplish the four-step program discussed in the Business Vignette in Chapter 3.

11. Some people believe the methodology presented in this chapter is too time-consuming and bureaucratic. What are the alternatives?

12. Discuss the relationship between this methodology and the task of reducing or removing the issues found in Chapter 1.

Assignments

1. Read the Schendler article in the list of readings. Identify the significant differences and similarities between the problems found by firms developing software for internal use and those developing applications for sale.

2. Study the Rackoff article in the references and prepare a report on the details of the SIS planning process at GTE.

ENDNOTES

[1] Jeffrey Rothfeder, "It's Late, Costly, Incompetent—But Try Firing a Computer System," *Business Week*, November 7, 1988, 164.

[2] See Note 1 above.

[3] See Note 1 above.

[4] Benn R. Konsynski, "Advances in Information System Design," *Journal of Management Information Systems* (Winter 1984-85): 5. Programming backlogs are found in other areas too. See Leslie Goff, "486 Rollout Means Next-Generation PCs," *MIS Week*, April 17, 1989, 26. The Intel 80286 is well supported with many applications, software for the 80386 is beginning to become available, but the 80486 is relatively unsupported. At this time there is a two-generation gap in software support for the 80486.

[5] Konsynski, "Advances in Information System Design," 7.

[6] Scott D. Palmer, "Software Maintenance," *Federal Computer Week*, December 5, 1988, 26. The General Accounting Office estimates that maintaining software costs the federal government 40 to 60 percent of its software budget. International Data Corp. found that enhancement maintenance consumes 50 percent of the maintenance budget in the federal sector. Fixing bugs requires 24 percent, and adapting programs to new hardware or system environments consumes 26 percent of federal maintenance expenditures.

[7] Palmer, "Software Maintenance," 26.

[8] Robert L. Glass, "Help! My Software Maintenance Is Out of Control," *Computerworld*, February 12, 1990, 87.

[9] Arnold S. Levine, "Survey Says People Problems Plaguing Software Maintenance," *Federal Computer Week*, October 16, 1989, 35. The survey found the most important problems to be attracting good people, lack of maintenance standards, and absence of current documentation.

[10] Martin Buss, "How To Rank Computer Projects," *Harvard Business Review* (January-February 1983): 118.

[11] Michael R. Mainelle and David R. Miller, "Strategic Planning for IS at British Rail," *Long Range Planning*, August 1988, 65. The method used at British Rail parallels that developed in this chapter.

[12] Edward Yourdon, *Modern Structured Analysis* (Englewood Cliffs, NJ: Yourdon Press, 1989). This book contains a thorough discussion of cost/benefits analysis in Appendix C. Some intangible benefits are realizable benefits that are hard to quantify and some are based on feelings or hunches that can't be quantified.

[13] See Carl L. Moore, Robert K. Jaedicke, and Lane K. Anderson, *Managerial Accounting*, 6th ed. (Cincinnati, OH: Southwestern Publishing Co., 1984), 347. Also see Yourdon, *Modern Structured Analysis*, 510.

[14] Nick Rackoff, Charles Wiseman, and Walter A. Ullrich, "Information Systems for Competitive Advantage: Implementation of a Planning Process," *MIS Quarterly* (December 1985): 285. The planning process includes instructing executives on competitive strategy and SIS, applying SIS concepts to actual cases, and reviewing the firm's competitive position. This is followed by brainstorming and discussing SIS opportunities, evaluating the opportunities, and developing detail for strategic systems planning. This article describes some of the difficulties associated with the application portfolio and, in particular, states that "the application growth rate in major data processing centers is anywhere from 25 to 40 percent per year."

REFERENCES AND READINGS

Buss, Martin D. J. "Penny-Wise Approach to Data Processing." *Harvard Business Review* (July-August 1981): 118.

Fjelstad, R. K., and W. T. Hamlen. "Application Program Maintenance Study: Report to Our Respondents." In G. Parikh and N. Zvegintozov, eds., *Tutorial on Software Maintenance*. Los Angeles: *IEEE Computer Society*, 1982, 13.

McFarlan, F. Warren. "Portfolio Approach to Information Systems." *Harvard Business Review* (September-October 1981): 142.

McNurlin, B. C., and R. H. Sprague. *Information Systems Management in Practice*, 2nd ed. Englewood Cliffs, NJ: Prentice-Hall, Inc., 1989.

Rackoff, Nick, Charles Wiseman, and Walter A. Ullrich. "Information Systems for Competitive Advantage: Implementation of a Planning Process." *MIS Quarterly* (December 1985): 285.

Schendler, Brenton. "How to Break the Software Logjam," *Fortune*, September 25, 1989, 100.

8 Managing Application Development

A Business Vignette

Introduction

The Application Development Problem
  Reasons for Development Difficulties

The Traditional Life Cycle Approach

Application Project Management
  Business Case Development

The Phase Review Process
  Phase Review Objectives
  Timing of Phase Reviews
  Phase Review Contents
  Who Are the Participants?
  Phases for Large Projects

Managing the Review Process

Resource Allocation and Control

Risk Analysis

A Business Vignette

Risk Reduction Actions

Successful Application Management

Summary

A Business Vignette

Review Questions

Discussion Questions

Assignments

References and Readings

Ramada, Inc., a Phoenix lodging firm, found itself caught in an expensive trap. On one hand, Ramada's DP shop found itself purchasing needed mainframe applications. On the other, it discovered that the cost of software support and maintenance was large and growing—and for some applications that were more than ten years old. New generations of hardware easily replaced the old, but old applications were difficult and expensive to replace.

So Ramada, along with a growing number of software-application users, moved to using computer-aided software engineering (CASE) tools on both personal computers and mainframes. At a division of I.E. DuPont de Nemours & Co., for example, PCs also help keep development costs down. By using a PC, DuPont developers avoid tying up the large mainframe systems.

Ramada projects cost savings averaging $40,000 per application from the use of CASE tools and PCs over eight years. It began its use of CASE with a mainframe-based code generator from Transform Logic Corp., a Scottsdale, AZ company. Not only could Ramada create large databases and on-line services with the Transform product; it could also create help screens for every field, screen, and transaction. Such details may have to be left to last when one creates a system using traditional coding by hand. Ramada found that the CASE application generated 95% of the code it needed for its applications; the remaining work could be handled by its programmers and analysts.

But Ramada wanted to reduce the amount of time spent on analysis and design done by people, so it procured Excelerator/IS, a product of Index Technology Corp., of Cambridge, MA. Excelerator runs on Ramada's PCs. Ramada found that creating the original of a data-flow diagram took about as long using the tool as drawing it by hand. But revisions, which formerly required as much time as originals, could now be produced in a matter of minutes. The result: dramatic savings in time. Ramada used the savings to improve communications between developers and users, and this further improved productivity.

When a company uses CASE tools, a change of mindset is required, from writing code to analyzing the problem, according to a spokesman in a Columbus, Ohio bank which employs software design applications. The emphasis on analysis reduces early design problems which are the most difficult and the most expensive to correct. Users at McDonnell Douglas claim this is a major advantage for CASE tools. The result is better systems at lower cost.

Ramada agrees. The shift in its emphasis results in 65% of development time being placed on analysis. This involves all users. Ultimately, Ramada finds that its development efforts have become much more productive. And the jobs formerly done by programmers can now become much more interesting and challenging as the tools permit the software designers to shift their attention from writing code to solving customers' problems and meeting their expectations.

INTRODUCTION

The previous chapter concentrated on the management of the application portfolio and related databases. The central problem with the portfolio is that of prioritizing scarce resources toward application maintenance, enhancement, or enlargement. This book emphasizes a management system to use with strategizing and planning systems developed earlier. This approach prioritizes resources; it focuses on doing the right things. The current chapter focuses on doing things right.

This chapter emphasizes application project management. It concentrates on the task of developing applications, one of several alternatives for application acquisition. For many applications, the preferred option is to perform maintenance, enhancement, or development activity within the organization. This option is best for large, unique applications or for important, strategic applications. Applications that require specialized, proprietary knowledge or that depend on exclusive, restricted databases must be maintained by the firm itself. Directing application development is a critical success factor for managers.

Technical considerations are important in application development; management considerations are critical. Weak or ineffective development managers are the most frequent and most expensive source of project difficulty. This chapter relies on traditional approaches to application development as a means for exploring these management issues.

The text emphasizes application project management. It concentrates on details important to application systems, such as application phase reviews, programming resource management, and risk analysis and control. The purpose of this emphasis is to develop a framework and a process for delivering application systems on time, within budget, and to the users' satisfaction. These objectives must be high-priority concerns for IT managers.

THE APPLICATION DEVELOPMENT PROBLEM

Application development continues to be a significant concern for most firms today. In fact, the difficulties of building applications seem to be increasing. For many firms, some of which were highlighted in the previous chapter, application development is a traumatic experience for developers and users. The trauma is not confined to the private sector. At a major university recently, the introduction of a phone registration system and its subsequent failure created major embarrassment for the administration. The breakdown caused an estimated 5700 students to wait in line for up to 10 hours to add classes. The administration said the failure was the result of a series of minor glitches that came together to create a gridlock. "We have just experienced the pain of new technology," commented the embarrassed president.

The problem seems to spare no one. The Interior Department's new payroll and personnel system is six months behind schedule. It is a planned adaptation of the FBI's undeveloped system using a fourth-generation language. The Office of Management and Budget identified 73 programs in 16 federal agencies that are at high risk.[2] Usually reliable firms like IBM, Lotus Development, and Ashton-Tate have all fallen victim to program development problems.

Reasons for Development Difficulties

Program development is one of the most difficult tasks for IT managers. The difficulties seem to fall into two categories: those deriving from programming itself, and those stemming from the firm or its management. Problems stemming from the programming discipline or from the program characteristics are:

1. Large program size

2. Increased program complexity

3. Measurement weaknesses

4. Weak theoretical foundations

The firm or its management is also a source of difficulties, such as those from:

1. Inadequate development tools
2. Improper use of tools
3. Weak management systems
4. Environmental factors

Programming itself has become more difficult for several reasons. Many of the small, easily written programs were completed years ago; current applications frequently are large and complex. For example, the first accounting programs or early billing systems consisted of perhaps ten or twenty thousand lines of code. Their cost was several hundred thousand dollars or less. The student registration system referred to earlier was under development for five years; the Allstate Insurance system is a $100 million project; and the space shuttle contains 25.6 million lines of code and represents a $1.2 billion investment. Here are some figures that suggest the size in lines of code (LOC) and cost of current application systems:[3]

Lotus 1-2-3, version 3	400,000 LOC	$22 million
1989 Lincoln Continental	83,500 LOC	$1.8 million
Citibank ATM	780,000 LOC	$13.2 million
Supermarket checkout scanner	90,000 LOC	$3 million

Many current systems are built around complex telecommunication nets. They interact frequently with many users in highly visible ways. Some are founded on older, less complicated, but relatively inflexible systems. Today's systems are larger and more complex than previous systems by orders of magnitude. While applications continue to increase in size and complexity, tools to build these difficult applications have not improved in capability and ease of use at the same rate.

In addition to shortened development cycles, firms expect increases in development productivity and improved application quality. Senior executives want to capture the benefits of new or improved applications sooner; consequently, they expect development managers to create applications in less time. Executives demand productivity improvements from their organizations, and they expect developers to build applications at reduced cost. They also expect higher quality applications. IT executives are searching for controllable development processes so that the results are more predictable and complexities can be managed effectively. Lastly, continuing maintenance expenses for error correction, enhancements, and changes must be minimized.

Because software development involves intellectual processes, software is flexible; it is very easy to change a line of code. It may be very difficult to understand the consequences of the change completely, however, particularly if the line of code is part of a large program. Large programs, therefore, tend to be rather inflexible and difficult to change. Advances in the theoretical foundations of software development and in tools for development lag behind the demand for applications by a large measure. Software development tools can be worthwhile, but their adoption and use are still limited. According to a study by Focus Research Systems, about 6 percent of IBM mainframe and compatible sites were using CASE tools and nearly 5 percent were planning to install tools during 1990. The greatest adoption of CASE tools, approximately 21 percent, is by the largest system users.[4]

Most management systems for application development utilize the notion that systems have life cycles. Embryonic concepts for new applications emerge, the ideas are developed, and systems are designed and implemented. They are maintained and enhanced, and are ultimately replaced. This systems development life cycle concept is the basis for studying the management issues and considerations in the traditional approach to application development. The life cycle approach to systems development is widely used, and its use is expected to continue.[5] The model was conceived to bring order to complex activities and to provide the basis for management intervention in systems development. This chapter outlines the management aspects of this subject, beginning with a discussion of the ingredients of the life cycle approach and culminating in some broad considerations of the overall process. The management issues surrounding the traditional approach to systems development are explored in detail.

THE TRADITIONAL LIFE CYCLE APPROACH

The life cycle approach to systems development divides this complex task into phases, each culminating in a management review. There are several valid reasons for using a partitioned or phased approach to development. Complex development activities are more easily understood and controlled in small increments. Skill requirements vary considerably over the life of the project and various skill groups are best managed if the project is subdivided. In addition, interactions between the development and user functions during development are much more easily managed incrementally. Finally, managers must evaluate progress and make decisions on an interim basis for maximum effectiveness.

Each phase has distinct characteristics; each contains unique development activities. Each phase has management activity, but they tend to be similar from phase to phase. It is normal for some nonmanagement activities within the life cycle to span several phases. This segmentation and overlapping will become clear in what follows.

The life cycle approach to systems development usually divides the project into five or more phases. The specific approach selected depends on the size of the project and the management system unique to the firm.[6] Table 8.1 outlines the phases as they are frequently employed in practice and as they are discussed in this chapter.

TABLE 8.1   Phases in the Development Life Cycle

1.   Initial investigation

2.   Requirements definition

3.   General design

4.   Development

5.   Installation

6.   Postinstallation activities

Each of the phases in Table 8.1 requires both unique and routine management activities. Specific types of management information are required during the phases and at the end of each phase. The activities and information requirements for each phase are presented in detail.

The techniques of application development are widely discussed in the literature. Many good texts discuss systems analysis, systems design, and system implementation.[7] There is much less information on *management systems* for systems design, development, and installation. This chapter concentrates extensively on the management of systems analysis, design, and installation activities because, as in most information technology endeavors, the ability to manage the work effectively is critically important.

APPLICATION PROJECT MANAGEMENT

The management of application development projects contains many elements common to the management of other types of projects but some elements are unique to programming projects. The success of application development managers, especially on large and complex programs, is relatively low. And, as pointed out in the Chapter 7 vignette, there have been some magnificent failures. It appears that there are important and significant differences between application development and other types of projects. These differences are critical for the management system employed by application developers.

Table 8.2 lists elements of the application project management process. The items in the table are necessary ingredients for success in managing application development. Development managers utilize these six elements as a framework for decision making. Although adopting these project management notions will not guarantee success in application development, not employing these techniques will greatly increase the probability of failure. Collectively, these actions and controls form the basis of a management system for application development.

TABLE 8.2   Elements of the Application Project Management Process

1. Business case development
2. Phase review process
3. Reviews management
4. Resource allocation and control
5. Risk analysis
6. Risk reduction actions

Business Case Development

The business case for an application program development activity illustrates and clarifies the costs and benefits of the project. The firm can expect to incur these costs and enjoy these benefits when the application is complete. The business case must address both tangible and intangible costs and benefits, providing a basis of comparison with the alternatives. One alternative must be the business-as-usual case. The tangible costs and benefits of the selected alternative, the remaining alternatives, and the business-as-usual case will be evaluated using one or more traditional financial analysis tools. In many cases, intangible factors are very important.

The department manager who owns the application is responsible for providing the business case. The IT development manager must propose the most cost-effective solution to the business problem, and this solution is used by the owner to articulate the business case. The business case is revised during the life cycle. It is an important constituent of the data on which the outcome of the phase review depends. The purpose of the business case analysis is to give executives vital information with which to make important decisions.

The first step in preparing the business case is to establish the objectives that the development activity is designed to meet. These objectives provide the base for future analyses; they are the foundation for subsequent decision making. They include a statement of the problem this application is targeted to solve. The objectives may describe an opportunity on which the organization intends to capitalize. The issues may be operational, tactical, or strategic and the goals of the system may yield tangible or intangible benefits. The second step is to perform an analysis of these benefits and the related costs.

The cost and benefit analysis must be performed for each alternative solution to the problem or opportunity and must also be completed for the business-as-usual case. It is important to understand the current situation in detail. Against this background the decision to proceed with alternative courses of action will be judged.

The analysis of the current system may not seem difficult, but it is substantially complicated if one includes the costs associated with lost opportunities or the cost of less than completely satisfactory customer service, for example. These intangibles must receive evaluation in both the current solution and the proposed alternatives if sound decisions are to result.

The financial evaluations must cover an extended period so that costs associated with the operation of the proposed solution can be assessed. This may mean that the analysis involves five years or more and that the time value of money becomes important. This also means that the analysis is subject to a wide variety of uncertainties and may require substantial amounts of judgment for interpretation. Under these circumstances, it may be best to make a range of projections of both costs and benefits. The decision makers can then make assessments considering the expected probabilities over the range of values.

In addition, the economic assessment must include an evaluation of all the tangible and intangible costs and benefits accumulated across all functions affected by the application over the appropriate period of time.

The estimate of development costs includes people-related expenses such as salary and benefits, hiring and training if required, occupancy and overhead costs, and costs associated with development activity. The development activity costs include hardware or computing costs, travel, and other items directly related to the development tasks. These costs will be found in the IT organization and in the user organizations. They must be summarized for the period starting at project inception and ending at application installation.

Operating cost calculations include IT and user people-related costs from the time of implementation to the end of the calculation period. Additionally, all costs associated with hardware and telecommunications systems, costs of purchased services and equipment, and all other cost items related to the operation of the system must be included. If the system is installed in phases, cost estimates need to reflect this phased approach accurately.

Upon completion of the cost analyses, it is possible to calculate the return on investment. The return on investment may be calculated via the payback method, which tells when the operating benefits exceed the development costs, the net present value method, or the internal rate of return method.

The net present value (NPV) method recognizes the time value of money over the life of the project. The NPV method accounts for the fact that money received today is worth more than money received at some future time, and that costs incurred today are more expensive than future costs. For most development projects, the NPV is the sum of the discounted net cash flows over the life of the project. The discount percentage is usually directly related to the firm's cost of capital. It may also be a function of time.

The internal rate of return (IRR) also includes the time value of money and the time span of the project. The IRR calculation yields a percentage rate which the firm receives for the money spent on the project. People from finance and accounting take part in these financial analyses to provide guidance in the details of the analysis. This gives confidence to the financial executives that appropriate financial considerations have been applied. Large projects require the concurrence of the controller in any case. Details of these financial analysis techniques can be found in most texts on financial management.[8]

If the analyses have been performed consistently and the considerations are mostly financial, the selection among the alternatives, including the business-as-usual case, will be straightforward. The alternative that yields the fastest payback, has the highest NPV, or yields the most positive IRR should be selected. However, in practice this situation almost never occurs because the intangible, not easily quantifiable factors generally are very important.

When purely financial considerations dominate the business case, some additional factors should be taken into account. For example, the payback method favors operational systems over tactical systems and tactical systems over strategic systems. The payback method will almost always exclude long-range investments; the use of this method tends to encourage short-range thinking. On the other hand, if the development costs are incurred over long periods of time, there are difficulties as well. There is a risk that technical obsolescence, changing business conditions, changes in management objectives, or other factors may make the system less desirable when installed than when originally conceived. A balance among these conflicting ideas can be achieved only through experience and judgment.

The most common situation is one in which intangible benefits are important or where long-range implications weigh heavily in the decision making. Strategic systems, those applications most highly sought by executive management, have both of these conditions by definition. In these circumstances the value of the strategic planning process and the resource allocation methodologies discussed earlier are very valuable. These processes and the thoughtfulness and carefully considered judgments made by senior executives over sustained periods of time all contribute to increasing the probability of success in what is a risky endeavor.

Firms that embark on large, complex application development projects incur risk. If the firm proceeds with the development project without the preparation embodied in the management systems and disciplines discussed earlier, it is incurring needless risk. This risk is accepted by the executives, their organizations, and the firm itself. Misunderstanding of the risk or thoughtless acceptance of it lies at the heart of the difficulties found in most application project runaways.

THE PHASE REVIEW PROCESS

Management control of important activities in the application development process is best accomplished when the work is divided into logical segments. The primary purpose of the life cycle approach, with its phases, is to ensure commitment and to understand and control the risk elements in the project.[9] The management control aspects of each phase have four dimensions: scope, content, resources, and schedules. The detailed activities of the phase and the information required for the phase review discussed subsequently reveal these dimensions. Each phase depends on other phases of the project, and the management process leads naturally from one phase to the next through the life cycle.

The phase review process is management oriented. It focuses on decision-making. The decision to continue, continue with modifications, or terminate is the essential part of the process. The decision usually is conditional and will be reviewed again at the next phase. That is, a subsequent review may uncover facts that lead to a different decision from that taken at previous reviews.

Phase reviews normally are conducted in meetings, but may be handled via correspondence in unusual circumstances. Each review should be documented in a phase review report, which forms part of the permanent record of the project. The documentation presented in the phase review process should develop from the workings of the management system employed to manage the project activities.

The management system must be well defined, organized, structured, and consistent across the phases of a project and between projects. It must ensure that the phase activity includes the ingredients in Table 8.3.

TABLE 8.3   Ingredients of a Phase Review

Project description

Well-defined goals, objectives, and benefits

Budgets and staffing plans

Specific tasks accomplished vs. those planned

Risk assessment

Process to track plans vs. actual accomplishments

Asset protection and business controls plans

User concurrences with objectives and plans

Phase reviews should produce documents and results useful for implementing subsequent phases. Implementation proceeds from one phase to the next providing the review results in a favorable outcome. The application system owner must concur that completed work meets specifications. Participants in the next phase must agree that they have all necessary inputs to continue.

It may appear that implementation progresses stepwise from phase to phase. Frequently, however, especially in large or complex applications, there is concurrent activity within and between phases. This does not eliminate the need for valid checkpoints that determine project status. Checkpoints required for tracking, measuring, or authorizing project continuation are more important when the project involves parallel activity.

It may not be possible to resolve all the issues at a given phase review. The decision to continue, continue with alterations, or terminate is a management judgment based on the circumstances and the information available at that point. The application owner may elect to proceed with an unresolved issue. However, this option should include an assessment of the risk involved. All open issues must be resolved promptly to ensure project integrity.

Phase Review Objectives

The objective of the phase review process is to provide the greatest probability of project success. This means that phase reviews must measure the accomplishment of agreed-upon objectives within the planned time and at the planned cost. They must provide a tool to assess activity status and to develop alternate action plans if required. They must provide a vehicle for reviewing plans and objectives for subsequent phases. All persons affected by the design, schedules, cost, data, and operating requirements of the project must have the opportunity to evaluate the status of the project at specific times during the process. The phase review process documents the decision reached by all affected managers.

Timing of Phase Reviews

Phase reviews should take place at the completion of each phase in the system development life cycle. Successful completion of the review indicates successful completion of the phase. When this process is applied to maintenance or enhancement activities that qualify as significant projects, phases and corresponding schedules must be identified. For large projects, with phases extending beyond six months or so, interim reviews should be held. Interim reviews should concentrate on plans versus actual accomplishments and expenditures and on the estimates to complete the phase.

If the system is large and complex, it is probably composed of multiple, concurrently developed subsystems. Phase reviews should be conducted for each subsystem; in addition, the entire project should undergo a comprehensive review at least once every six months. The purpose of this review is to coordinate the subsystem reviews and to assess overall project status. Each review must establish, alter, or confirm schedules for subsequent phase reviews.

Phase Review Contents

The activities of Phase 1, initial investigation, begin with an idea for a new application or for an enhancement to an existing application. These embryonic concepts usually (but not always) arise in the application-using department. Strategic systems, for example, may originate through the strategizing or planning processes as discussed earlier. Systems analysts perform a preliminary review of the existing system and prepare a preliminary definition of the new system requirements. A preliminary concept for the system is developed, and alternatives are prepared. This is followed by a feasibility evaluation, and plans and schedules for the Phase 1 review are then developed. Table 8.4 itemizes the management information requirements for the Phase 1 review.

TABLE 8.4 Phase 1 Review — Management Information Requirements

Statement of need and estimate of benefits

Schedule and cost commitments for Phase 2

Preliminary project schedule

Preliminary total resource requirements

Project dependencies

Analysis of risk

Agreed-upon project scope

Plans for Phase 2

Phase 2, requirements definition, consist of modeling the existing physical system and deriving a logical equivalent to which the new system requirements are added. This results in a logical model of the new system. The global technical design is created from the model. Updated costs and benefits are developed for the project and system control and auditability requirements are established.[10] System performance criteria are established at this time. The Phase 3 plan is complete, and the Phase 2 review is scheduled. The management information requirements for the Phase 2 review are shown in Table 8.5.

TABLE 8.5 Phase 2 Review — Management Information Requirements

Documented statement of requirements

Refined benefits commitment

Schedule and cost commitments for Phase 3

Refined project schedule

Refined total resource requirements

Updated analysis of dependencies

Analysis of risk

Agreed-upon project requirements and scope

Plans for Phase 3

Activities of Phase 3, general design, consist of developing external and internal system specifications. The system software specifications are refined and utility program requirements are specified. At this time, hardware requirements are completed and system architecture definitions are finalized. During Phase 3, system control and auditability requirements are completed and user documentation and training are planned. The plans for Phase 4 are developed and the Phase 3 review is scheduled. The management information requirements for Phase 3 are shown in Table 8.6

TABLE 8.6   Phase 3 Review — Management Information Requirements

Finalized general design

Final benefits commitment

Schedule and cost commitments for Phase 4

Committed project schedules through Phase 5

Committed costs through Phase 5

Resolution of remaining dependencies

Analysis of risk

Preliminary installation plan

Plans for Phase 4

During Phase 4, development, the activities consist largely of program design, building, and unit testing. File and data conversion strategies are developed during Phase 4 and program modules are written. Program module testing is completed during this phase and system installation is planned. The development team begins user training and develops user documentation. Plans for Phase 5 are developed and the Phase 4 review is scheduled. Phase 4 management information requirements are presented in Table 8.7.

The activities of Phase 5, installation, are critical. During Phase 5, user training and user documentation are completed. User acceptance testing is passed during this phase and file conversion is completed. The system is installed and operation is ready to begin. The Phase 5 review is scheduled. Upon satisfactory completion of Phase 5, the strategy for beginning use of the new system and phasing out the old system is implemented. The management information requirements for Phase 5 are shown in Table 8.8

TABLE 8.7   Phase 4 Review — Management Information Requirements

Finalized installation plan

Satisfactory completion of program test

Schedule and cost commitments for Phase 5

Committed project schedules through Phase 5

Reaffirmed commitment to system benefits

Commitment to system operational costs

Analysis of risk

Final installation plan

Plans for Phase 5

TABLE 8.8   Phase 5 Review — Management Information Requirements

Satisfactory completion of system test

User acceptance document signed

Finalized application business case

Reaffirmed commitment to system benefits

Commitment to system operational costs

Analysis of risk

Plans for Phase 6

Phase 6, postinstallation activities, consists mainly of the operation and maintenance of the new system. This is the time to evaluate the effectiveness of the life cycle management system and to review the management techniques applied during the development process. The system is evaluated against the original specifications and objectives. Managers conduct an analysis of programming and implementation effectiveness and review requested enhancements. The Phase 6 review is scheduled.

The Phase 6 review evaluates the results of previous phase reviews and makes plans for incorporating what was learned into subsequent projects. Phase 6 is important to IT application development management because it is introspective and reinforces sound management techniques. The IT organization learns through experience just as individuals do. The Phase 6 review provides checks among the strategy, the various plans, and the actual realized implementation. These checks are an essential part of management control.

There is a significant amount of management information implied by the outline above. This information enables managers to assess the progress of the programming project and make judgments regarding its future. Without information of this type and a setting in which to evaluate it, managers take on unnecessary risk and may face the prospect of operating out of control. Note that Phase 6 is mainly a review of the quality of the management process. This review is introspective. It assists management in refining the process for the organization's future use.

Who Are the Participants?

The owner of the application system and data, key user managers, and representatives of all functions affected by or affecting the project must participate in and evaluate the results of phase reviews. Representatives of other senior executives should participate if the project is of significant potential benefit to the firm. Usually it is not difficult to gain the attention of executives when large amounts of money are involved. IT managers or their representatives normally orchestrate the review process.

Phases for Large Projects

Exceptionally large or complex projects may require special attention, which is best accomplished through modifications to the phase review schedule. The terms "large" and "complex" are relative to the size of the organization and to its skills. The decision to invoke special treatment must be individually determined for each organization. If the proposed project is the largest or most complex ever considered by the firm, then it must be given special handling. Other projects may qualify for special treatment as well. Special consideration consists of additions to the phased process, which are illustrated in Table 8.9. For major projects, the phases consist of more events.

The modified phased approach presented in Table 8.9 divides the design and development phases discussed earlier and creates a project with eight phases. The review that concludes Phases 3A and 3B evaluates the detailed external design and the detailed internal design. Likewise, the program design can be evaluated separately from the programming and testing. This additional perspective greatly assists in maintaining project control. Further subdivisions can be made if conditions warrant. In addition, project managers must define appropriate intermediate checkpoints within each phase to assure proper controls.

Phase reviews are essential to control project activities. They relate directly to the system development life cycle, and they force thoughtful and organized decision making. Managing phase reviews is a critical IT activity.

TABLE 8.9   Expanded Phase Review Process

Phase 1	Initial investigation
Phase 2	Requirements definition
Phase 3	General design
Phase 3A	Detailed external design
Phase 3B	Detailed internal design
Phase 4	Development
Phase 4A	Detailed program design
Phase 4B	Program and test
Phase 5	Installation
Phase 6	Postinstallation

MANAGING THE REVIEW PROCESS

The review process is management-oriented and it can lead to sound decision making. Each phase review must contain unambiguous documentation with which all parties concur. As a minimum, the documentation must describe the scope, content, resources, and schedules of the work effort. In addition to these items, it is advisable to produce a clear statement of the assumptions and dependencies involved in each phase and the complete plan. At some point in the project schedule prior to successful project completion, the assumptions must turn into facts and the dependencies must be formalized explicitly.

Every effort must be devoted to resolving issues and nonconcurrences prior to or during the phase review. The emphasis during the review should be on the items of scope, content, resources, and schedules. Changes to any of these major items should receive special attention. Managers who have a strong tendency to search for the underlying reasons for changes and variances at each step are less likely to get into trouble later on.

A management system for issue tracking and reporting must be developed. Each issue must be assigned to an individual for prompt resolution within an agreed-upon schedule. Subsequent phase reviews must document the fact that all open issues have been resolved satisfactorily. Unresolved issues inevitably lead to an unsatisfactory review; they can become sufficient cause to terminate the project.

The results of the phase review must be documented and placed in the management file on the project. A summary of the information should be prepared and distributed promptly to all concerned parties and to appropriate members of management.

Project planning describes the ebb and flow of skills from Phase 1 through Phase 6. Analyst and user activity is high during the beginning of the project and is high again just before and during installation. Programming activity is low at project definition. It peaks during implementation and tapers off after installation. Similarly, computer operators, technical writers, database administrators, trainers, and others all have patterns of deployment over the life cycle. Managers need to track and control these resources just as they do physical or monetary resources. Life cycle resource management and control are fundamental to the success of application program development.

Information on the deployment of resources by skill type is part of the application development plan. The resource plan must be available during the life of the project; it is given additional detail at each phase review. Resources applied by skill type versus the plan must be tracked by the management team during each phase. For example, during Phase 3 analyst effort declines and programmer activity increases as the application moves toward implementation. Deviations from the plan during this phase in these skills is a signal for management to attack the underlying cause. Simply analyzing total resources will not reveal the degree of detail that management needs.

Stephen Keider identifies six tasks that are most likely to cause failure if mismanaged and presents seven early danger signs.[11] Through interviews with 100 MIS professionals he learned what they thought was the single most important cause of system failure. The results are:

Reason for failure	No. of responses
Lack of project plan	23
Inadequate definition of project scope	22
Lack of communication with end users	14
Insufficient personnel resources and associated training	11
Lack of communication within project team	8
Inaccurate estimate	8
Miscellaneous	14

Some projects fail because of technology or design problems, but most difficulties are controllable by the project manager.

An essential and vital part of application project management and of each phase review is the analysis of risk. Software projects fail for a variety of reasons, but most of the reasons for failure can be traced to inadequacies in the project management system. Management system failures are greatly reduced by using the principles developed in this chapter. But there is one additional technique that will significantly improve chances of success. This additional element is the analysis of risk.[12]

In the application development process there are analogies to the familiar economic concepts of leading, coincident, and lagging indicators. Some project managers are relatively uninformed and are observing lagging indicators. For example, some development managers will find out one day that the program they developed has failed because the user community has never embraced the program, or has ceased using it because it lacks function. These managers will discover the failure after the fact.

Most programming managers are better informed. They have an exhaustive list of project metrics that they track on a continuous basis. These indicators relate to budget, schedule, function, or to a host of other relevant items. The managers' careful monitoring of these indicators allows them to know when their project fails at the very moment it does. They are the first to know, because they are employing coincident indicators. This is certainly better than using lagging indicators, but it is not nearly good enough. The idea, after all, is to succeed, not fail. What is needed is a set of leading indicators, indicators to alert managers in advance that their projects are headed for trouble unless they take action. The analysis of risk provides these indicators.

What is the analysis of risk? How will this analysis work to provide early warning for project managers? How will project managers be alerted to impending difficulties? To answer these questions, managers need to search for the sources of programming project difficulties. Next, they must develop a set of quantified measures to describe the extent of risk to which they are subjected in each of these areas. Additionally, managers must track these risk measures to discern their trends over the life of the project.

The major sources of risk for an application programming project can be grouped into five categories with their approximate weights. Table 8.10 lists these sources of risk and their approximate weights.

Each category can be further subdivided into quantifiable and measurable items that collectively provide a basis for the assessment of project risk. First, let's discuss the ingredients of these six categories and then develop a rationale for quantifying them.

TABLE 8.10   Major Sources of Risk

	Source of risk	Weight
1.	User activity	20
2.	Programming and management skill	10
3.	Application characteristics	30
4.	Project importance and commitment	20
5.	Hardware requirements	10
6.	System software requirements	10

Active user involvement in any programming project is vital for the success of that project. If support from the user community is weak or missing, the project is in jeopardy. User activity can be measured by considering the quality and quantity of requirements definition and the depth and breadth of user involvement in the prerequirements activity. User involvement throughout the postrequirements activity and the user's knowledge level of the proposed system and its relationship to other user systems are good measures of activity, as are the extent of user training efforts and the success of these efforts.

The knowledge, skill, and experience levels of the system implementors and of the project manager are extremely important for success. Having insufficient numbers of implementors or having implementors with insufficient skills to perform the tasks indicates that corrective action should be taken. For instance, if the system requires new telecommunications software and technicians are not sufficiently experienced or trained in this area, management should be concerned.

A third and obvious area of concern is the nature of the project itself. The items of importance are the size or scope of the system under development, its duration and complexity, and the output or deliverables of the project. The project logistics and the extent or degree to which development takes place over a wide geographic area are other significant factors. For instance, application programs developed across several locations are much riskier than those developed at one site. Additional items in this category include the sophistication of project control techniques employed and the extent to which the development team is homogeneous and used to working together. This latter item should include the effects of subcontract programming, for example.

Project importance and commitment can be measured by the aggressiveness of the schedule, the number of management agreements needed for implementation, and the extent of management commitment to the project. Each factor adds a measurable element to the risk assessment for the total project.

System hardware forms a fifth area of risk. If new or unfamiliar hardware is required for the system, or if the system has unusually stringent performance requirements, risk is introduced. Programs developed to run within the capacity of existing hardware systems incur little or no risk in this area. Performance specifications for the system and capacity planning for the hardware system are important.

The last item of concern is the operating system software needed for application development and implementation. If the application uses routine system software and is developed using familiar languages, risk is minimized. Programs using new and unfamiliar languages are much riskier. Programs using prereleased or relatively untested software packages are even riskier.

There may be more risks. For example, if part of the application is purchased, there may be additional risks associated with the vendor. On the other hand, depending on the capability of the firm's programming staff, purchasing all or part of the application may reduce risks to the firm. Another example of an external source of risk is in the use of third-party telecommunications systems. This dependency brings with it additional risk considerations, which must be analyzed and quantified.

A useful but somewhat arbitrary weighting or quantifying of these risk elements is shown as a guide for the project manager. Managers may want to adjust the values to suit their individual situations. However, it is important that the methodology remain constant during the life of the project. Because absolute values of risk cannot be measured accurately, the change in the value of risk is much more valid and useful. It is worthwhile to keep a history of risk measures to establish trends within the organization and across projects and project managers. A more detailed breakdown of the weights for each category is given in Table 8.11.

TABLE 8.11   Detailed Risk Items

1.0	User activity	
1.1	Quality and quantity of requirements definition	6
1.2	Prerequirements planning activity	3
1.3	Postrequirements activity	4
1.4	Knowledge and understanding of proposed system	4
1.5	User training	3

(Continued)

TABLE 8.11   Continued

2.0	Programming and management skill	
2.1	Implementor experience, skill, and ability	5
2.2	Management experience, skill, and ability	5
3.0	Application characteristics	
3.1	Size, scope, and complexity of the system	6
3.2	Duration and complexity of the project	5
3.3	Project deliverables	3
3.4	Project logistics	5
3.5	Project control techniques	7
3.6	Organization considerations	4
4.0	Project importance and commitment	
4.1	Aggressiveness of the schedule	4
4.2	Number of managers involved	8
4.3	Extent of management commitment	8
5.0	Hardware requirements	
5.1	New hardware	6
5.2	Stringent performance requirements	4
6.0	System software requirements	
6.1	Operating system and utilities software	5
6.2	Language requirements	5
Total		100

The 20 items must be scored and evaluated prior to each phase review and preferably more frequently. An item with no risk would be scored zero; an item having a great deal of risk would be given the highest score possible for that item. The weights should be consistent from scoring to scoring in order to determine risk trends. Precision is much preferred to accuracy in this analysis. The absolute level of risk is valuable in determining whether the project should continue, continue with modifications, or be terminated. For example, if at Phase 1 the risk totals 75, most prudent managers would seriously consider terminating the project. For the risk to equal 75, many of the individual items would have to be at high risk. Therefore, the entire project will be high risk at this time. On the other hand, if the total risk at Phase 1 is 20 or lower, this would be an indication of low or manageable risk. Numbers between these ranges require a detailed review and analysis at the phase review prior to decision making.

As the program proceeds toward implementation, the initial risk should decline. (A program successfully installed and operating has zero risk!) Programs proceeding smoothly shed risk from Phase 1 to Phase 6. Increases in total risk over the life of the development project are danger flags. More detailed analysis is required. Yet, if the total risk is declining, the project may still be headed for trouble. An example would be a situation in which one of the items previously having a zero or low value suddenly increased in value. If, during Phase 3, Item 3.2, duration and complexity, increased from a value of 1 to a value of 5 because it was discovered the program was much more complex than previously predicted, that could be cause for serious concern. The concern will be present even though other items decline in value and the total declines.

This analysis provides leading indicators of project success.[13] Management commits to deliver a product at some future time, at some cost, and with a stated function. This commitment is made with an understanding of the risks involved at the time the commitment is made. It is further understood that risks would be mitigated during the course of development in order to meet the commitments. If analysis over time indicates unfavorable risk trends, management must take corrective action to meet the previous commitments. Analysis of the sort developed above permits management to take corrective action before commitments are missed. The consequences of proceeding with the project in the face of high or rapidly rising risk can be severe.

Seemingly easy tasks, such as switching to a new computer system, can be troublesome even for firms experienced in computerization. The experience of Sun Microsystems in the next Business Vignette illustrates the point.

A Business Vignette

Sun Microsystems Apologizes In Letter For Late Payments[†]
Suppliers of Computer Maker Are Told Faulty System Caused Tardiness in Past

Sun Microsystems Inc. sent a form letter to its 6,000 suppliers apologizing for failing to pay its bills on time.

Late payment of bills is the latest problem to surface as a result of the desktop-computer maker's much publicized switch to a new system for providing its management with information.

In June, Sun, based in Mountain View, California, disclosed that problems with its new computer system for getting information to management had resulted in lost customer orders and an inability to track inventory. At the time, Scott McNealy, Sun's chief executive officer, said the firm's information system played "a significant part" in a slowdown of Sun's business, but that it was "screaming now."

"I am aware of the problem and inconvenience this may have caused you and your company," wrote Terence Lenaghan, Sun's corporate controller. "I would also like to assure you that any delays that you might have experienced in receiving payments are in no way reflective of a deterioration of Sun's financial status."

A spokeswoman for Sun said the company was late in sending out the letter and that it was paying suppliers on time now. "The letter really refers to a past period of time," said Carol Broadbent.

Ms. Broadbent said that Sun "had extra-high payables at the end of June," but has since corrected the problem. In August, Sun acknowledged that the information system still hampered it, requiring the company to, among other things, finish some of its accounting work by hand.

Partly because of the problems, Sun reported a loss of $20.3 million on revenues of $431.2 million for its fourth quarter ended June 30.

[†] Reprinted by permission of *The Wall Street Journal*, ©Dow Jones & Company, Inc. 1989. All Rights Reserved Worldwide.

RISK REDUCTION ACTIONS

There is much value in having some quantification of the risks inherent in the project. Prudent managers can take actions to reduce or mitigate the risks and take advantage of risk trends. It may not be possible to eliminate completely all the risk during the project, but it certainly is possible to focus attention on the areas of risk and to take action to manage them. Management action may consist of deployment of special resources; institution of special control techniques; or utilization of alternatives having less risk. Project managers have many resources at their disposal during the life of a project with which to manage problems. They need risk assessment tools to alert them to impending problems.

For instance, if user training lags behind schedule, risk for the project will rise. Alert project managers will recognize this trend and search for the underlying causes. They know that training users in the operation of the application is very important and that poor or late training jeopardizes the successful installation of the application. In the midst of development activities it is easy to conclude that training will catch up later when there is more time. Aggressive managers will resist this tempting thought and will take action to eliminate this risk item now.

Problem management is a major part of a project manager's job. Risk analysis is an analytical tool that yields early warning of impending difficulties. Alerted to future difficulties, IT managers can initiate action to cope with the difficulties when they are small and manageable. It is an indispensable part of the project management system for successful application development.

SUCCESSFUL APPLICATION MANAGEMENT

The ingredients of successful application management flow from a well-designed and smoothly functioning management system. The application management system is a controlled process capable of yielding predictable results and of dealing with increased complexity. Its products meet all technical specifications and are surprise-free regarding function. The resulting applications meet schedule and budget agreements and satisfy the conditions expressed in the business case regarding operating costs and realizable benefits. The purpose of the review process and the other efforts contained within it is to ensure that these goals are attained. Successful application management yields products that contribute important assets to the organization.

The application management process defined in this chapter applies not only to programs but to user and program documentation too. Programs, program documentation, and user documentation are products of the development process and are managed alike. The documentation products must be subjected to critical scrutiny at the phase review just as the emerging application is.

Business managers in most firms today are intensely interested in improving productivity and enhancing performance. Application project managers must concentrate on productivity too, and they must be able to demonstrate productivity improvements. The thoughtful, organized, businesslike approach to the development process detailed in this chapter is the basis for productivity improvements. Additional ingredients also are required. Ramada's consistent use of development tools illustrates one of these ingredients. Several others will be discussed in later chapters.

The U.S. Congress is also concerned about these issues and is demanding changes in procedures and improvements in development and acquisition processes, as shown in the next vignette.

Software development continues to be a high-risk activity. Program complexity is increasing in many cases, and some easily stated problems contain obscure, partially hidden difficulties. Software developers do not have appropriate tools in many cases, and software project management techniques must be substantially improved. As organizations attempt increasingly complex projects, managers must explore new support technology and must pay special attention to managment control techniques. Techniques developed in this chapter are valuable and important to success.

SUMMARY

Managing application development projects is a difficult task because systems are becoming larger, more complex, and more strategic in nature. Application development involves more resources, occurs over a longer time period, and is inherently riskier than it was a decade ago. *Ad hoc* management techniques and routine project management methodologies need to be augmented by disciplined processes. These processes must be specifically designed to cope with the difficulties of application development and with the associated increased risk.

The phased development approach divides the project into manageable tasks. It requires a management review process to focus on fundamental project issues. The foremost issue is the validity of the project itself; this concern is addressed through sound business case analysis. The business case is reviewed periodically throughout the life cycle of the application to ensure continued project viability. Without this essential foundation for the project, all efforts to develop a successful application ultimately will lead to failure.

The phase review process includes a careful reevaluation of the business case. It focuses on the control issues of scope, content, resources, and schedules. It reviews the assessment of risk at the end of each phase. The phase review is a decision point in the life of the project. The decision takes into account the relevant inputs from all parties involved in or affected by the development effort. It is designed to illuminate all important issues relating to the continuation of the project and to resolve them to management's satisfaction. Failure to accomplish these objectives is sufficient reason to terminate the project.

Risk analysis is an important management tool because it yields leading indicators of project difficulties. Therefore, it provides management with an opportunity to resolve problems when they are small and easily handled. Failure to take advantage of these early warnings frequently leads to major problems later in the cycle. Major problems late in the life cycle of an application lead to extreme distress for project managers and for all who depend on the successful conclusion of the development effort.

Successful management of application development requires a management system tuned to the needs of programming projects. It demands a rigorous, thoughtful process and candid, open communication among all participants. The participants must pay attention to detail and must demand excellent follow-up on the resolution of the many issues that will continue to surface. The basis for much of this process lies in the antecedent activities of strategy development, strategic and tactical planning, technology assessment, and application portfolio asset management. Given a firm foundation in these activities and the rigorous employment of project management tools, techniques, and processes, the risks involved can be substantially mitigated and chances of success greatly increased.

The management system for applications builds on previously developed management processes. This means that application development contributes to the business goals in a positive and well-understood manner. The firm and its managers know that IT's development activities are congruent with the firm's objectives. Strategy development and strategic planning, along with resource prioritization in tactical and operational plans, are effective processes. If projects are well managed, the firm's executives will be likely to hold realistic expectations of the IT function. The management system for applications is valuable because it focuses on the IT manager's critical success factors.

A Business Vignette

Rep. Murtha Criticizes Software Failings at Cheyenne Mountain[†]

Pittsburgh—An Air Force four-star general and a congressman cited space and missile defense programs at Cheyenne Mountain, Colorado, as an example of computer software problems causing long delays and drastic cost over-runs in a critical defense system.

Their remarks were addressed to an attentive audience of 1,000 computer software executives at the annual TriAda convention here last week. "The costs of Cheyenne Mountain improvements are going to double from their original estimate, and we still don't have a workable system at America's most important installation involving air, missile and space defense . . ." because of software application inadequacies, said Rep. John Murtha, chair-man of the House Appropriations Defense Subcommittee, the body that rules on virtually every penny spent on U.S. military outlays.

"Cheyenne Mountain's missile warning system is one example of falling behind in software development," agreed Gen. Bernard P. Randolph, com-mander of the Air Force Systems Command.

"The C-17 has been delayed by a year because of software problems. Both the Air Force and the Congress told the prime contractor there was a problem with software, but he didn't do anything about it," Murtha said.

In an era of DOD cutbacks, Murtha said the high-tech industry shouldn't be surprised that "money for software and computers is going to be reduced substantially in the budget," which he said contains $30 billion in requests for spending in the field. Randolph noted that advances in software develop-ment have trailed far behind those of computers themselves.

"Hardware productivity has improved by a ratio of 1 million to one in the past 25 years, and software has improved only 17 to one. And the time for studying the problem is over. The Department of Defense has had 17 studies of it in the past few years. You just can't sit still and define the prob-lem, you have to begin resolving it," he added.

Review Questions

1. What are the benefits of using CASE tools for the firms discussed in the first vignette?

2. What are the goals of successful application development? Why is this area a critical success factor for the IT manager?

3. Why is local development the only reasonable alternative for many applications?

4. Why does the life cycle approach divide the project into phases?

5. What are the phases in a typical life cycle?

6. What activities take place in each phase? What information is required?

7. Why is Phase 6 important?

8. Do the system development life cycle and the phase review process apply to the maintenance of large applications? Under what conditions are these ideas most important?

9. What are the six essential elements of a sound project management system for application development?

10. What are the similarities and differences between managing a computer system application and managing the construction of an office building?

11. What are the ingredients of a computer-system application business case?

12. Why is it becoming more difficult to develop and assess application business cases?

13. Why are phase reviews an indispensable part of the management system for application development?

14. What are the objectives of a phase review?

15. How often should phase reviews be held?

16. How should the phase review process be modified for very large applications?

17. What role does documentation play in the review process?

18. Describe the risk analysis process. What are the elements that are reviewed in risk analysis?

19. Why is risk analysis described as a leading indicator?

20. What is the connection between critical success factors and application development management?

Discussion Questions

1. Who should participate in systems analysis and design? How does the degree of involvement for analysts and programmers change from inception of the idea to implementation?

2. Discuss briefly what a systems analyst does in each of the six phases. What does the user manager do at the completion of each phase?

3. Discuss the phase review discipline as applied to very large and complex applications.

4. Describe the management system that must be in place in order to ensure project integrity from phase to phase.

5. How is resource control maintained in application development project management?

6. Describe some of the intangibles in the application business case and discuss why they are important.

7. What are some of the intangible issues likely to be present in the development of a new on-line customer order entry system?

8. What would be the major risk factors for the replacement of the current payroll program with a new program incorporating the latest tax changes to be implemented at year end? How would these risk factors tend to vary over the development cycle?

9. Where do the risks reside in the development of a totally new strategic information system?

10. Describe the management actions that form the main ingredients of the management system needed for successful application development.

11. What elements of the management system developed in this chapter would be most valuable to those firms in the vignette in Chapter 7?

12. What antecedent activities are necessary in the firm for application development projects to proceed successfully?

Assignments

1. Outline the agenda for the Phase 3 review of an internal application program. Name the positions that should be present and identify the order in which they speak. If the program under development is going to be developed as a product of the firm and Phase 3 precedes product announcement, what additional considerations will be important?

2. Select a popular CASE tool and review its specifications. Prepare a report on the management aids found in the tool and the problems they solve for application project managers. Prepare your report for presentation to your class.

[1] Janet Mason, "Corporate Programmers Lead the Way with CASE," *PC Week*, September 12, 1988, 50.

[2] Carolyn Duffy, "OMB Tags 73 Systems in Trouble," *Federal Computer Week*, December 11, 1989, 1. The difficulties are in system acquisition and system security. The departments of Commerce, Defense, Education, Housing and Urban Development, Labor, and State are involved.

[3] Brenton Schlender, "How To Break the Software Logjam," *Fortune*, September 25, 1989, 100.

[4] John Mahnke, "CASE Momentum on a Roll," *MIS Week*, January 2, 1989, 29.

[5] Charles R. Necco, Carl L. Gordon, and Nancy W. Tsai, "Systems Analysis and Design: Current Practices," *MIS Quarterly*, December, 1987, 461. The authors believe the life cycle approach will benefit from structured development and increased use of tools, and it will be influenced by prototyping methodologies.

[6] There is little agreement on the number of phases into which the systems life cycle should be divided. However, the management principles apply regardless of the number of phases.

[7] Some representative texts on analysis and design are listed in the references. See, for example: Aron; Kendall and Kendall; Powers, Cheney, and Crow; Wetherbe; and Yourdon (1989).

[8] For example, see Paul M. Fischer and Werner G. Frank, *Cost Accounting* (Cincinnati, OH: South-Western Publishing Co., 1985), 219-226.

[9] Joel D. Aron, *The Program Development Process: The Programming Team* (Reading, MA: Addison-Wesley, 1983), 340-343.

[10] System control and auditability features are critically important to applications programs. They are discussed extensively in Chapter 16.

[11] Stephen P. Keider, "Managing Systems Development Projects," *Journal Of Information Systems Management* (Summer 1984): 33.

[12] The analysis of risk presented here is derived from the author's experience at IBM. Other organizations may use similar analyses.

[13] Aron, *The Program Development Process*, 343-348, presents risk assessment in a somewhat different manner.

REFERENCES AND READINGS

Aron, Joel D. *The Program Development Process: The Programming Team.* Reading, MA: Addison-Wesley, 1983.

Bachman, Charles W. "A Personal Chronicle: Creating Better Information Systems, with Some Guiding Principles." *IEEE Transactions on Knowledge and Data Engineering* (March 1989): 17.

Brooks, Frederick P. *The Mythical Man Month—Essays on Software Engineering,* Reading, MA: Addison-Wesley, 1974.

Fischer, Paul M., and Werner G. Frank. *Cost Accounting.* Cincinnati, OH: South-Western Publishing Co., 1985.

Heckel, Paul. *The Elements of Friendly Software Design.* New York: Warner Books, Inc., 1984.

Keider, Stephen P. "Managing Systems Development Projects." *Journal of Information Systems Management* (Summer 1984): 33.

Kendall, Kenneth E., and Julie E. Kendall, *Systems Analysis And Design.* Englewood Cliffs, NJ: Prentice-Hall, 1988.

Lucas, Henry C., Jr. *The Analysis, Design, and Implementation of Information Systems.* New York: McGraw-Hill Book Company, 1985.

Mason, Janet. "Corporate Programmers Lead the Way with CASE." *PC Week,* September 12, 1988, 50.

Powers, Michael J., Paul H. Cheney, and Galen Crow. *Structured System Development,* 2nd ed. Boston, MA: boyd & fraser Publishing Co., 1990.

Wetherbe, James C. *Systems Analysis and Design: Traditional, Structured, and Advanced Concepts and Techniques,* 2nd ed. St. Paul, MN: West Publishing Co., 1984.

Yourdon, Edward. *Managing the System Life Cycle,* New York: Yourdon Press, 1982.

Yourdon, Edward. *Managing the Structured Techniques.* New York: Yourdon Press, 1986.

Yourdon, Edward. *Modern Structured Analysis.* Englewood Cliffs, NJ: Yourdon Press, 1989.

9

Alternatives to Traditional Development

A Business Vignette

Introduction

Advances in Tools and Techniques
   Fourth-Generation Languages
   CASE Methodology
   The Object Paradigm
   Prototyping

Subcontract Development

Purchased Applications
   Advantages of Purchased Software
   Disadvantages of Purchased Applications

Additional Alternatives

A Business Vignette

Managing the Alternatives

Summary

Review Questions

Discussion Questions

Assignments

References and Readings

Appendix

A Business Vignette

Unusual partnerships are being formed to defray the cost of new technologies and to recoup investments in existing systems; profits, however, are still elusive.

IS Shops Form Alliances as Development Costs Rise[1]

When a systems audit last year at Kidder, Peabody Inc. declared the firm's trading systems out of date and its data largely unusable, the new IS management faced a tough choice. Kidder, which has just been acquired by General Electric, could either spend six years and upwards of $100 million developing new systems, or it could acquire technology and systems from Wall Street rival First Boston Corp.

In an unusual move, Kidder chose the latter course. It purchased First Boston's technology and agreed to co-develop a version for retail trading. "We are going to do this for a fraction of what it would cost to do it ourselves, and in half the time," explains VP of Information Systems Robert J. McKinney. "That was a huge factor."

With multimillion dollar price tags now common in major systems development, IS organizations are re-evaluating how they develop or acquire their most crucial systems. Increasingly, they are open to trading them, and technology, with others—even competitors—to defray the staggering costs. Many more seek to recoup investments by licensing their software directly or by creating subsidiaries to market systems and expertise.

"We can't afford to keep doing what we've been doing: everyone building essentially the same systems," maintains Joseph P. Castellano, Managing Director for Information Systems at Drexel Burnham Lambert, Inc. "The biggest challenge we all face is reducing costs. Sharing the development, buying packages built by others—vendors or competitors—is one way to do it."

The $60 million cost of a global trading system was all the motivation First Boston needed to consider reselling its technology, according to Eugene F. Bedell, VP of Information Services. "When everyone agreed this system had to be built, one of the first questions was, 'How do we pay for it?'" Bedell recalls. Having taken the plunge with Kidder, First Boston now is exploring several ways of selling the fruits of its development.

For San Francisco-based construction and engineering firm Bechtel Group Inc., and Peoria, Illinois-based Caterpillar Inc., a large equipment manufacturing company, it was not the recouping of monies in particular systems,

---

but reversals in their traditional businesses that led each to form IS subsidiaries this year. Falling demand for nuclear power plant construction forced Bechtel to look inward for ways to boost falling revenues; IS was one area that offered opportunities.

Similarly, Caterpillar's $2.5 billion sales drop in the early 1980s "was a catalyst" in its decision to consider IS as a revenue source, says William B. Heming, a manager in the company's venture capital operation. "We had a lot of resources in the computer area that weren't being fully used," he says.

"Faced with intangible advantage or cold, hard cash to recoup part of their investment, a lot of companies would take the latter," says Thomas H. Davenport, director of research at IS consultants Index Group Inc., Cambridge, Mass. The economics of building major systems is relaxing old taboos about dealing with competitors. When it comes to sharing technology, "senior management is open to it," declares Drexel's Castellano.

First Boston's sale of technology to Kidder, for instance, precludes only those portions of the trading system that deal with internal accounting and with decisions on when to enter or leave a securities market. Minus the analytical and accounting portions, the global trading system "will help Kidder Peabody do business more efficiently, but won't help them take any business away from us," argues First Boston's Bedell. "There is nothing proprietary in the functions we're doing . . . [They have] all been done for decades."

Castellano agrees. "No one wants to lose control of his systems, but lots of people think many of these do not represent a proprietary edge. Let's face it; egos aside, most firms in our business are competing without any proprietary edge at all."

Forest products giant Weyerhaeuser Corp., of Tacoma, Washington, has run an IS service venture for three years. During that time, the company has refused to sell IS services for competitive reasons only once, according to J. George Pikas, director of sales and marketing for Weyerhaeuser Information Systems.

Still, the issue of competitiveness remains a touchy one. When Bechtel Group recently formed its own IS venture, it also entertained questions about selling IS to competitors. Information and Technology VP Bill Howard says, "With most of the software, it's really as much a matter of how you used it as of anything else. We think we know how to use it, and obviously we will have access to it earlier."

Yet, for all the enthusiasm behind companies turning to IS ventures to boost revenues, there is plenty of caution. Caterpillar Information Services and WIS operate near break-even or in the red, according to executives.

Others, such as Hartford Insurance Co., have found attempts to sell IS expertise a money-losing experience. They have abandoned the venture. Bert Walker, president of Caterpillar Information Services division, says the division confines itself to working as a subcontractor to larger systems integrators. "In this business, the key is marketing. Companies like ours jump in and find 70% to 80% of the critical factor to success is the ability to market, where they have little capability and no expertise. Any people who think what they're building for their own use is marketable at a commercial level has a surprise coming to them. It doesn't work that way," asserts Walker.

INTRODUCTION

The preceding two chapters concentrated on the application portfolio as a major asset to the firm. The text also developed management methods for prioritizing the scarce application development resources. For those programs most appropriately developed by employees within the firm, the preceding chapter discussed project management practices and techniques calculated to increase the success of local development.

Building on this foundation, Chapter 9 explores additional application-development tools and techniques and considers additional acquisition alternatives. This chapter discusses the topics of prototyping and object-oriented programming. It also considers the merits of purchased applications. It explores subcontract development and the employment of service bureau organizations. Joint development activity through the formation of alliances also is discussed.

This chapter concentrates on management systems issues important to managing the alternatives. The approach in this chapter is not only to build on the traditional alternatives, but also to present additional alternatives. These alternatives are valuable in managing the complete portfolio asset. The final alternative, that of end-user computing, is covered in detail in Chapter 10.

ADVANCES IN TOOLS AND TECHNIQUES

Fourth-Generation Languages

During the past 40 years, computer programming has advanced through several stages or generations of technology, generally identified by the types of languages used in the programming task. The first computers were programmed in machine language, sometimes referred to as the first generation. Machine language used the language of the hardware, namely, the binary number system. This was followed closely by the second generation, or assembly language. Assembly language replaced operation codes and addresses with easier to understand natural language–like terms or mnemonics. In each case, one line of code resulted in one machine instruction.

The invention of third-generation languages was a major step forward. These are more English-like; they generate several machine instructions from each language statement. Programs such as COBOL, FORTRAN, BASIC, PL/1, and others are examples of third-generation languages. There are a hundred billion or more lines of code running on computers today that were created in these languages. More code is being written every day with these popular programming tools.

Programmer productivity, beyond that attainable with third-generation tools, is desperately needed. In addition, the profession needs tools that are easier to learn and to use. This will permit more people to participate in program creation more effectively. Fourth-generation languages, used with development tools supporting documentation, library functions, and other needs of the developers, promise to satisfy these needs in part. Indeed, James Martin defines fourth-generation languages by their ability to improve productivity. "A language should not be called fourth generation unless its users obtain results in one-tenth of the time with COBOL, or less," states Martin.[2]

A large number of languages are considered to be fourth-generation languages. They can be classified as shown in Table 9.1.

TABLE 9.1   Types of Fourth-Generation Languages

Database query and update

Report generators

Screen and graphics design

Application generators

Application languages

General-purpose languages

Many products and varieties of languages fall into the categories in Table 9.1. Some examples of fourth-generation languages are database query languages such as SQL or QUERY BY EXAMPLE, information-retrieval and analysis languages such as STAIRS or SAS, report generators like NOMAD or RPG, and application generators like MAPPER,

FOCUS, and ADF. Many more language tools (perhaps a hundred or so) are said to have fourth-generation characteristics. Most of these languages or programming tools do not have the general capabilities possessed by the third-generation tools we are accustomed to using. They are powerful, but more specialized; easier to use, but less flexible. Usually several fourth-generation languages are required to satisfy the needs of most programming departments. In any case, their use improves productivity and quality. Table 9.2 tabulates the characteristics of fourth-generation languages.

TABLE 9.2   Characteristics of Fourth-Generation Languages

Advantages	Disadvantages
Can be learned easily	Low performance in large systems
Reduces programming time	Possible slow response
Improves productivity	Inefficient use of memory
Improves program quality	
Reduces maintanence effort	
Problem oriented	

Fourth-generation languages usually consume more machine resources in operation than their predecessors. This means that their use must be restricted to tasks for which they are most suited. Very large programs supporting many simultaneous on-line users are probably best done in conventional languages. Fourth-generation languages are built to optimize their use by programmers. They are not designed to optimize computer performance. Therefore, transaction processing systems associated with very large databases may perform poorly if programmed in fourth-generation languages.

Fourth-generation languages reduce program development time at the expense of computer resources. This is generally a favorable trade-off since the trend of programmer costs is upward and the costs of computing power are declining. Some tools called cross-compilers translate programs written in fourth-generation languages into lower-level languages. The purpose of these tools is to combine the advantages of both systems of languages. This improves the performance of the development *and* the operational processes.

Programmer productivity improves considerably with these languages because they are easy to use and are more powerful than their predecessors for many applications. Most applications in most firms today can be developed using fourth-generation languages. Some firms have converted to one or more 4GLs in total. They have captured improvements in productivity and in program quality. Reliable reports of programmer productivity gains of more than 20-to-1 and significant reductions in expenses are attributed to fourth-generation languages. For example, Kawasaki stopped using COBOL three years ago and now has more than 800 new functions and programs written in Pro-IV. All new on-line applications are also developed in Pro-IV.[3] Arco Coal has eliminated COBOL programs from its portfolio and replaced them with programs written in Focus. This has been accomplished while reducing the programming staff.[4]

Many seasoned programmers resist these new languages because the new technology makes their skills, learned over a long time, obsolete. It requires retraining. Some programmers regard fourth-generation languages as technologically unsophisticated, suitable only for end-users. Managers frequently feel more comfortable with third-generation languages because they fear the introduction of another language. They are unsure if the compatibility issues can be solved and they dread the transition period. Some programming departments and their managers remain firmly anchored in the past, seeking comfort from the lure of the familiar.

In some instances, fourth-generation languages are used as direct replacements for earlier languages. Tools to support the development process are frequently considered a separate technology. In the future, however, the distinction between languages and tools will disappear. Developers will perform their tasks at programmer workbenches, unable to distinguish one support mechanism from another. This will truly be the era of computer-assisted software engineering.

CASE Methodology

The application of fourth-generation languages through additional programmer-support tools adds further potential for productivity enhancements. The tools, or programmer workbenches, are the heart of what is called computer-assisted software engineering, or CASE. These automated functions represent the results of trying to use computer technology for the benefit for computer professionals, programmers, analysts, programmer technicians, programmer librarians, and so on. The tools support the development process from requirements definition through maintenance. Support functions include analysis and design, code generation, test case development, and many forms of documentation. CASE tools are characterized by on-line workstations containing functions and features to assist professionals in virtually every aspect of software development.

Eight or ten firms market fourth-generation languages with CASE support. Major computer manufacturers are also addressing this type of programmer support. Prices for CASE programs vary widely depending on the function but range from $8000 to $100,000 or more. A list of CASE vendors can be found in the appendix to this chapter.

Sophisticated workbench technology utilizes a network of workstation devices. Networking assists in the complex task of communication between the members of the development team. These networks support timely and accurate information exchange. CASE technology uses graphics extensively. It provides a controlled process for managing the versions and releases of the developing product, including the documentation. CASE provides an automated system for managing the test case library and for assisting with program validation. In some instances, the tools contain code-generation or code-rebuilding capability so that old programs can be rebuilt or refurbished rapidly and productively.

Some tools are designed to rebuild current programs. File descriptions, database configurations, and source code are placed in the systems design database where they are enhanced and updated, or migrated to new languages or data management systems. The process includes both reverse-engineering and forward-engineering.[5]

There are several types of CASE tools. Some support requirements planning, analysis, and design. They also have documentation tools for these tasks. Rapid production of design graphics and system documentation is a feature of these tools. These upper CASE or front-end tools assist in the early phases of the systems development life cycle. Lower CASE or back-end tools provide code-generation capability, test-case development support, and assistance in database development. They help produce documentation of the programmed application. Some development-support tools assist in project management. These tools provide support to life cycle administration. They develop and communicate project metrics and support the programming environment.

General-purpose products, or products that support the complete life cycle, are called Integrated or I-CASE tools. I-CASE tools are used in conjunction with a development methodology, usually the systems development life cycle. They may, however, also be employed with other methodologies. CASE systems support many of the common languages. They operate on personal workstations or are supported from workstations connected to mainframes.

Modern CASE tools assist in the task of project management by collecting development statistics, displaying status reports, and communicating among and between members of the team and the project managers. The U.S. Department of Defense recognizes the importance of CASE tools. In the definition and analysis of software requirements for DoD, developers are directed by DoD standard 2167A to use systematic and well-documented methods. Many CASE tools help satisfy this requirement.[6]

CASE is an important technology for several reasons. Users believe that CASE improves the quality of design and assists greatly in developing system documentation. CASE improves communication among developers and between developers and users. Although some developers resist new tools or methodologies, improved tools will reduce this resistance and will encourage programmers to adopt proven methodologies. CASE tools that are not integrated with languages or databases are somewhat inconvenient to use. Integrated tools will remove this objection.

Management control is vital in application program development. A controlled environment is a prerequisite to improved quality. High-quality development is also productive development. CASE tools used with proven project management systems and with improved programming methodologies such as the object paradigm have great potential to improve program quality and maintainability. The phase-review approach, valid business case analysis, and risk analysis focus on the management process. Metrics developed from these management processes and augmented by development metrics from the CASE tools permit the project to remain under control during its life cycle.

The growth and development of automated tools supporting highly productive languages will accelerate significantly in the future. The same is true for the deployment of these tools and techniques in application development. Many specialized firms are developing CASE tools, but major computer manufacturers are also involved. Digital Equipment Corporation is providing DECdesign V1 and IBM is developing AD/Cycle, which includes 35 third-party suppliers who will package their products under the AD/Cycle umbrella.

Businesses are demanding significant improvements in programming productivity. CASE tools and fourth-generation languages have proven track records. For example, programmer productivity improved from 65 to 400 lines of code per day in two years at Con Edison and quality also increased, reducing future maintenance costs. Con Edison uses an application generator to turn design specifications into COBOL code. BDM International, a large systems integrator, reduced costs nearly $5 million on a fixed price contract for the Air Force. Error rates have declined by 75 percent. Souvran Financial Corp. saved $1.2 million on its first four projects using a CASE tool and has adopted the tool as a company standard. Souvran anticipates maintenance savings since programs developed with CASE contained fewer lines of code. There are many other examples. These improvements can come about only through the application of automation to the programming task. The introduction of new languages, advanced tools, and new techniques is partly a technical problem. However, it is mostly a people problem. Application-development departments have little choice but to embrace these tools. Astute IT managers will use their influence and management skills to encourage early adoption and effective utilization.

The Object Paradigm

Object-oriented programming technology is gaining popularity in the academic world, in business, and in industry. The beginnings of the technology were in Norway in the 1960s, where scientists developed a language called Simula. A research team at Xerox continued this effort and developed an object-oriented language called Smalltalk. This is widely used today, particularly in academic institutions. Object technology consists of object-oriented design, development, and databases. Object-oriented programming and object-oriented knowledge representation are also part of the technology.

Object programming approaches a problem from a different level of abstraction than does conventional programming. Conventional languages separate code and data; object languages bring the two together in a self-contained entity called an object. With each object is some code appropriate for its use. These codes are called methods. Objects that share common methods and attributes are called classes. For example, a file may be defined as an object and the methods appropriate to the object might be copy, display, edit, and delete. The object and the methods common to it form a single entity. Another file may have the same methods as the previous file, and the two files would belong to the same class. There can be many members of any class.

One of the most powerful concepts of object technology is that of inheritance. New object classes can be defined as descendants of previously defined classes. The new classes inherit the methods of their ancestors but these methods can be altered by adding new methods or redefining previous methods. For example, one could define a new object in the file class called output. Output would inherit methods of copy, display, edit, and delete. For use in an application system, programmers may remove the method called copy and the method called edit. The object programmer may add the method called print. Methods appropriate to output would then be display, print, and delete. The possibilities are unlimited.

Object technology is especially useful in forming user interfaces; screen applications such as menus, displays, and windows; and text, video, and voice databases. It isolates the effects of change and permits expansion of features in a modular manner. Object technology provides a way of dealing with complexity through abstraction. It will be useful for operating systems, programming languages, databases, and user applications.

The object paradigm is already widely used. Many developers are using C++, a version of C with object-oriented capabilities. C++ is available on Sun and Apollo workstations and on 80386-based systems. It is also available on Digital Equipment Corp. VAX platforms and others. IBM's OS/400 provides object capability, and object capability is available in the operating system for NeXT, Inc.'s workstation.

There are many new developments in this rapidly emerging technology, and it is being applied widely by computer manufacturers, application developers, and government agencies. A new international organization, the Object Management Group, has been formed to develop a common applications environment through object-oriented standards. Its mission is to establish standards enabling the rapid and organized development of object programming on a worldwide basis. Object technology may be to the 1990s what structured techniques were to the 1980s.[7]

Prototyping

Many systems incur problems from the very beginning of their life cycle caused by the extreme difficulty of the specification process. Problems originating in the early stages of system design are difficult and expensive to correct later on. For many applications, the hardest part of system design may in fact be specifying the problem. In many situations it is preferable to experiment a little with the problem to obtain a more realistic feel for the solution domain.

In many cases it is unrealistic for the development team, the analysts, and the users to reach agreement early in the cycle on the precise final specifications. On the one hand, developers want to freeze the specifications so that they can begin implementation. On the other hand, the users want to reserve some flexibility because they may not be positive about their requirements. The ability to experiment with the design and to prototype the solution prior to final commitment is very desirable. It represents a compromise solution to the conflicting demands of developers and users.

Prototyping demands rapid implementation of many small alterations, with correspondingly rapid feedback on the merits of the changes.[8] Prototyping is best performed in an environment of high automation where the tools support these needs. Third-generation languages operating in batch mode are not suitable. Prototyping is made feasible by fourth-generation languages supported by CASE workbenches. Figure 9.1 illustrates the prototyping process.

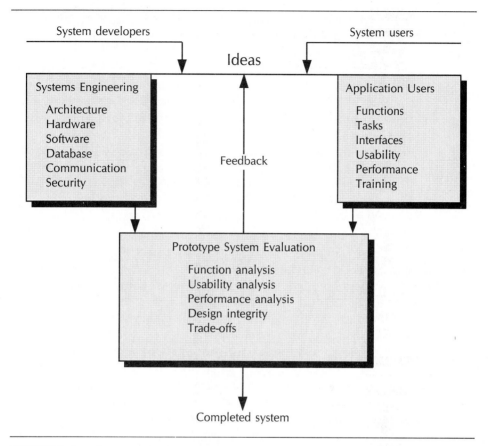

FIGURE 9.1　The Prototyping Process

System developers and system users generate ideas for the new system. An elementary model is devised for evaluation. Developers are concerned with hardware, software, architecture, and communication technology. Users are devising human/machine interfaces, functions, and usability criteria. Their ideas are integrated; the model is evaluated; new ideas are generated; and the process is repeated until the problem and most solution parameters are specified. The experimentation continues until both parties are comfortable with the prototype.

The advantages of prototyping are that system specifications can be changed as a result of early experience with a running model. The final system will then closely meet user needs and expectations. A disadvantage is that the prototyping effort is difficult to manage (when is the prototype finished?). Another disadvantage is that users may adopt the system prior to completion or integration with other systems or databases. It may be more difficult to obtain final documentation on a prototyped system, as well.

Prototyping activity can continue until the system is fully developed and operational. In this instance, the prototype model is refined until it completely satisfies the user requirements. With CASE technology, most of the documentation and other supporting material is available. The system is then ready for production.

An alternative approach employs a combination of the life cycle and prototyping methods. Prototyping is used to develop the specifications and remove some of the uncertainty of the initial stages. Prototyping refines the user interfaces and solidifies the architecture. When these crucial tasks are finished, the life cycle methodology, beginning at Phase 2, is employed to complete the remaining stages. This is particularly attractive for large programs with complex human interfaces. Prototyping is very useful for developing critical specifications; the life cycle methodology disciplines the final development process. The advantages of each are combined to form a superior process.

A manufacturer of large computer printers uses this approach. Much of the user interface on these devices consists of backlighted panels for operator messages and touch panels for operator response. The printer contains five microprocessors to manage the interface and control printer operations. All the interfaces were prototyped in software and hardware, and the specifications were developed and approved. Following this, both the hardware and the programming continued with formal development processes. The product was introduced on an accelerated schedule, at minimum development cost, and with high confidence that the human factors would be satisfactory.

Managers have several alternatives for developing required additions to the application portfolio. The traditional life cycle can be used for applications that are easily and unambiguously specified. It is a valid and useful approach for applications that demand careful control because of risk factors in the business case or the application itself.

Prototyping offers many advantages. It can solve several important and vexing problems. The usefulness of this approach can be extended by combining it with others. In all cases, high-technology languages and tools are mandatory.

The choice among the alternatives must account for development time and cost and operational and maintenance cost. Many of these considerations are directly dependent on programmer productivity, program quality, and user satisfaction. A careful financial analysis should accompany the decision to use one method in preference to or in combination with another. Because of the steadily declining cost of computer hardware and the increasing costs of programming, the economics are becoming increasingly favorable for the prototyping methodology.

SUBCONTRACT DEVELOPMENT

Not all programs requiring development must be completed within the firm. Some application development can be off-loaded to a firm specializing in this type of work. This activity is called *subcontract development*. The firm seeking to acquire the application subcontracts with another to perform some of the development activity for a fee. Several advantages to this form of program acquisition are listed in Table 9.3.

TABLE 9.3   Advantages of Subcontracting

Balances resources and possibly skills

Trades money for people

Assists in managing staffing

Useful for obtaining experience

One advantage in taking this path to program acquisition is that it balances resources within the firm. Subcontract development trades money for application developers. It also reduces the firm's requirements for people, as compared with local development. It permits skill imbalances within the firm to be resolved. Subcontracting allows a firm to manage its long-term staffing more effectively because the subcontract tasks are a workload buffer for the firm's permanent staff.

Subcontracting may also be used to obtain skills on a temporary basis for specialized tasks, again reducing the need for permanent staffing. Because the firm doing the subcontract work may have both skills and experience in the general problem area, the results may be superior to those of alternative methods. Taking all of these factors into account may also result in lower application acquisition costs to the firm.

Subcontracting also has some disadvantages, as shown in Table 9.4.

Subcontracting is bureaucratic because it involves contracts, reviews, reports, and more formal supervision than local development. It is therefore much more management-intensive than in-house development. The task of analysis and requirements definition for subcontractors must be especially thorough and well documented. (The contract terms and conditions are based on this documentation.) Alterations to the work scope or content after work commences usually result in financial penalties and schedule changes. Unless the contract makes provisions for a prototyping approach, what is contracted for is what is delivered.

TABLE 9.4   Disadvantages of Subcontracting

Creates bureaucracy

Requires management attention

Requires rigorous requirements analysis

May expose confidential information

The act of subcontracting also results in diffusion or dissemination of information. If the system development involves confidential information, nondisclosure agreements may be required. The risk of exposure or breach of confidentiality may be greater with a subcontractor than with the firm's regular employees.

Conflicting objectives must be understood and worked out with both parties. For example, the firm wants the job completed on schedule, to specifications, and within budget. The contractor wants to maximize profits. A cost-plus contract could encourage the contractor to increase the job size or scope and lengthen the schedule. A fixed-fee contract may incline the contractor to perform as little work as possible to meet the contract terms. The contract terms and conditions must explicitly recognize this conflict. Also, the firm's management system must operate to enforce the contract in fairness to both parties.

An analysis of risk along the lines discussed earlier is mandatory prior to entering into a development contract. Periodic risk analyses must continue during the life of the contract as well. Subcontract development places considerable burdens on the firm's managers and on the IT management system.

PURCHASED APPLICATIONS

Increasingly, businesses are turning to external sources for some of their application programs. The trend toward purchasing applications has been underway for several decades, but it has accelerated considerably since the introduction of personal computers. Indeed, the wide availability of highly useful applications has been the major driving force behind the explosive growth of personal computing. Although purchased applications are widely associated with PCs, they are becoming more common in the minicomputer arena and in mainframe installations as well. For instance, the IBM AS/400 has a portfolio of more than 2500 applications. Most of these applications have been developed in collaboration with IBM customers. They can be purchased with the hardware.

Advantages of Purchased Software

As we learned earlier, there are many applications in most installations that must be constructed and maintained by the resident programming staff. Some applications, however, can be performed satisfactorily by purchased software. The decision to build or buy is many-faceted. Some portions of the decision rest heavily on the advantages and disadvantages of purchasing application programs. The advantages of purchased applications are numerous. They are outlined in Table 9.5.

TABLE 9.5   Advantages of Purchased Software

Early availability

Well-known function

Known and verifiable quality

Inspectable documentation

Lower total cost

Availability of maintenance

Periodic updates

Education and training

In contrast with the usually extended development cycles in-house, purchased applications can be installed and utilized relatively promptly. Since the development time in the application life cycle may be as large as 50 percent of the total time, this time savings can be considerable. This usually translates into substantial benefits in the business case. The benefits from the application also occur earlier and improve the cost/benefits analysis. Thus, in many instances, purchased applications are more affordable.

Purchased applications contain functions that are usually well defined or that can be determined easily. This functional certainty removes some of the risk, as compared with locally developed programs. Purchasers can be relatively certain of the functional capability of the program before spending much money. The functional capability of the purchased application can be learned from the vendor's documentation, through test runs of the program, or from other purchasers. This kind of information can be objective. It also removes some of the biases associated with functional questions surrounding in-house development.

Through discussions with prior users, industry experience, and pre-purchase application tests, a good understanding of the quality of the code and of the documentation can be obtained. Again, this reduces risk. Documentation forms a major portion of the application product and is a source of concern with locally developed programs. This concern is twofold. Good programmers are not necessarily good writers; their documentation may be of low quality. The documentation may not be available when it is needed because, if anything slips schedule, it's likely to be the documentation. Having both the code and the documentation available for inspection eliminates these risks.

Not only does early availability of the application improve the business case, but purchased applications usually have a lower total cost. This results from the vendor's ability to spread application development costs over a large number of customer purchases. In contrast with those of hardware, per unit software manufacturing costs are very low. The primary manufacturing cost is that of documentation reproduction and distribution. Therefore, there are large economies of scale in purchased applications.[9]

Many widely available applications have a continuing stream of enhancements available to prior customers at reduced prices. These updates, built on previous versions, add functional improvements to the application. They can support new hardware features or functions, or provide integration with applications from the same or other vendors. These enhancements are designed to improve or broaden the product and to increase its value to customers. Usually these enhancements are available at relatively modest prices. Perhaps not all of the added function is usable by each customer, but the part that is usually has a very favorable cost/benefit ratio.

Lastly, user training is simplified with purchased applications. Many popular applications are supported by outstanding training materials: training manuals, reference manuals, user aids, help lines, user groups, and in some cases vendor or third-party classes. Because of the popularity of some applications, this training can be obtained at modest cost. And, in contrast to most locally developed training material, vendor-supplied training aids available with popular applications tend to be of high quality.

Disadvantages of Purchased Applications

Given the impressive array of advantages offered by purchased applications, it is easy to understand why they have become so popular and grown so rapidly. However, there are some important disadvantages as well. Management needs to analyze both the advantages and disadvantages and understand them in detail so that intelligent decisions can be reached. Table 9.6 lists some disadvantages.

TABLE 9.6   Disadvantages of Purchased Applications

Functional deficiencies

Program and database interactions

Difficulties of modification

Management style considerations

The principal disadvantage of purchased programs relates to functionality. In some instances, applications cannot be purchased to perform the functions desired by the firm. Good examples of this can be found in the area of strategic systems. Because of the nature of the business and the need for confidentiality, or because of unique and proprietary information or processes, purchased applications may not present a good alternative to local development. Most strategic systems must be developed with the resident programming staff.

Some groups of application programs are so interrelated and so highly dependent on common databases that integrating a purchased application may be practically impossible. For instance, the array of financial application programs may depend so heavily on locally defined databases that the integration of a purchased general ledger program is difficult or impossible. The cost of integration will exceed the development savings in some cases. It may be infeasible to replace the entire set of financial and accounting applications for the same reasons. As the degree of interaction among and between application program sets increases, the insertion of an application procured from outside the firm becomes increasingly difficult.

Modifications to a purchased program may or may not be easily performed, depending on several factors. These include the availability of the source code; the nature of the source language itself; and the availability of program logic manuals or other forms of program documentation. There are other detracting factors such as the need for test-case development. If the source code is not available, or if the programming documentation is incomplete or unavailable, program modification may be impossible or impractical. In any event, a comprehensive cost/benefit analysis must be performed to determine the financial reasonableness of proposed modifications.

In some instances, it may be necessary or appropriate to surround the application with customized programs that bridge to the present application set. Doing this may be preferred to modifying or customizing code within the purchased application. In addition, the need to preserve compatibility with future releases is important. Any modifications to the application may need to be replicated in the next release. The vendor usually feels no responsibility to migrate your custom code in the next release. The financial considerations involved in this contingency must be factored into the business case for the proposed application.

An additional disadvantage can be that the function of the application may be deficient, incomplete, or implemented in a manner foreign to the firm's business operation. Application programs frequently implement the management system employed within the firm. Sometimes they include functions derived from management style considerations. To a large degree, customized application programs represent "how we do things around here." It is not likely that a popular, widely distributed application will include these nuances. The adoption of a purchased application may require modification to either the purchased application or to the management system. Resources required to perform these modifications detract from the business case for the application.

In most firms, the costs of modifying the management system to accommodate an application program are never evaluated even though they may be small. Many managers believe that programs are easily changed and that management systems are not. Some managers believe that programmers are employed to support management or the administration. Some managers are unwilling to consider the pros and cons of the issue. This type of myopic thinking has led many organizations to spend large sums developing customized systems, the primary functions of which are common in the industry.[10] For many reasons, the issue of functionality looms large with purchased applications.

Not all purchased software is of high quality, has great documentation, or is supported by updates and quality maintenance. And some purchased applications come with no training or educational support. There is no substitute for a thorough investigation of all aspects of the product prior to purchase. The reputation of the vendor is important. And it may well be worth the additional cost, if any, included in the price. Good applications tend to be long-lived. The association between your firm and your vendor will endure for years if it is mutually beneficial.

The issues related to purchasing applications for the firm's portfolio are many and complex. This is especially true if the application replaces or supplements a current application and uses the firm's traditional databases as sources and sinks of information. The issues are somewhat simpler if the proposed application is "stand-alone" or if the application forms a new area of automation for the organization. Given all these considerations, purchased applications are becoming popular. This is because they tend to be financially attractive and because they represent a viable way to reduce the application backlog through the expenditure of funds.

Software is big business. IBM is the largest supplier of software, but thousands of firms are in the business. U.S. vendors annually sell more than $22 billion worth of software worldwide and about $15 billion domestically.[11]

In the microcomputer world, the applications and the hardware on which they run are generally considered together from a business and a financial perspective. The purchase of applications in support of a microcomputer strategy is usually a given condition for the firm. As micros grow in capability and as automation of business processes continues, purchased applications will be an extremely important factor. New businesses, and many businesses worldwide not currently highly automated, will implement strategies centered on purchased applications. Some firms will opt for programmerless environments, and many that do will succeed. For those that succeed, the application programming problem will be a nonissue.

ADDITIONAL ALTERNATIVES

Other alternatives to in-house application development are evolving. One of the most promising is the formation of alliances. In the business vignette, for instance, Kidder, Peabody and First Boston agreed to codevelop a retail trading system for a fraction of the cost that each would have incurred separately. Large computer manufacturers have created alliances with retailers and with telecommunication firms. And some have created many partnerships with businesses to bring software solutions to their customers. These alliances also help in the marketing and distribution of other products.

Alliances are being formed between large and small businesses to further their strategic interests. These alliances are founded on strategy and implemented with structural alterations. They are useful for many endeavors, but they are becoming increasingly popular in computer technology. These relatively new approaches to development can be very effective when the circumstances are favorable. When a satisfactory alliance partner can be found, there is the potential of reduced costs, improved schedules, and increased function for each of the partners. Successful product-development alliances offer the potential for revenue and profit for both partners.

Successful joint development and marketing efforts have been accomplished between Northwestern National Life and Infodata Systems Inc.; Avon Beauty Group and IMI Systems Inc.; and Hilton Canada and Control Key Corporation. Some joint ventures are developed to solve mutual problems without the intent of selling the resulting product. For example, Security Pacific and five other banks are cooperating on an imaging technology development project. If successful, the project will eliminate millions of pieces of paper and will reduce data processing costs.[12] These and other considerations are driving the trend toward alliances and joint ventures.

The following vignette about a health-care sector venture illustrates an alliance between two large organizations designed to market software and hardware.

Yet another alternative for reducing expenses and off-loading program development work is presented by the IS service bureau. These firms provide systems and operating environments capable of processing routine applications for their clients. For example, payroll is an application that is frequently processed at a service bureau. The firm provides payroll input information to the service bureau. The service bureau processes the payroll, mails the checks or makes direct deposits, completes the payroll register, and returns it with other reports to the client firm.

The advantages of using a service bureau are that it off-loads work and responsibility and reduces in-house computer requirements. The service bureau makes program changes required by law or regulation and keeps the payroll functionally modern. For example, it will update the payroll program with new federal withholding tax changes or keep the program in compliance with changes in state regulations. The cost of this service is spread over many clients, reducing expenses for all.

Service bureaus eliminate application development and maintenance costs for some applications. They also reduce hardware capacity requirements. The use of service bureaus is appropriate for many operational applications that have little strategic advantage. The discovery of these applications and the consideration of service bureau processing frequently arises during the prioritization process discussed in Chapter 7. They are a viable alternative through which to trade money for people and computer capacity. The notion of using service bureaus is the first step in considering out-sourcing, an important current trend that will be discussed in detail later.

Health-Care Sector Venture Formed by IBM, Baxter[13]

IBM and Baxter Healthcare said they formed a joint venture to sell computer hardware, software, and other services to the health-care industry. The move should let the companies enhance their already large positions in that market, which consultants say total more then $4 billion a year in the U.S. They add that various technologies crucial to health-care applications are coming together, so the field could enjoy a growth spurt.

Mark Gross, national director of health-care information services at Ernst & Young, the accounting firm, said that Baxter has produced some excellent software for running hospital operations but hasn't made much money at it. IBM hasn't been as big a player in health-care information services, but Mr. Gross said he thought the joint venture could make a compelling argument that it could provide hospitals with all the software, hardware and services they needed. At the moment, the market is so fragmented that hospitals must typically shop around.

Jay Toole, world-wide director of health care consulting at Andersen Consulting, said he thinks that improvements in image-processing, networks, and workstations are coming together quickly and could change the way hospitals are run.

Image-processing has improved to the point where patients' records can be routed electronically throughout a hospital, or to doctors' offices outside the hospital if a patient is being treated there. Items like X-rays and CAT scans are tougher to route electronically, but that's becoming feasible, too, with the increasing prevalence and speed of networks.

IBM and Baxter said they think sales of computer equipment and services should also benefit greatly from the pressure that government and companies are putting on hospitals to cut costs by becoming more efficient.

MANAGING THE ALTERNATIVES

Chapter 7 examined the resource prioritization problem inherent in maintaining the application portfolio. The text explored some management practices that help resolve the difficulties in a rational manner. The alternatives available for portfolio acquisition are presented in this chapter and the preceding one. Given the range of alternatives, what methodology can be employed in the selection process? How can managers make best choices among the alternatives?

In many firms, the prioritization methodology reveals that programming talent is the constrained resource and that money is a less constraining factor. This realization frequently arises during the final discussion on prioritization when it becomes obvious that the programming staff cannot respond to all requested work on the portfolio.

A sequence of steps leading to optimum selection among the alternatives is summarized in the following questions:

1. Which applications can be processed at a service bureau, saving people resources and computer resources?

2. Which applications needing development or replacement can be purchased to save programming resources and time?

3. Are there ways in which the firm can enter into agreements with others, either on a contractual or a joint development basis, to utilize the firm's resources more optimally?

4. Is it possible to utilize sophisticated development tools and techniques to improve programming productivity?

5. What alternatives are there to increase the human resources applied to application development? (End-user computing is discussed in the next chapter.)

Given this array of alternatives and the thought processes discussed in earlier parts of this text, the firm is likely to achieve a balanced approach to the prioritization problem. Most likely, those programs with low strategic but high operational value will be candidates for a service bureau. Applications of high strategic value or potential strategic value are most appropriate for in-house development. These highly valuable assets must be managed carefully. They probably involve proprietary information, and they usually reside at or near the top of the prioritized list.

For the remaining applications, the choices must be made in a manner that applies available resources to optimize the firm's goals and objectives. The strategizing and planning processes discussed earlier are critically important foundations for the application program management system. IT management must ensure that the environment supports and encourages high productivity through the use of advanced tools, techniques, and management systems.

SUMMARY

Application acquisition offers IT and user managers a variety of opportunities for optimizing the firm's resources. It provides opportunities to develop important strengths for the firm. Managers must respond to these opportunities in a disciplined manner. This response begins with a clear vision of the application portfolio's contribution to the firm's success. This vision of the application portfolio resource is developed and enhanced through careful strategic planning. Tactical and operational planning fine-tunes the allocation of resources to strengthen the portfolio.

The portfolio of applications is augmented by using traditional life cycle methodologies or by employing alternative approaches. Although all the alternatives offer opportunities for gain, they are all accompanied by risk elements. These risks must be thoroughly understood and they must be mitigated in some manner. Advanced tools and system development techniques are available, but managers must have technical and people management skills to implement and use them. Advanced tools shorten development cycles, offer great productivity improvements, improve product quality, and increase user satisfaction. Achieving these ends is a high priority for IT managers. Indeed, these objectives are critical success factors for IT managers and user managers.

The exploration of alternative approaches to system maintenance, enhancement, and acquisition is causing firms to reassess their programming development and computer operation functions. Purchased applications for large and small systems are rapidly gaining in popularity. Developing applications for sale is a major industry. Firms recognize the economic value of the alternatives, and they are more willing to consider them. Many firms are very reluctant to take on the commitments and long-term liabilities associated with a large permanent programming staff. Many others are reconsidering their strategies of IT self-sufficiency and are actively pursuing alternatives. In many cases, CIOs are leading these dramatic new directions.

Review Questions

1. What unique management problems does joint development, as exemplified by the Kidder/First Boston effort, surface?

2. Where are the leverage points in using fourth-generation languages in conjunction with CASE tools?

3. Under what circumstances can prototyping used in conjunction with normal life cycle development be highly effective?

4. What are the advantages of prototyping in connection with the use of CASE tools?

5. What are the factors accelerating the trend toward purchased applications? Do you think these factors will increase or decrease in importance in the future, and why?

6. What are the disadvantages of purchasing application programs? Are these disadvantages more or less important for well-established information technology departments?

7. What are the risks in using purchased applications? How can these risks be quantified and minimized?

8. What is the role of purchased applications in connection with the explosive growth of personal computers?

9. In what ways does quality enter into the purchase decision?

10. What are the advantages and disadvantages of using subcontract development?

11. How does subcontracting application development assist in balancing resources?

12. What are the advantages and disadvantages of using a service bureau for some of the firm's applications?

13. Discuss the pros and cons of joint application development.

14. What is the relationship of the topics in this chapter to the topics in the previous chapter?

Discussion Questions

1. Information systems service business units have been established by many firms over the years. What are the potential opportunities and pitfalls with such endeavors?

2. Discuss the implications of sharing software between competitors, especially considering Bill Howard's statement, "With most of the software, it's really as much a matter of how you used it as of anything else."

3. Using the techniques of risk analysis discussed earlier as a starting point, identify the elements of risk inherent in subcontract development. List these risk elements in order of importance as you perceive them.

4. What trends in the information-processing field are favorable to service bureau firms? What trends are unfavorable?

5. Draw a flowchart of the questioning process discussed in the section on managing the alternatives.

6. Discuss the relationship of the material in this chapter to the section on critical success factors in Chapter 1.

7. Why is programmer productivity such an important issue today?

8. Discuss the significance of the object paradigm.

9. How can the analysis leading up to Figure 7.4 assist in the task of managing the alternatives?

Assignments

1. Using library resources, perform an analysis of two firms in the industry that build application programs for sale.

2. Obtain descriptive material on two fourth-generation languages. Compare and contrast their capabilities and limitations. For what class of problems is each language most suitable?

ENDNOTES

[1] Reprinted from "IS Shops Form Alliances as Development Costs Rise," *DATAMATION*, September 1, 1988, © 1988 by Cahners Publishing Company.

[2] James Martin, *Application Development Without Programmers* (Englewood Cliffs, NJ: Prentice-Hall, Inc., 1982), 28.

[3] John Kador, "Kawasaki Prefers 4GL over COBOL for On-Line Development," *MIS Week*, October 16, 1989, 22.

[4] Ben Chao, personal communication.

[5] Mike Feuche, "New CASE Reshapes Software," *MIS Week*, February 1, 1988, 35.

[6] Alexander J. Polack, *Federal Computer Week*, October 30, 1989, 19.

[7] Object technology is taught at Brigham Young University, UT Austin, Baylor University, Dartmouth, University of Illinois, University of Colorado, and others.

[8] Edward Yourdon, *Modern Structured Analysis* (Englewood Cliffs, NJ: Prentice-Hall, Inc., 1989), 97-100. This text presents another discussion of the prototyping life cycle.

[9] Lotus 1-2-3, version 3, cost $7 million to develop and Lotus Development Corporation spent an additional $15 million on testing and quality control. This $22 million product can be purchased for much less than one-hundredth of one percent of the development cost alone. This is an extreme case, but it illustrates how economies of scale act to the purchaser's advantage.

[10] Examples of this phenomenon are widespread. Tens of thousands of programmers are developing and maintaining unique ledger systems, payroll programs, manufacturing applications, and inventory control programs. These unique programs offer no competitive advantage, but they do maintain the firm's culture. They are terribly expensive. In some cases, these development activities are late, over budget, and sources of embarrassment for their firms.

[11] George Briggs, "U.S. Software Spending Grows by 12 Percent," *MIS Week*, March 13, 1989, 31. The largest noncomputer manufacturer in the software business is Computer Associates International, Inc. with annual revenues exceeding $1 billion.

[12] Preston Gralla, "Something Ventured, Something Gained," *CIO Magazine*, May 1988, 12.

[13] Reprinted by permission of *The Wall Street Journal*, ©Dow Jones & Company, Inc., 1989. All Rights Reserved Worldwide.

REFERENCES AND READINGS

Borovits, Israel. *Management of Computer Operations*. Englewood Cliffs, NJ: Prentice-Hall, Inc., 1984.

Gralla, Preston. "Something Ventured, Something Gained." *CIO Magazine*, May 1988, 12.

"IS Shops Form Alliances as Development Costs Rise." *Datamation*, September 1, 1988, 19.

Kendall, Kenneth E., and Julie E. Kendall. *Systems Analysis and Design*. Englewood Cliffs, NJ: Prentice-Hall, Inc., 1988.

Martin, James. *Application Development Without Programmers*. Englewood Cliffs, NJ: Prentice-Hall, Inc., 1982.

Martin, James, and Carma McClure. *Structured Techniques: The Basis for CASE*. Englewood Cliffs, NJ: Prentice-Hall, Inc., 1988.

Wojtkowski, W. Gregory, and Wita Wojtkowski. *Applications Software: Programming with Fourth-Generation Languages*. Boston, MA: boyd & fraser Publishing Co., 1990.

Yourdon, Edward. *Modern Structured Analysis*. Englewood Cliffs, NJ: Prentice-Hall, Inc., 1989.

Many vendors supply a variety of software supporting the CASE environment. Some CASE tools and vendors are listed below.

Aims Plus	AIMS Plus, Inc. 1701 Directors Blvd. Austin, TX 75234
Teamwork/SA	Cadre Technologies Inc. 222 Richmond St. Providence, RI 02923
Pacbase	CGI Systems Inc. One Blue Hill Plaza P.O. Box 1645 Pearl River, NY 10965
PC/Workshop	Computer Corp. of America Four Cambridge Center Cambridge, MA 02142
Developer Workstation	DBMS, Inc. 1717 Park Street Naperville, IL 60540
Excelerator	Index Technology 101 Main Street Cambridge, MA 02142
Information Engineering Workbench	Knowledgeware 3340 Peachtree Road NE Atlanta, GA 30326
Structured Architect	Meta Systems 315 E. Eisenhower Parkway Suite 200 Ann Arbor, MI 48104
DesignAid	Nastec Corp. 24681 N.E. Highway Southfield, MI 48075
MacBubbles	StarSys, Inc. 11113 Norlee Drive Silver Spring, MD 20902
CASE	Tektronics P.O. Box 500 Beaverton, OR 97077

10 *Successful End-User Computing*

A Business Vignette
Introduction
Why Adopt End-User Computing?
What Are the Issues?
Organization Changes
   The Workstation Store
   The Information Center
Policy Considerations
Downsizing
   The Attributes of Downsizing
   What To Downsize
Office Automation
Planning for Office Automation
   Stages of Growth
Implementation Considerations
   Coordination
   Some Possible Difficulties
People Considerations
   Change Management
Summary
Review Questions
Discussion Questions
Assignment: A Case Discussion
Case Discussion Questions
References and Readings

Echlin's users were fearful of the plan to replace the mainframe computer with LAN-based PCs because they anticipated reduced function. The reality of the implementation is that their applications are more effective on networked PCs than they were on the central mainframe.

The auto-parts manufacturer Echlin, of Branford, CT, saved nearly $800,000 in 1989 by pulling the plug on its mainframe. The company put in place a local area network (LAN) over three years, using microcomputers as stations on the LAN, which then took over all data processing tasks at the Echlin corporate headquarters. During the conversion from mainframe to PCs, Echlin was able to improve its end-user applications and make them more effective.

The company had a well-established MIS department which handled all personnel and financial applications as well as general ledger and accounts payable. It also managed applications for financial planning and reporting; marketing; and employee incentives. To cover the information requirements of the growing firm, Echlin projected a budget that would reach nearly $2 million by 1989, about double the figure for 1985.

The mainframe that once ran Echlin's headquarters operation has been sold. In its place is a network with four Novell file servers and about 60 PCs. The compatible PCs vary in size depending on the user's application. The network also supports personal applications such as word processors and spreadsheets. The applications have been rewritten in PC/Focus, from Information Builders, Inc. This arrangement does not have the storage capacity of the original computer, but—and this is a key factor—that mainframe required a Gbyte of storage for its own operating system. Total network and computing capacity has grown to keep up with the growth in Echlin's business and information processing requirements.

What could go wrong in moving to a more efficient and cost-effective management information system? First, the end users of the applications were not happy. They knew the old mainframe, and they were content with it. They resisted the change that was being forced on them. And part of their resistance come from the fact that the MIS organization was driving the change and was planning to reduce its size! More resistance came from the threat of the unfamiliar.

The idea for downsizing the MIS operation originated at the top of the organization and had to be carefully managed during implementation. Echlin MIS managers made sure that no part of the new system was put in place without the approval of the users, who were trained and supported as they learned new applications and procedures. The Echlin managers also made sure that the end users were convinced they would receive benefits from the new system. In some cases this meant that money had to be spent rewriting mainframe applications to user specifications because suitable packages were not available. In those cases where no improvement could be made, MIS gave users identical application tools.

Resistance to the conversion from mainframe applications to LAN-based PC applications was found within the professional MIS staff as well. Turnover among mainframe analysts and programmers was nearly 100 percent during the three-year conversion period.

The new system cost $750,000 over three years; the cost savings have exceeded projections by a wide margin.[2]

INTRODUCTION

Earlier chapters discussed the difficulties associated with managing the application portfolio and the related databases. Several promising methods for prioritizing the development backlog and for managing the portfolio were presented. The text presented several alternative approaches to development and explored the option of purchasing commercially developed applications. The text discussed the advantages and disadvantages of these alternatives; it reviewed situations in which one or more of these options would be attractive to the firm. This chapter turns our attention to one remaining alternative having great potential for coping with some of the difficulties mentioned earlier.

The situations described in the previous three chapters are frequently found in organizations that have installed end-user computing. Capitalizing on advances in telecommunications technology and workstation products, firms are introducing capability that permits and encourages end users to participate in the application development process.[3] This participation increases the effort applied to application development. Also, it tends to reduce user frustration with the application development group. The potential benefits of end-user computing are encouraging firms to invest in the technology.

Firms are stepping up their funding for end-user computing and are devoting a larger share of IT resources supporting end users. Studies show that IT organizations are spending approximately 25 percent of their budgets on end users, and user organizations are spending a significant amount of their own money on information systems and services. In total, about a third of the firm's expenditures for information technology is spent by or for end users.[4] Resource deployment of this magnitude significantly influences organizations and their people.

End-user computing places special demands on employees, managers, and organizations, and implementation of end-user computing raises many important issues. If firms are to attain the considerable benefits of end-user computing, critical organizational and political barriers must be removed. Problem prevention and rapid problem resolution are essential.

This chapter discusses the opportunities and issues surrounding end-user computing; it presents an organized approach toward addressing them. The introduction of organizational changes such as a workstation store and an information center are presented as part of the approach. The chapter introduces office systems and discusses the important considerations relating to office automation.

WHY ADOPT END-USER COMPUTING?

End-user computing is being widely implemented in business, industry, and government organizations because it is attractive for several compelling reasons. Table 10.1 itemizes the important factors encouraging the trend toward end-user computing.

TABLE 10.1   Factors Favoring End-User Computing
_____

Lengthy development backlogs

Increasing programming costs

Availability of new end-user tools

Declining workstation costs

Availability of workstation applications

Advances in telecommunications

Growing base of skilled users
_____

Major driving factors for end-user computing are the growing backlog of work facing application development groups and the increasing costs of programmers and program development.[5] Abundant evidence demonstrates that users of IT services can participate actively and productively in application maintenance, enhancement, and development. And user participation is advantageous to everyone. New development tools and techniques provide important advantages to end users. The tools enable users to be successful program developers.

Significant reductions in personal workstation hardware costs and the growing availability of application programs for these workstations have motivated the trend toward personal or end-user computing. As the business vignette illustrated, Echlin capitalized on the declining cost of personal workstation hardware and the availability of inexpensive and highly functional networks. Echlin eliminated the mainframe computer entirely in its headquarters operation. Hardware and software that is inexpensive, reliable, and available at the individual workplace provides the tools to make end-user computing feasible. Economic considerations permit and foster widespread adoption of personal computing.

Advances in telecommunications, such as local area networks and sophisticated communications support to mainframes, have fostered program and data sharing. Workstation users have benefited from these advances. Currently most workstations are independent, stand-alone units, but the rapid growth in local area networks indicates that interconnections will grow rapidly in the near future. Networking will increase the utility of workstations and will bring the power of mainframes to the employee's workplace. Distributed data processing is becoming commonplace. Having all the firm's employees working together through networked individual workstations is the latest trend in parallel processing. Parallel processing at the level of the firm is becoming the norm.

The number of skilled individual users of personal computing applications is large and growing. Skilled users can be found within the ranks of professional, technical, and office workers and within the managerial population. The growth of this skilled population within the firm will accelerate rapidly upon the adoption of end-user computing. Numerous precedents for end-user computing have been established by firms that have implemented the technology. All these factors serve to provide additional incentive to those who may still be contemplating the technology.

The adoption of end-user computing has significant consequences for the firm, for the organizations within it, and for many of the employees and managers. This technology introduction is a microcosm of the larger phenomenon of electronic data processing itself. Stages of growth, for example, have been identified as an important concept to end-user computing.[6] End-user computing is worth studying, not only because it is highly relevant to today's world, but because it causes us to reflect upon the application of many principles that have been discussed thus far.

The implementation of end-user computing precipitates dramatic changes within the firm. The resulting changes occur within a short period, perhaps three years or so. For this reason, the management team must have a solid knowledge of management principles to install end-user computing successfully. End-user computing has high potential benefit for the firm, but it carries significant risk. Fortunately, the risks can be substantially mitigated by skillful management practices. What are the risks, and what can be done to reduce or eliminate them? What management practices, if skillfully employed, remove most of the risk?

WHAT ARE THE ISSUES?

The answers to these questions can be found by considering eight potential problem areas and by introducing management practices and organizational changes fashioned to cope with these difficulties. Each problem area requires strategic, tactical, and operational analysis for optimum resolution. Corporate culture, political considerations, and company policy also enter into the analysis of these issues.[7] Table 10.2 identifies the issues of end-user computing.

TABLE 10.2   The Issues of End-User Computing

Software and application issues

Hardware compatibility and maintenance

Telecommunications concerns

Data and database issues

Business controls

Financial concerns

Issues relating to people

Political, cultural, and policy issues

Compatibility among commonly used applications significantly aids the effective installation and application of end-user computing. Application software compatibility will simplify networking, training, and hardware installation. Installation and utilization tasks are simplified and benefits are recognized sooner if software compatibility is attained within the installation. Standards and guidelines for user-developed application programs assist in achieving program compatibility. User training programs for standard applications are mandatory. Training programs are simplified and are more effective in organizations that concern themselves with application compatibility.

Hardware implementation is also more effective if compatibility exists among the workstation devices. Hardware compatibility will simplify networking, software installation, and application portability.[8] Plans to provide hardware maintenance for user equipment with minimum disruption to their activities are essential. Also, it is reasonable to expect that hardware upgrades can be obtained conveniently and at minimum cost. Finally, a policy for workstation ownership must be established.

Communications issues include the physical network architecture, network software, and network policy questions. Important policies regarding the utilization of outside databases and dial-in capability must be established. Network management and maintenance must be transparent to end users. IT must take the lead in managing these technical and policy items so that users perceive them as part of a seamless environment. The firm's senior executives will provide numerous inputs to answer policy questions and resolve policy issues.

Data and databases raise many questions for end-user computing. Question such as what data resides where, who owns the data, and how databases are controlled must be answered unambiguously.[9] The use of personal computers and their access to large databases greatly increases risks for the organization. Serious damage or loss can result from improper or nonexistent backup procedures. Also, data integrity issues must be resolved with well-conceived controls when uploading and downloading in networked environments. Proper resolution of these concerns is vital because end-user computing thrives on availability and security of information assets in a distributed environment.

In addition to the protection of information assets, security for application programs must also be provided.[10] Programs that maintain important records or control valuable assets require special attention. The firm's managers and the application owners must carefully consider the security risks and decide whether to distribute these programs to individual workstations. Programming processes and procedures must ensure that acceptable levels of programming quality are maintained. Program documentation must be managed through the application and enforcement of documentation standards. End users are more likely to omit documentation than professional programmers; they must be trained and counseled in documentation disciplines. In addition, user-generated auditability features are generally required and must be installed in user-developed applications. Providing network security is an IT responsibility.

As technology dispersion brings micros and minis to user departments, responsibilities formerly handled by IT also migrate to user managers. Disaster recovery planning and management is one of these responsibilities. With end-user computing, department managers own and operate major information processing resources; they must take precautions against a variety of potential problems. Frequently, user managers do not understand what recovery planning is or why it is needed. The central IT organization must provide training and assistance on this important subject. Providing training and guidance on this and other matters is part of IT's staff responsibilities.

Maintaining positive financial returns as end-user computing develops demands special attention.[11] The conditions most likely to achieve satisfactory results include an aggressive attitude toward benefits accounting and specific attention to cost containment. In many cases, intangible benefits are prominent in the justification. Both tangible and intangible benefits should be recorded. Excessive expenditures for hardware and software upgrades, coupled with incomplete benefits recording, will lead to serious difficulties later.

The cost of decentralizing end-user support is significant, and the business case partly depends on whether reductions in centralized computing can be achieved. In the Echlin situation, for example, the central mainframe system was sold and helped account for part of the cost reduction. In addition, some costs are organizational in nature and are difficult to measure. In most cases, more than 50 percent of the total cost consists of nonhardware items.

KPMG Peat Marwick reports that total end-user computing expenditures in 1989 were about $130 million,[12] broken down as follows:

Hardware	$ 57.9	million
Software	17.1	million
Training	6.1	million
Support	22.4	million
Data	4.7	million
Communications	4.9	million
User time	17.0	million
Total	130.1	million

User training is likely to be understated due to the common practice of charging training costs back to user departments. Data and communications costs may also be understated because these expenses are probably absorbed within the IT budget. Software and support accounts for about 30 percent of the total cost. Managers can anticipate that financial reviews will inevitably occur. The organization should plan on dealing with this potential problem at the outset.

Although all the conditions above are mandatory for success, the most important and demanding topic involves solving the people-related issues. Substantial training is required to convert nonprogrammers or novice programmers into productive and skillful end users. Skill migration takes place among the employees as they add information technology to their professional job skills. Not all individuals will make this transition smoothly. Some employees will take the lead quickly and easily and will set the example for others.[13]

Employee transitions require skillful handling on the part of managers. In some cases, employees experience traumatic job dislocations. In all cases, managers must be especially skillful in handling the disruptions. Management training should be stressed; it should not be confined to first-line managers.

Political issues are important because end-user computing changes the power structure in the firm. Individual managers are concerned about what power shifts mean to them, and emotionalism frequently intrudes on rational decision making. Strategy development, long- and short-range planning, and application portfolio management are effective tools for dealing with political issues. These management systems put decision making on a business basis and help diffuse political concerns.

The issues are not all equally important at all times during implementation. In the early installation phases, user support is highly important. Good support encourages early adoption by hesitant users. It smooths the transitions for early adopters. When adoption is well along, financial considerations, data management, and business controls become increasingly important.[14] During implementation, the organization is undergoing substantial and permanent change. The firm's leadership at all levels must manage and shape the transition in accordance with long-term goals and objectives.

Given the magnitude of the issues and their pervasiveness, how will the management team cope? What actions can they take to deal with the potential problems, and what must happen for the firm to capitalize on the potential benefits?

ORGANIZATIONAL CHANGES

Structural changes within the IT organization will provide many opportunities to deal with the tasks noted above. IT must define support levels, perhaps through service level agreements, and must assign support tasks. Implementation of user support includes defining management processes and allocating resources in support services.[15] This can be performed most effectively by establishing two new entities within the firm. These organizational units must receive assistance from IT, but their staffing generally includes non-IT members as well. These new units are the workstation store and the information center.

The Workstation Store

The workstation store usually reports within the IT organization. Its mission is to perform many vital support functions for end users.

One important support function performed by the store is the purchasing of workstations and software. In order to save time and money for the firm, the store purchases workstations in volume quantities and at discount prices for resale to user managers. The discounts may amount to 30 or 35 percent of the quantity-of-one unit price. The store also purchases and distributes licensed applications with the hardware. In addition to reducing costs, this approach provides a control mechanism to ensure hardware compatibility for the firm. In some cases, it may be advantageous to secure compatible hardware from more than one vendor in order to satisfy the firm's requirements. The store can also ensure favorable prices for software by securing site licenses. Software compatibility can be achieved by this approach as well.

Workstations may not actually be sold to user managers, as suggested above, but hardware and software costs may be summarized and reported by organizational unit. Whether the costs are funded at a high level in the firm or distributed to the using department is a policy matter that must be resolved. The distribution method makes the cost justification more credible but increases the administrative overhead. In any case, careful accounting for the costs and the benefits is mandatory.

The store is also responsible for obtaining and providing central maintenance for all workstations and for making substitute equipment available while maintenance is being performed. Maintenance support for end users is vital. Users want to get their jobs accomplished; they prefer not to be responsible for the maintenance of their equipment. The workstation should be handled in the same manner as the telephone—if it fails, someone must correct the problem promptly.

The store should manage upgrades and migration of hardware and software. The store obtains approved hardware and software upgrades and makes them available to users with justified needs. These functions are important to users and valuable to the firm. They must be performed efficiently and effectively.

The store also performs other important functions. It gathers information for IT from sales and inquiries regarding user requirements. IT can keep track of the pace of end-user adoption and can obtain trend information through sales analysis. This information is valuable as IT develops further plans for end-user computing. The workstation store also obtains benefits information from users at time of purchase. Users are requested to submit this information when taking delivery of equipment or software. The benefits data, with other cost and benefits information, is analyzed by IT and is used to provide financial justification for end-user computing on a continuing basis.

As the utilization of workstations advances, there will be demands for both hardware and software improvements. Satisfying this demand by the store is accomplished in much the same way as the initial products were obtained and distributed. As a condition for obtaining upgrades, customers should provide an approved benefits statement that can be reconciled with the cost of the additional equipment or programs.

In addition to being a control point in the implementation of procurement policies, the store can serve very effectively as an internal control mechanism for physical assets. The store's record-keeping function is important as a business control. When units are distributed, records are maintained that show the equipment location, the requesting manager's name, and the equipment configuration. This data is required for effective inventory control.

Information routinely gathered from users while providing services is valuable to the organization for understanding trends and establishing user preferences. This information becomes part of the IT planning data and serves as a leading indicator of future demand.

The workstation store provides a variety of valuable services to users, and it removes some barriers to successful technology introduction. Volume purchases result in significant cost savings and eliminate the potential incompatibility problems of individual purchases. The store centrally distributes common applications. This permits effective control of licensed products and encourages widespread utilization of these tools. The store maintains a buffer supply of hardware for use by customers as replacements for failing components. This practice keeps interruptions at the workplace to a minimum and manages hardware maintenance in a manner nearly transparent to the end user.

The workstation store is not an optional entity. Effective implementation of end-user computing requires this unit to function efficiently and with an eye toward outstanding customer service.

The Information Center

Several additional activities demand careful attention if the implementation of end-user computing is to proceed smoothly and reach a successful conclusion. These functions are performed by an organization called the information center.[16] This unit usually reports within the IT structure but requires very close interaction with the user groups. The information center is designed to facilitate end-user computing and performs many important functions, enumerated in Table 10.3.

TABLE 10.3 Functions of the Information Center

Conduct or provide user training

Provide development assistance

Evaluate new applications

Distribute user information

Collect trend information

Perform problem determination

Gather planning information

The information center conducts or provides training for users on workstation hardware, software, and procedures. KPMG Peat Marwick found that nearly three-fourths of the companies it surveyed had formal training programs for end users and that most firms utilized a mix of internal trainers, contract trainers, and outside classes.[17] The information center is responsible for this training activity. It also provides assistance to end users in the program development process and trains users in business controls and recovery management techniques.

The center evaluates newly available application programs and other software; it distributes information on user-generated programs. Data collection for determining trends in user software requirements is an important function of the information center. The information center serves as a first-level clearinghouse for the software requirements of end-user computing. Trend information gathered by the center is valuable to IT and to the firm for establishing future computing requirements and for shaping strategies and plans.

The center provides guidance and assistance to users on software and hardware migration and upgrades. If properly staffed and managed it will be a force for preventing duplicative development activity. When the center is operating effectively, users will seek advice and direction prior to undertaking significant activity. The center provides coordination among the user groups as part of its staff responsibility.

The center also assists users by maintaining a telephone help line service and by performing initial user-problem determination. The center is responsible for customer support. This includes answering questions, even those that seem trivial to the experts. The center is a place to turn when things go astray, as they occasionally will. When other sources of help are not available, the information center is the user's safety net. Adopting a helpful attitude toward end users is the key to success for this group.

The information center plays a central role in end-user computing. It complements the workstation store. It solves or helps avoid problems likely to be encountered by the firm during the transition. In addition to the important task of employee training, the center must train managers on development processes and control issues. Training is a continuing activity for employees and managers. Employee proficiency with new tools and management skills in user departments are learned traits: The importance of training activities looms large.

Successful introduction and implementation of end-user computing requires the skillful employment of management techniques and the adoption of structural changes. These management actions are intended to reduce risk and to pave the way for capturing the benefits.[18] The firm will benefit substantially from this technology and will experience significant change in the process. Not only will the organizations within the firm be flexible and more responsive, but the individuals within the organizations will also be better equipped to assume more responsibility and to become more productive. Therein lies the payoff.

POLICY CONSIDERATIONS

The firm must make a number of policy decisions during the adoption and implementation of end-user computing. Some policies smooth the way for technology adoption and encourage end-user computing; others affect computing costs or bear on fundamental concerns such as business controls, security, or employees. These policy matters are usually important to the firm in the long term and they warrant attention by senior IT executives and others.

One such issue requiring attention at the policy level is hardware and software compatibility. The firm must decide whether to adopt one or more of the many popular application solutions available for end users. Adoption of one reduces costs and makes training easier, but this strategy may reduce functionality. For example, should the firm adopt one word processing application or should more than one be permitted? Most firms limit the choice to one to ensure easy document interchange and to reduce training and cross-training. Since there are many applications and many choices within each application the decisions on application policy are usually not easy. But allowing the issue to resolve itself by default is not responsible and will lead to problems later.

Selection of an application policy paves the way for a policy on hardware compatibility. Again, restricting the hardware options available to users reduces costs, improves maintenance, and makes migration to future systems easier. Most firms limit the hardware options to one or two popular brands, but they may purchase compatible models or clones from several manufacturers. Some firms maintain a small advanced technology group for the purpose of testing new hardware and software in order to be sure that new technology developments are understood when planning for the future.

Policies regarding ownership and control of end-user hardware, applications, and data must be developed by the firm. Individual or local ownership of these items improves security and reduces risk of loss or damage. Local or individual control may retard sharing and usage of these assets with negative consequences to the organization. Executive policy must choose among conflicting factors.

DOWNSIZING

In many organizations today, executives are looking for ways to reduce costs, increase responsiveness, and improve flexibility. Information technology enables executives to achieve these objectives through decentralized operations with centralized control. Distributed data processing linked to centralized control processors through advanced networks is the enabling technology.

The trend toward distributed data processing is not new; it started in the 1960s with the development of minicomputers and departmental computing. More recently, the trend has accelerated as increasingly powerful microcomputers appeared in the workplace. Today's workstations contain several megabytes of primary memory; they have 100 or more megabytes of secondary disk storage and they operate at 20 MIPS or faster. Local-area networks connect these powerful workstations together and to high-speed printers, very large datastores, and huge mainframes. The LANs are connected through gateways to other networks making additional capabilities available to the workstations. Distributed data processing has never been more attractive.

Competitive pressures demand rapid responses from business executives. Many business managers prefer local control of their vital and important application systems because they relish the flexibility gained by this control. In addition, they want to be responsible for their costs as they respond to changing market conditions. They prefer to manage their own information system. Accomplishing local control involves taking applications off the mainframe and implementing them at individual or departmental workstations. This process is called downsizing. The business vignette illustrated downsizing at Echlin.

The Attributes of Downsizing

Downsizing has advantages and disadvantages, and not all applications are appropriate candidates for consideration. Table 10.4 protrays the advantages and disadvantages of downsizing.

TABLE 10.4   Advantages and Disadvantages of Downsizing

Pros	Cons
Greater user control	Weakened central control
Increased flexibility	Hidden costs
Decentralized costs	Increased user skill demands
Lower costs	User management distraction
Improved responsiveness	Database disintegration
Fosters purchased systems	Discourages common systems
Reduces IT workload	Increases user workload
Encourages innovation	Encourages parochialism

Downsizing provides local control and flexibility to user organizations and improves their internal responsiveness. Downsizing frequently reduces costs, in some cases significantly. Some users report that downsizing reduces application costs by 80 to 90 percent.[19] Generally, mainframe application systems cost more to develop or enhance; mainframes cost more per MIP than personal workstations; and usually mainframe systems require more overhead. Costs incurred in user departments are experienced directly, but IT costs are allocated or charged via a billing mechanism. Users generally prefer to control their costs directly.

Downsized applications have hidden costs such as department attention and overhead; user training and operations costs; and costs derived from reverse economies of scale. For example, problem, change, and recovery management is much less efficient and costs more with decentralized systems. Indeed, user departments sometimes skip these essentials. They are implicitly trading higher risks for lower costs.

Downsizing encourages user innovation because there is little historical or cultural background to overcome. Traditions in IT may hinder innovative ideas unnecessarily. For example, IT organizations are prone to develop applications locally because that is what programmers do, but user organizations may opt for purchased applications because they purchase many other necessary items. They are predisposed to purchase software too.

Downsizing increases user workload and raises demands for special skills. User managers must cope with these demands and must increase their management skill levels. Managing an application portfolio and running a small data center are nontrivial tasks. These activities require talents not usually found in operational departments. Discharging these responsibilities detracts from the main function of the department and diverts managers' attention from their primary tasks. Some managers refuse to take on these added functions and resist downsizing.

Datastores and large integrated databases are a deterent to downsizing. Some databases are highly integrated and are difficult to separate into distributable units. Downsizing increases the data management problem and may lead to redundant data elements, increased storage costs, and possible asynchronous conditions. Both application portfolio management and downsizing are critically sensitive to database conditions and architecture. Firms with highly developed information architectures suffer less from these difficulties than others. But, as we learned in the first chapter, information architecture tops the list of issues. Most firms require major improvements in this area.

There are some additional disadvantages for the firm. Decentralized control of applications may lead to inefficiencies, particularly if corporate planning is poorly executed. This may mean, for example, that duplicate development occurs, or that corporate goals are deemphasized in favor of unit goals. Parochialism tends to favor unit performance at the expense of firm performance. Firms with highly disciplined strategizing and planning processes are much better prepared to capitalize on downsizing opportunities. Disciplined management processes pay dividends in many areas.

What To Downsize

Not all organizations should adopt downsizing, and not all applications are candidates for downsizing. Table 10.5 portrays applications suitable for downsizing and those that are not.

TABLE 10.5 **Downsizing Applications**

Candidate Applications	Unsuitable Applications
Small systems	Large systems
Single department applications	Enterprise applications
Isolated databases	Integrated databases
Stable applications	Evolving applications
Purchased applications	Customized applications
Technically simple systems	Technically advanced systems
Low-risk applications	High-risk systems

Small, stable, relatively simple applications with isolated databases are ideal downsizing candidates. Applications that are evolving or growing in size and importance are questionable candidates. Technically simple applications are preferred to technically sophisticated systems. Suitable applications are characterized by low risk; they can be managed easily by the using department. The number of people interacting with these applications is usually small and downsizing is likely to be successful. This type of system is an ideal first candidate for downsizing.

Large, enterprisewide applications involving many users interacting with integrated databases are best left on the mainframe. Evolving applications customized to the firm's requirements should not be disturbed. If the application relies on specialized or advanced technology it is best left in the custody of IT professionals. High-risk systems or applications are unlikely downsizing candidates.

The management system developed in Chapter 7 to handle the firm's applications portfolio is valuable for establishing the rational for downsizing. Business goals and objectives are more important than political or emotional reasons when making downsizing decisions.[20] The portfolio management system focuses on the firm's business objectives and on the objectives for individual functions and departments. It puts decision making regarding downsizing at the proper level in the firm and reduces the emotionalism and parochialism that can surround these decisions.

Although some applications have been dispersed to user organizations for operation, much of the maintenance and enhancement activity remains the responsibility of IT. One study revealed that in manufacturing firms, 75 percent of requests for new applications were handled by the traditional development organization, 17 percent by the information center and users together, and only 8 percent by users alone. The corresponding figures for service firms were 67 percent, 27 percent, and 6 percent.[21] Simple query applications are the most popular with most end users.

End-user computing is distributing information technology activities but it is also expanding the scope of IT. IT must provide training in system development and operation; develop information architectures for distributed systems; and provide system development tools and other support for users.

Downsizing is the ultimate expression of end-user computing. Downsized applications running on powerful workstations operated and maintained by departmental personnel which serve critical business functions represent distributed data processing at its best. Timely exchange of vital control information with corporate central systems through advanced telecommunication networks completes the architecture. Executives obtain reduced costs, improved responsiveness, and increased flexibility while retaining essential controls. For those firms prepared to capitalize on downsizing opportunities, the rewards are significant.

OFFICE AUTOMATION

In contrast with traditional data processing activities in the firm, operations within the office environment have usually been localized. Most firms receive information, produce documents, and store material within or adjacent to the office operation itself. With the exception of the mail room and the central telephone switching system, information handling activities for the office environment were traditionally decentralized. The introduction of electronic data processing applications to the office environment has provided both the means and the incentives to centralize some office activities. In particular, telecommunications systems have fostered trends toward centralized office operations.

Modern office systems are a specialized example of end-user computing supporting office workers. Office workstations are interconnected with local area networks and the office LANs are joined to broader networks within the firm itself. Specialized office applications such as document production, information distribution, and filing systems serve office workers. But the trend toward increasing connectivity and general-purpose applications is blurring the distinction between office workers and others. Knowledge workers in all parts of the firm are adopting and using these tools. Telecommunication and computer technology is being used to export traditional office capability to all functions within the firm.

We turn our attention to office systems because the study of office automation illustrates many of the opportunities and issues discussed earlier. Electronic office systems, like end-user computing, offers important advantages to modern organizations but requires skillful management during introduction and implementation. Many firms have implemented some form of electronic office system; thus there is a body of experience upon which to build.

Experience with office systems demonstrates many important financial and operational advantages to the firm. Office systems improve productivity and increase responsiveness to business needs. Larger volumes of information are handled with greater speed and improved accuracy, which increases performance and improves profitability. For example, document production systems such as word processing programs or publishing systems coupled with electronic mail systems speed the flow of information to all parts of the firm and even to customers or suppliers.

Not only do electronic office systems allow office workers to do things better, but they support workers in their quest to do better things. Office applications provide improved tools for better and more timely analysis and synthesis of data. They build foundations for more effective information integration. For example, knowledge workers employ spreadsheet techniques or statistical analyses to routine business data and incorporate the results into final documents without the intervention of paper. The entire process is electronic.

Modern office systems provide improved support to the business environment through automation. Electronic keyboards and displays enhance the process of converting ideas into text; graphics composition tools and graphic display devices permit rapid integration of words and pictures. High-quality laser printers provide hard-copy presentation materials. Electronic media can be used to store, index, and retrieve the material and associated reference materials. Electronic communications or E-mail permits and encourages the sharing of ideas and information. It greatly improves the dissemination of business information.

Technology trends and business demands are speeding the adoption of many new and varied means of communication. Facsimile transmission, video conferencing, cellular phones, electronic data interchange (EDI), and other knowledge worker communication tools are changing the office environment forever. For some workers the rapid pace of change creates stress and anxiety. But for others, improved tools increase their productivity and lead to skill upgrades and higher job levels. If managed properly, office automation can bring increased job satisfaction and enhanced quality to the work environment.

PLANNING FOR OFFICE AUTOMATION

Planning for office automation is a nearly continuous process because the introduction and implementation of new tools for knowledge workers seems to be a never-ending task. Many planning questions address issues such as what technology to introduce next; where to begin the introduction; and when and at what speed changes should occur. The firm's routine technology planning process serves to answer these questions if the proper background material on the current and future office environment is available.

An important first step in planning for the office of the future is to understand the existing office environment. Is the firm office-intensive such as an insurance company, or are office operations less vital as might be the case in a warehousing operation? Office planners should catalog the strengths and weaknesses of the present office environment for their firm. They should develop areas of opportunity for the firm. Their plan should leverage the strengths, capitalize on the opportunities, and reduce the areas of weakness.

Stages of Growth

Managers responsible for implementing office automation should recall the stages of growth developed by Nolan; they apply to office systems. In the first stage, initiation of the technology occurs and early proponents begin experimentation and start developing applications. For example, an innovator may acquire and install a publishing system and begin to produce documents for use within the department. As coworkers begin to see the advantages of the new tools, they adopt the technology and its application spreads. The second stage has begun. In this example, desktop publishing would continue to develop until everyone who had a need for the capability would have access to it and would be qualified to use it.

During the time that widespread adoption and use of an office technology takes place, the issues of control, data administration, and integration emerge. These issues will be resolved as the firm marches toward a state of comparative maturity. For most office technologies this stage will be attained over a period of years—one, two, or up to ten or more. For example, the implementation of a new application package for office workers who are currently using office workstations may only take a few months. The implementation of video conferencing, on the other hand, has been underway for a decade or more in some firms.

Chapter 4 pointed out that planners must use several methodologies if they are to be successful for most technologies. Except for firms with low degrees of office automation, critical success factors or business system planning methodologies must be considered during the planning process. For many firms, office automation is both highly diffused and highly infused, and eclectic planning approaches must be employed. Sullivan states that firms in this position are in a complex environment. Essentially, they have entered the information age.[22] In general this means that planning must be overt, systematic, and well developed. Office system planners must be proactive and must have strong strategic perspectives.

IMPLEMENTATION CONSIDERATIONS

When the decision to introduce a new office automation tool has been made and implementation plans are being formulated, action should be taken to develop the preliminary program goals and objectives. As with application development, detailed results are difficult to predict and alternative conclusions usually result. At this point a feasibility study can provide a sound basis for establishing a plan to achieve organizational objectives.[23] For instance, if the plan called for the installation of facsimile machines at appropriate places in the firm it would be wise to test the plan at one location and determine feasibility. The test would also help develop a more detailed business case for the plan. The prototypical installation serves as a test for the technology and reinforces planning details for subsequent installations.

The feasibility study must also analyze all supporting functions presently being performed within the targeted areas. The introduction of new office tools alters methods and procedures. The feasibility study is the basis for determining what changes in present procedures are necessary or desirable. The results also help refine office automation planning. An analysis of the feasibility study results will generally involve most of the functions and processes of administrative and office operations. The installation of office automation usually results in difficult-to-forsee consequences, and the feasibility study or prototyping approach provides confidence.

Expectations must be carefully managed by the planning and implementation teams. Prototyping efforts must proceed with the understanding that senior managers support the effort because they have confidence that the work has potential to improve the firm's operations. Individuals involved in the installation must be prepared and trained to accept change and to cope with disruptions to routine workflow. Managers must tolerate the infrequent difficulties and must permit the technolgy to be implemented according to plan without undue pressure or displays of overconfidence.

The prototype installation or feasibility study helps managers solidify the plan to introduce the technology throughout the selected functional areas. The study also focuses on current applications and assists in developing cost and benefits information about the new technology. Other results from the prototype include confirmation of software and hardware choices; physical installation preparation; and finalized practices and procedures surrounding the new technology.

Successful introduction of new office technology is highly dependent on human factors considerations.[24] As with end-user computing, hardware and software costs are important. But costs are less important than factors that allow the new tools to be installed easily or that permit them to be used easily and conveniently. It is very important that the software be easy to use. Easy-to-use software is characterized by smooth interactive capability and well-constructed option menus. Well-designed systems usually use a keyboard, a mouse, and a graphics screen to allow users to alternate between products or functions. Some current windowing systems have splendid architecture making them easy to use. One window can be used for tutorials and help sessions, thus giving the user the ability to learn through use.

Other environmental factors are also important. The physical environment including lighting, seating, and dimensioning must be carefully reviewed in order to reduce physical stress. Facilities planning must include sufficient space. But planning must also address items such as noise levels; temperature and humidity; and asthetics such as color and office appearance. Careful attention to these human factors will reduce resistance to change, speed the implementation, and improve morale.

Coordination

An effective office systems strategy should be governed by a formal charter that describes the mission and responsibilities of the implementing

---

managers. Corporate policy statements covering such areas as hardware and software selection, return on investment, cost estimating procedures, and resource allocation should guide the office automation effort. It may be appropriate to establish a steering committee to gain senior management involvement and commitment to office automation. The committee can help link the office systems plan to business plans, goals, and objectives. The information center should procure and administer the training. IC people act as internal consultants providing advice, assistance, and guidance for office automation projects just as they do for end-user computing projects.

Some Possible Difficulties

There are many office systems technologies being developed today; many manufacturers and vendors are claiming sizable increases in productivity for their systems. As with end-user computing, office managers must keep a watchful eye on costs and real and intangible benefits. In some cases, office systems have not matched user needs and the planned productivity gains did not materialize.

There are many risks in placing sophisticated information technology in the workplace. Many of these risks can be averted by using the information center and the workstation store in the manner described earlier. Issues such as business controls and data security and control must be resolved. The information technology organization must provide guidance on these issues; IT must ensure consistent treatment of these topics across the firm. On many of these matters, the IT manager has staff responsibility for office systems just as for end-user computing. IT will play a vital role in the successful implementation of office automation. The mutual interaction of the office systems implementors and the IT staff will assist considerably in ensuring office systems success.

PEOPLE CONSIDERATIONS

Office automation may alter people's behavior and their attitudes toward their jobs. Roles and organizational reporting structures may also change. For instance, in many firms that have installed integrated office systems, secretaries report to an administrative manager rather than to individual executives. This has taken place because office systems bring consistency to the job and make job rotation and personnel reassignment more effective. For many secretaries and for many executives this was a critical change calling for adjustments on everyone's part. Not only does the electronic office have a more direct effect on traditional reporting relationships such as secretary to principal, but it calls into question traditional loyalties as well. For example, loyalty to the administrative function may become stronger than loyalty to the person served by the secretary.

Today's integrated office is far more complex than the office of 20 years ago. It is more difficult to manage because it potentially affects many more functions in the business environment. The new office environment is greatly impacted by changes in technology, administrative reorganizations, and the nature of office work itself. Therefore, the firm must plan to address the human concerns of office automation. The plan should be centered around employee communication: Keep employees informed about system changes; explain changes in policies and procedures; obtain employee input on planned changes; and respond to employee grievances promptly. It is almost impossible to overcommunicate in a rapidly changing work environment.

Change Management

Change management during office automation can be divided into three phases. During the first phase, when the prototype installation is underway and the early, more innovative employees are being trained, progress bulletins should be released and managers should hold information meetings for adopters. The system and its impact should be explained. Expectations of the system should be established. Information center personnel should be available to answer employee questions; managers must respond to concerns. In particular, issues that may impact morale should be addressed promptly.

In the second phase, when employees are beginning to use the new tools, experienced employees should be assigned to assist beginners. Additional staffing should be available to offset the reduced output of employees just learning the system. Training materials must explain how each employee fits in with the system and how the system fits into the plan for the office environment. Some users will experience frustrations with the system. Managers must be especially sensitive to these employees. They must take steps to reduce or eliminate the problems. Managers should provide frequent positive feedback to employees as they make progress.

During the final phase, employees should be competent in the use of the new technology; they should be sufficiently skilled to recommend or implement improvements to applications on their own initiative. Managers should recognize good ideas and should encourage their adoption.

Some employees find difficulty in adapting to changing work patterns and to the structural changes that usually accompany them. Success for these employees can be achieved by clearly explaining the reasons for change, by planning for change as it affects them personally, and by allowing them to appreciate the benefits of change. Both managers and employees must be flexible in order to respond to new opportunities. Successful change management copes with the many personal anxieties, reduces apprehension, or perhaps even turns it into enthusiasm.

Managers should recognize that changes can be implemented in the areas of technology, organizational structure, management style or technique, and organizational culture. Almost all changes, technological or organizational, result in changes in employees' attitudes or behavior patterns. When technology creates jobs with higher skill requirements, traditional occupations are upgraded, modified, or eliminated.

When managers make changes that affect employee attitude or behavior, the success or failure of the change may well depend on whether workers perceive personal rewards or benefits. It may also depend on whether employees have the capability to learn new skills and whether they in fact want to learn them. The scope of communication and the quality of orientation and training play an important role in managing changes successfully.

Several points are worth emphasizing. The impact of change will not be enthusiastically welcomed by all individuals—for some people change carries threat with it. The challenge comes from the need to start over again learning new skills. Attentive managers will cope with this difficulty through effective people management practices such as one-on-one communication, coaching, counseling, and training. The most important task for the manager is to recognize that each individual struggles with change in his or her own way. Successful managers deal with the problems on an individual basis.

SUMMARY

End-user computing is one of the most important developments in the history of information technology. For the first time since the invention of the digital computer, immediate and convenient access to large computers is available to most professional employees in modern firms. There are many reasons to take advantage of these developments.

To capitalize on the technology of end-user computing, the firm's managers must select and implement software and hardware systems. Telecommunications systems will need to be implemented to support the new environment. In addition, managers must deal successfully with financial issues; they must resolve business controls problems; data management and data ownership issues must be settled; and, most importantly, managers must deal with people concerns skillfully.

The implementation of the information center and the workstation store will provide organizational resources to deal with many of the issues. These resources will allow the IT organization to engage with end users in a meaningful and productive manner. These organizations form the IT manager's staff support to end-user computing.

The adoption of end-user computing by the firm will cause the IT organization itself to undergo significant change. The information technology manager's job will substantially increase to include considerable staff responsibility. The role of IT in the firm will expand; this translates into more responsibility for the people in IT as well. Several new and exciting jobs will be created in the workstation store and the information center. Employees in the user organizations and in IT should be given the opportunity to fill these jobs. With a plan for job rotation, many people can benefit from these opportunities.

Office systems, as an example of the application of end-user workstations, provide many of the challenges to management that they will experience with more sophisticated end-user applications. Careful attention to the areas of risk and to the effective functioning of the support organizations will help contain or reduce implementation problems. These actions coupled with sound people management techniques will pave the way for success.

End-user computing has high potential payoff. It includes significant risks of various forms. If managed properly it can be a win-win situation for all involved. The firm gains substantial benefits; the management team has greatly improved resources for dealing with the problems facing it; and the employees have new, more valuable skills and are more productive. Although the effort is great, the benefits can significantly exceed the costs.

Review Questions

1. What is the most important lesson learned from the Echlin business vignette?

2. What actions did Echlin MIS managers take to reduce users' fears of downsized systems?

3. What factors are encouraging the trend toward end-user computing?

4. What role do advances in telecommunications play in the trend toward end-user computing?

5. What are the issues or problems that must be solved in order for end-user computing to be successful in a firm?

6. Why is compatibility such an important consideration in the implementation of end-user computing?

7. What are some of the questions surrounding the issue of databases?

8. End-user computing causes some problems faced by owners and operators of mainframes to appear in user departments. What are some of these new problems for user managers?

9. What percentage of end-user computing costs are hardware and software related?

10. What is the most important issue of end-user computing? Why?

11. Describe how the issues of end-user computing vary in importance over time.

12. What are the functions of the workstation store? How does the workstation store assist in maintaining sound business controls for the firm?

13. What are the responsibilities of the information center?

14. Where do the workstation store and the information center report in the firm? What is the source of staffing for these organizations?

15. What is the linkage between the store, the information center, and the firm's planning process?

16. What are some of the policy issues that accompany the thrust toward end-user computing?

17. What do we mean by downsizing? Why is it important today?

18. What are the advantages and disadvantages of downsizing?

19. What are the characteristics of an application that should not be downsized?

20. Today's office environment is undergoing rapid change. What are the technology drivers behind these changes?

21. What are the advantages of prototyping new technologies in the office environment?

22. What are the phases of change management during the implementation of office automation?

23. Why is it important that managers deal with employees on an individual basis when implementing changes in the workplace?

Discussion Questions

1. The implementation of end-user computing raises the issue of expectations. What steps can be taken to ensure realistic expectations?

2. The text presents a number of factors motivating the trend toward end-user computing. Which of these factors relates to the firm's competitive posture in the industry and why?

3. In earlier chapters we discussed advances in hardware and telecommunications. How do these advances increase or decrease the risks associated with end-user computing?

4. Many end-user computing installations do not have well-developed telecommunications networks supporting them. What are some of the consequences of this? Does this increase or decrease the probability of success? Why?

5. The firm may need to make some policy statements concerning the use of both purchased applications and user-developed applications. What policy questions may need to be resolved regarding these programs?

6. End-user computing increases some of the management responsibilities in those departments using it, especially regarding data and business controls issues. What are some additional considerations that management needs to deal with in these areas?

7. What alternatives are available to the firm regarding the question of who should own the workstation hardware and purchased software?

8. What are the pros and cons of funding the workstation hardware and purchased software centrally at the corporate level versus at the individual department level?

9. What financial advantages can be captured by an effective workstation store? Can you make some estimates and quantify these financial benefits?

10. Elaborate on the notion that the workstation store and the information center provide a "leading indicator" function for the IT organization. How does this relate to planning for the IT organization?

11. What benefits can be achieved for the organization and for some of its people by the staffing opportunities presented in the workstation store and information center?

12. Discuss the strategic implications of downsizing.

13. Discuss the advantages and disadvantages of downsizing the general ledger application.

14. What is the importance of securing the involvement of senior management during the feasibility study of office systems?

15. How might the implementation of office systems impact the traditional manager/secretary relationship?

16. What actions can be taken by managers to ease the transition for employees into the automated office environment? What approaches might be most successful with the laggards?

17. Upon successful implementation of end-user computing, the responsibilities of the IT manager will include a greatly expanded staff role. What skills are useful in discharging these new responsibilities? What measures of success can the manager employ in these endeavors?

18. Effective end-user computing and office systems permit the "logical office" concept, i.e., the office is where the workstation is. What opportunities does this concept present?

19. The logical office could be in the employees' homes. What are the advantages and disadvantages of the office-at-home concept?

20. What issues does the concept of the office at home raise for the organization?

Assignment: A Case Discussion

The SUNEXPLOR corporation is a manufacturer of durable equipment for the petroleum industry and has carved out a niche in the world market for its products. The firm is very successful. It attributes its success to a productive technology laboratory, low-cost manufacturing operations, and in part to its aggressive approach toward the utilization of information processing technology. The firm is very concerned about business controls, cost effectiveness, productivity, and cost controls.

SUNEXPLOR is located in a major southwestern city and employs approximately 2000 individuals. About 1000 employees are engaged in production line activities and the remainder are mostly professional employees. The firm's organization chart displays the major functional units.

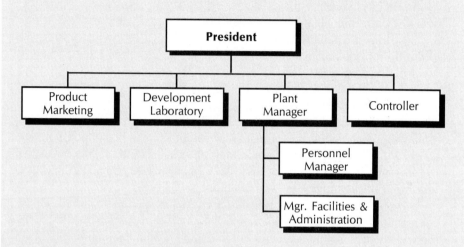

The executives:

    President — Ray Sweeney

    Product marketing — Jim Blake

    Development lab — Hal Williams

    Plant manager — Ed Gunderson

    Controller — Pete Burns

    Personnel manager — George Adamson

    Facilities manager — Joyce Groves

The information systems manager, Larry Swanson, reports to Joyce Groves. He has a staff of 50 people and operates four CPUs each having approximately 20 MIPS capacity. The computing capacity is utilized by all areas of the organization; however, the backlog of applications is growing rapidly, primarily as a result of the successful introduction of a new line of oil well service equipment and the need to develop and build follow-on equipment.

The original cost of the computing equipment presently installed was about $10,000,000. Swanson's annual budget is approximately $2.5

million, not including depreciation on the equipment. The IS organization manages the computer facility and the application programming group skillfully. It has a reputation for being aggressive in the introduction of new information technology.

The growing backlog of application development work has reduced the level of user satisfaction; some customers of IS are threatening to purchase their own equipment and develop programs to satisfy their unique requirements. Some users are engaged in programming activity using the computer center equipment and are requesting authorization to access major application data files. Swanson is concerned about the situation and has met with his staff in order to develop a plan to cope with the problem as he sees it.

The plan finally adopted by the IS organization is to seek approval to install about 1000 personal computers over the next two years in the offices of the professional employees. The PCs are intended to communicate with each other and with the main computing center via coaxial cables, control units, and appropriate software. The plan has been discussed with some of the users, and they generally support the need for more computing capacity.

The president holds a regularly scheduled staff meeting each Tuesday morning. Swanson has arranged for one hour on the agenda next Tuesday to present his plan.

Case Discussion Questions

1. What factors concerning information technology make the above scenario highly plausible?

2. If Swanson's proposal is adopted by Blake and his staff, what might this mean for:

   a. the enterprise

   b. the people in the enterprise

   c. the Swanson organization

3. Assuming the adoption of the Swanson proposal, what are the issues that need to be dealt with in the following areas:

   a. hardware

   b. software

    c. communications

    d. finances

    e. data management

    f. business controls

    g. support activities

4. During the two-year installation period, what trend information should Larry Swanson be accumulating in order to continue into the next stage of growth?

5. Making estimates of some of the parameters, construct the financial plan for the installation of end-user computing at SUNEXPLOR.

6. What role will Joyce Groves fulfill during the staff meeting and during the next two years?

ENDNOTES

[1] Mike Feuche, "Echlin Pulls Mainframe Plug and Installs Networked PCs," *MIS Week*, June 19, 1989, 1. Echlin introduced networked PCs at headquarters. The firm is decentralized into about 50 divisions, most of which do their own data processing.

[2] Not all firms are replacing mainframes with microcomputers. Hannaford Bros., an 80-unit, $1.2 billion supermarket operation, relies on minicomputers and credits them for cost reductions that helped raise profits above the industry average. See Mike Feuche, "Minis Win Out over PCs at Supermarket Chain," *MIS Week*, October 23, 1989, 17.

[3] A KPMG Peat Marwick study predicted that 80 percent of all applications will be written by end users by the end of the 1990s. "The Shape of Things," *MIS Week*, February 5, 1990, 36.

[4] Chester Frankfeldt, "The Dispersion of IS," *CIO*, November 1989, 12.

[5] E. R. McLean, "End Users as Application Developers," *MIS Quarterly*, December 1979, 37.

[6] Sid L. Huff, Malcolm C. Munro, and Barbara H. Martin, "Growth Stages of End-User Computing," *Communications of the ACM* (May 1988): 542. The authors developed a five-stage growth model from the six-stage model of Nolan *et al*.

[7] Peter G. W. Keen and Lynda A. Woodman, "What To Do with All Those Micros," *Harvard Business Review* (September-October 1984): 142.

[8] J. Daniel Cougar, "E Pluribus Computum," *Harvard Business Review* (September-October 1986): 87. This excellent article describes how to obtain high returns from investments in end-user computing.

[9] Lawrence S. Corman, "Data Integrity and Security of the Corporate Data Base: The Dilemma of End-User Computing," *Data Base* (Fall-Winter 1988): 1.

¹⁰ Joel S. Zimmerman, "PC Security: So What's New?" *Datamation*, November 1, 1985, 86.

¹¹ Lee L. Gremillion and Philip J. Pyburn, "Justifying Decision Support and Office Automation Systems," *Journal of Management Information Systems* (Summer 1985): 5. The authors recommend the portfolio approach for justifying management support systems and office automation in contrast to the traditional return on investment approaches.

¹² "The Shape of Things," *MIS Week*, November 27, 1989, 35. For additional insight into cost structures see Strassmann in the references.

¹³ James C. Brancheau and James C. Wetherbe, "Understanding Innovation Diffusion Helps Boost Acceptance Rates of New Technology," *Chief Information Officer Journal* (Fall 1989): 23.

¹⁴ J. C. Henderson and M. E. Treacy, "Managing End-User Computing for Competitive Advantage," *Sloan Management Review* (Winter 1986): 3.

¹⁵ Robert L. Leitheiser and James C. Wetherbe, "Service Support Levels: An Organized Approach to End-User Computing," *MIS Quarterly* (December 1986): 337.

¹⁶ For an alternative to the information center approach see Gerrity and Rockart in the references.

¹⁷ "The Shape of Things," 36.

¹⁸ William J. Doll and Gholamreza Torkzadeh, "The Measurement of End-User Computing Satisfaction," *MIS Quarterly* (June 1988): 259. This article describes a valid methodology for measuring end-user satisfaction.

¹⁹ Kathleen Milymuka, "Honey, I Shrunk the Mainframe!" *CIO*, September 1989, 36.

²⁰ John Mahnke, "How To Size Up Downsizing Opportunities," *MIS Week*, December 4, 1989, 18.

²¹ Chester Frankfeldt, "The Dispersion of IS," *CIO*, November 1989, 12.

²² Cornelius H. Sullivan, "Systems Planning in the Information Age," *Sloan Management Review* (Winter 1985): 4.

²³ James H. Green, *Automating Your Office—How To Do It, How To Justify It* (New York: McGraw-Hill Book Company, 1984), 140-157.

²⁴ Barbara S. Fischer, *Office Systems Integration* (New York: Quorum Books, 1987), 103–112.

REFERENCES AND READINGS

Alavi, Maryam, and Ira R. Weiss. "Managing the Risks Associated with End-User Computing." *Journal of Management Information Systems* (Vol. 2, Winter, 1985-1986): 5.

Carr, Houston H., *Managing End-User Computing*. Englewood Cliffs, NJ: Prentice-Hall, Inc., 1988.

Cougar, J. Daniel. "E Pluribus Computum." *Harvard Business Review* (September-October 1986): 87.

Gerrity, Thomas P., and John F. Rockart. "End-User Computing: Are You a Leader or a Laggard?" *Sloan Mangement Review* (Summer 1986): 25.

Henderson, John C., and Michael E. Treacy. "Managing End-User Computing for Competitive Advantage." *Sloan Management Review* (Winter 1986): 3.

O'Donnell, D. J., and S. T. March. "End-User Computing Environments—Finding a Balance Between Productivity and Control." *Information and Management* (September 1987): 77.

Poppel, Harvey L. "Who Needs the Office of the Future?" *Harvard Business Review* (November-December 1982): 146.

Pyburn, Phillip J. "Managing Personal Computer Use: The Role of Corporate Management Information Systems." *Journal of Management Information Systems* (Winter 1986-87): 49.

Rivard, Suzanne, and Sid L. Huff. "An Empirical Study of Users as Application Developers." *Information and Management* (February 1985): 89.

Rockart, John F., and Lauren S. Flannery. "The Management of End-User Computing." *Communications of the ACM* (October 1983): 777.

Strassmann, Paul A. "The Real Cost of Office Automation." *Datamation*, February 1, 1985, 82.

Swider, Gaile A. "Ten Pitfalls of Information Center Implementation." *Journal of Information Systems Management* (Winter 1988): 22.

Withington, F. G. "Coping with Computer Proliferation." *Harvard Business Review* (May-June 1980): 152.

Part
Four

11 Developing and Managing Customer
Expectations

12 Problem, Change, and Recovery
Management

13 Managing Production Operations

14 Network Management

Tactical and Operational Considerations

Business operations are critically dependent on the effective and efficient operation of information and telecommunications systems. A disciplined management approach to the routine operation of business systems is a key success factor for IT managers. An approach which deals effectively with numerous operational issues and potential problems in a systematic manner is required in this complex arena. Part Four begins with the chapter Developing and Managing Customer Expectations and continues with Problem, Change, and Recovery Management; Managing Production Operations; and Network Management.

A systematic, disciplined approach for handling operational issues enables IT managers to attain high levels of customer responsiveness from their information systems.

11

Developing and Managing Customer Expectations

A Business Vignette

Introduction

Tactical and Operational Concerns

Expectations

The Disciplined Approach

Service-Level Agreements

What the SLA Includes

   Schedule and Availability

   Timing

   Workload Forecasts

   Measurements of Satisfaction

The Role of User-Satisfaction Surveys

Additional Considerations

Congruence of Expectations and Performance

Summary

Review Questions

Discussion Questions

Assignments

References and Readings

Appendix

Security Pacific Corporation reduced IS expenses, expanded its business, and improved IS performance.

An IS Management Triumph[1]

John Singleton began experimenting with service-level agreements and defining his Management by Results program when he was vice president for operations and data processing at Maryland National Bank in Baltimore. When he moved to Security Pacific as IS Chief, he refined and implemented these concepts with outstanding results. His accomplishments paved the way for promotion to vice chairman of Security Pacific and chairman and CEO of Security Pacific Automation Co.[2]

Upon joining Security Pacific, Singleton was given the assignment of controlling IS spending then growing at an annual rate of nearly 30 percent. Within a year he had reduced the increase to about 10 percent, more in line with rates for IS organizations of this size. He accomplished this reduction by implementing his Management by Results program and by centralizing the IS organization, contrary to the current trend toward downsizing and dispersing IS activities.

Singleton believes that centralized IS organizations have significant cost advantages over downsized operations and maintain IS control over IS activities. In addition, centralized IS operations offer bigger jobs to IS professionals and allow the firm to retain critical IS management talent. Decentralized operations fragment the job and create isolated pools of talent. This encourages aggressive managers to look outside the firm for advancement opportunities. But to be successful in large companies, centralized IS operations must be responsive to the needs of diverse and remote business units. This is where the Management by Results program comes in.

The program has four related components: IS strategic planning, service level agreements between IS and users, IS performance tracking and reporting, and personal performance appraisal and salary programs. The goals of Management by Results are to link the IS business plan to the user's plan, understand exactly what the customer needs are and contract with users to fulfill their needs, report accomplishments and results to users and IS people, and tie individual accomplishments to salary compensation. The program ensures that IS people and the IS organization are committed to achieving company and user objectives.

Security Pacific Automation prepares an annual strategic plan tuned to user goals and uses a steering committee composed of board members of the parent organization to oversee IS activities. For example, Singleton reviews budget requests with the steering committee using IS achievements obtained through the measurement process prior to going to the full board. This approach ensures top level agreement with IS strategies and plans and virtually guarantees strong linkage between IS strategies and the firm strategies.

Service-level agreements (contracts between IS service providers and users) are at the heart of the Management by Results system. IS works with each user to establish exactly what the user needs and how the user measures delivered performance. IS and users agree on performance standards, performance calculations, and methods for reporting unsatisfactory results. The agreements also link IS costs and performance because Singleton believes cost effectiveness is a common goal. There is a balance between performance, quality, and good customer relations that must be achieved in the agreements he believes. When finalized, the agreement is signed by IS managers and users. There are more than 500 service-level agreements in place at Security Pacific.

Security Pacific Automation achieves high credibility with users by measuring and reporting all IS activities. The program sets standards and permits users to evaluate IS performance regularly and frequently—sometimes daily. Objectivity is improved through user evaluations of performance. Singleton attained excellent or above average ratings from 94 percent of his users during the first three quarters of 1988, a tribute to the program's effectiveness.

Individual performance appraisal and salary advancement are tied to IS performance, thus ensuring personal commitment to achieving results. Measurement techniques similar to those in the service agreements allow managers to track individual performance. Employees are measured on managing service levels, budgets, personnel, and new business development. They negotiate their measurements with their managers and obtain a basis for understanding their contributions to the business. Singleton believes this puts responsibility for career management more directly with employees.

Security Pacific Automation bids for customer business through proposals generated by IS employees, and competes with proposals from outside vendors for services or equipment. Largely because of its cost competitiveness and commitment to results through service-level agreements, the Automation Corp. is taking business away from outside vendors. Additionally, when IS underspends its budget or makes a profit on outside sales, it returns the excess to its internal customers. This strategy reinforces and cements relations between IS and its customers.

The Management by Results system has yielded high returns at Security Pacific. Substantial staff reductions were implemented and expenses dropped five percent over two years even though the company itself was growing by acquisition. During the past several years, Security Pacific purchased The Oregon Bank and The Arizona Bank financed partially by cost containment at Security Pacific Automation. Electronic transactions grew by 20 percent in three years. Later Security Pacific capitalized on its IS effectiveness by absorbing the data processing departments of three out-of-state banks having combined budgets of $100 million.

When Singleton joined Security Pacific he found that unhappy users were his most serious concern. They believed that IS costs were unreasonably high and that quality performance was lacking. IS also had a reputation for not finishing projects on time. In short, IS credibility with users was low and they wanted improvements quickly. If Singleton had not been able to correct the situation within six months to a year, users were prepared to dismantle the IS operation, carve up the pieces among themselves, and take charge of their own destiny.

Singleton moved rapidly but his program encountered resistance from IS managers and users. Some managers believed their responsibilities were not measureable and users resisted centralization efforts until the benefits were visible. IS people who could not adapt left the company voluntarily or were terminated. But Singleton demonstrated that centralized IS organizations can be flexible, responsive, and effective. He believes that IS must lead the way in reengineering the business and in destroying the old ways of doing things.

For his accomplishements, Singleton received the 1988 Information Systems Award for Executive Leadership from the John E. Anderson Graduate School of Management at UCLA. The accounting firm of Peat Marwick Mitchell & Co. found Security Pacific's IS expenses to be about 10 percent lower than comparable operations. IBM declared Security Pacific to be one of only two firms that measures and manages IS strategic resources effectively. Indeed, under Singleton, Security Pacific Automation Corp. is truly an IS management triumph.

INTRODUCTION

The acquisition, development, and enhancement of application programs and the management of the associated databases raise many long-term strategic issues. Certainly the applications raise tactical issues as well, but strategic concerns are dominant. However, the operation and use of the applications and databases raise mostly tactical and short-term concerns.

Tactical and operational concerns arising in the production operation portion of the IT organization are the focus of this chapter. Production operations concerns itself with the routine implementation and execution of the application programs supporting the firm's mission. Some of these applications are on-line nearly continuously. They operate in a highly interactive manner as they support hundreds, perhaps thousands of users. Other systems operate daily, weekly, or at month-end according to a predetermined schedule. The nature of this scheduled production work is operational and tactical. Because the firm's essential business activities vitally depend on the IT production operation, the applications must operate effectively. This activity is one of the critical success factors for the IT organization.

The beginning of the disciplined approach to production operations management will be developed in this chapter by considering the notion of service levels and service-level agreements. This chapter formulates the process of reaching agreement with IT customers on service levels. It describes the components of a complete service-level agreement and outlines methods for acquiring measures of customer satisfaction. The central themes in this chapter are the management and satisfaction of a customer's operational expectations.

TACTICAL AND OPERATIONAL CONCERNS

The operation of the computer center lends itself to a disciplined management approach. Careful management of an array of disciplines permits the organization to achieve success in this critical area. Difficulties in computer center operations can be quickly observed by a large number of the firm's employees. Failure of the central processing unit supporting the office system, for example, will be noticed promptly from the executive suite to the remote warehouse. Production operations is a vital and highly visible function.

In contrast to production operations, an error or an omission in the strategic planning process, while perhaps more devastating to the firm in the long term, may not become apparent for months or perhaps years. Production operations managers' lives are dominated by what is happening today and what will happen next week or next month. It is mandatory that they have a management system that provides the tools essential to their success. More than other managers, they need to rely on procedures and disciplines to cope with the wide variety of challenges that may come their way.

The first chapters in this text stressed the need for IT managers to meet expectations held by senior executives in the firm. The fulfillment of executives' expectations, whether or not they are realistic, is the principal standard against which IT managers are measured. We also learned that expectations are established through a variety of means. Some expectations are developed from sources external to the firm: from meetings with vendors, articles in the trade press, and meetings of business associations. These externally developed expectations are reinforced in the minds of executives by a number of internal factors, such as the amount of resources devoted to information technology. Expectations may not always be soundly based; in some cases, unreasonable expectations are formulated by the IT organization itself. It is essential that the organization utilize a management system designed to cope with the process of setting and meeting realistic expectations.

Tools, techniques, and processes designed to deal with expectations related to the application portfolio were discussed in the previous part. These processes considered the tactical period, but they had strategic implications. This chapter concerns itself with tools, techniques, and processes for developing and managing customer expectations in the operation of these applications, processes that apply principally to the short term. The concepts are formalized processes for achieving service-level agreements between the IT organization and all its clients. These agreements establish acceptable levels of service and include mechanisms for illustrating the degree to which service levels are attained.

One objective of the service-level agreement process is to ensure that the entire firm has a clear knowledge of what is expected from the computer center operation. The process must also provide an obvious means for recording and publicizing service levels delivered. Objectives always change over time, and some are not completely fulfilled all the time. Given this reality, the goal of this chapter is to present a means to focus on how well the organization is working to achieve its objectives. Clear visibility on this issue will terminate the sometimes endless discussions within the firm concerning what the measurements really are. The process being developed establishes service-level agreements and reports service levels achieved.[3]

THE DISCIPLINED APPROACH

Service-level agreements (SLAs) are the foundation of a series of management processes collectively called "the disciplines."[4] A discipline is

---

a management process consisting of procedures, tools, and people organized to deal with an important facet of the production operation within the IT organization. The word *discipline* is used because frequently it is easier in the short range to avoid the subject; discipline is required to maintain operational control. Figure 11.1 depicts the relationships among the processes that comprise the disciplines of production operations.

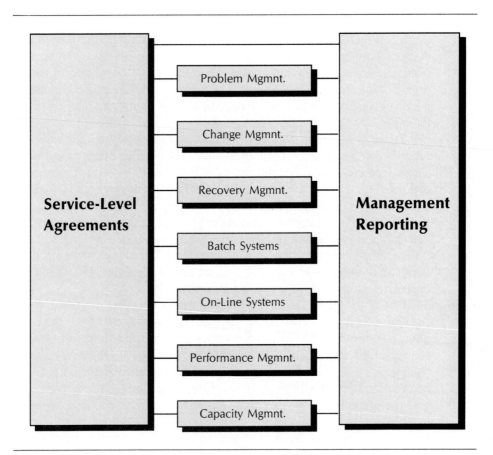

Service-Level Agreements

- Problem Mgmnt.
- Change Mgmnt.
- Recovery Mgmnt.
- Batch Systems
- On-Line Systems
- Performance Mgmnt.
- Capacity Mgmnt.

Management Reporting

FIGURE 11.1   The Disciplines of Production Operations

Computer center performance can best be judged against specific, quantifiable service criteria. Negotiated service agreements establish user-oriented performance criteria so that performance assessments can be made. Service control and service management depend critically on these criteria.

Computer center performance is made visible to management through specific reports. Service-level agreements involve all client organizations directly, but management reports are focused on specific organizations. Management reports are an essential tool for production operations managers because they organize and present the results of the intermediate disciplines. Managers in user organizations depend on them also. The reports provide information to users on the status of the service they are receiving, the actions being taken to correct any problems they may be having, and progress being made on improving system performance or capacity.

In order to provide good service, operations managers must control all other aspects of the computer center. The items that must be considered are:

1. Control of problems. Problems, defects, or faults in operation lead to missed service levels and expenditure of additional resources.

2. Control of changes to the system. Change control is vital because mismanaged changes lead to problems and reduced service.

3. Recovery from service disruptions. Managers must plan to recover from the inevitable faults or defects in system performance that occur.

4. Batch and on-line operations. Current batch and on-line workload must be scheduled and monitored and the results delivered to user organizations.

5. Performance analysis. System throughput must be maintained at the level that was planned to meet service agreements.

6. Capacity planning. Capacity can be planned correctly only when the preceding disciplines are operating effectively.

All of these topics are essential to the successful operation of a computer center. They are considered in more detail in subsequent chapters. Collectively, these processes or disciplines provide the tools and techniques required for successful management of computer operations.

SERVICE-LEVEL AGREEMENTS

The service-level discipline is a management process for establishing and defining levels of service provided by IT. The objective of the IT organization is to deliver service at established levels to client organizations in the firm. Service planning is the basis for tactical planning in the operations arm of IT. The process that establishes service levels involves a significant degree of negotiation between organizations. Users, their managers, and the firm's senior managers must be involved.[5] In most cases it is an iterative process because of the dynamic nature of the business, changing user requirements, and new technology. Additionally, the process becomes evolutionary as expectations are developed and leveled.

The process culminates in a mutually acceptable agreement that documents service levels to be delivered to each IT client.[6] The purpose of the agreement is to ensure understanding by the client organization using the service and the IT organization providing the service. The agreement discusses all aspects of service expected by the client. During the negotiation phase, the balance between service levels and the costs of providing service needs to be explored in detail. Both parties must strike an effective balance between these issues for the agreement to be totally satisfactory. A properly constructed service-level agreement establishes a cost-effective means for client organizations to use IT services.

The agreement must outline methods to measure and report service levels objectively. The cost of providing service to the user organization must be justified by the responsibilities discharged by the client organization on the firm's behalf. For example, the costs of providing improved on-line response to computer-aided design system users must equal or exceed the value of the improved productivity resulting from the response improvements. Much of the cost/benefits resolution should have occurred during the tactical planning process.

Although the initiative for establishing service-level agreements must be taken by the IT organization, client line managers and IT managers must negotiate the agreement. In especially complicated and difficult cases the negotiation may occur between the CIO and other executive managers. A higher level of attention is required when disagreements cannot be resolved at lower levels because of valid differences of opinion or insufficient breadth of vision. Frequently senior executives are better positioned to define required service levels and justify additional expenses. This is especially likely in cases where costs and/or benefits are intangible or difficult to quantify.

Services provided by IT must be expressed in terms meaningful to client managers. For instance, response times measured at the user's terminal are more meaningful to clients than CPU seconds measured by IT. Turnaround time (the time from job submission to job completion) must be measured at the user's workstation, not at the mainframe.

All client organizations must be included in the SLA process, and nearly all computer operation services should be considered. Applications that are infrequently executed or run on an as-required basis, or that place low demands on the resources of the organization, need not be specifically included in the agreement. Applications widely utilized in the firm require special attention. An earlier chapter stated that each application requires an owner-manager who takes charge of the application asset and discharges ownership responsibilities. One of the owner's responsibilities is to negotiate service levels with IT managers on behalf of all application users. It is the owner's responsibility to use cost-effective trade-offs to achieve a balanced agreement with IT management. The most effective service-level agreements are achieved when both managers exercise corporate statesmanship and keep the interests of the firm foremost in their minds during the negotiation process.

WHAT THE SLA INCLUDES

The service-level agreement normally begins with administrative information.[7] This includes the date the agreement was established, the agreement duration, and the expected date for renegotiation. It may specify unusual conditions, such as significant workload changes, which will mandate renegotiation. The agreement includes key measures of service required by the client organization and service levels to be delivered and measured by IT. Resources and associated costs required by IT to deliver the service are included.[8] A mechanism to report actual service levels delivered is also described in the agreement. Table 11.1 lists the elements of a Service-Level Agreement.

TABLE 11.1 Service-Level Agreement Contents

Effective date of agreement

Duration of agreement

Type of service provided

Measures of service

    Availability

    Amount of service

    Performance

    Reliability

Resources needed or costs charged

Reporting mechanism

Signatures

Negotiation of service-level agreements occurs when the operational plan is being prepared. At this time, resources for the coming year are being allocated and near-term requirements for IT services are becoming clear. The effective date of the agreement usually coincides with this portion of the planning cycle. For services that are relatively stable in operation, the duration of the agreement will probably be one year. Payroll or general ledger applications are examples of such systems. For applications of this type, the agreement will be reinstated during the planning cycle one year later.

Applications that have volatile demands or growing processing volumes may require more frequent SLA renegotiation. For example, a new strategic system experiencing great implementation success may have rapidly growing demand for IT services. It may require a renegotiated service-level agreement every six months. The agreement should spell out the circumstances that trigger renegotiation. Figure 11.2 illustrates these concepts in a partially completed SLA.

Figure 11.2 illustrates the administrative information for an SLA between the IT organization and the personnel function in a firm. The agreement notes that personnel will be charged for the actual service used according to rates established by the firm's controller. (This is not a fixed price contract.) The manner in which firms recover IT costs vary widely. This illustrates one of the possibilities. The subject of IT charging methods is covered in detail in Chapter 15.

The agreement is negotiated in mid-October, but does not take effect until the beginning of the next year. In this case, the negotiation took place during the planning cycle. It becomes effective with the new year.

SERVICE-LEVEL AGREEMENT

The purpose of this agreement is to document our understanding of service levels provided by _IT Computer Operations_ to _Personnel Function_. This agreement also indicates the charges to be expected for the service described herein and the source and amounts of funds reserved for these services. This is not a fixed price agreement. Charges will be made for services delivered at rates established by the controller.

The date of this agreement is _October 15, 1990_.

Unless renegotiated, this agreement is in effect for 12 months beginning _January 1, 1991_.

FIGURE 11.2   SLA Administrative Information

SLAs must specify the type of service required. The service may refer to batch operation on a regularly scheduled basis, batch runs on demand, or complex on-line transaction processing applications. The IT organization should negotiate the agreements by class of service since capacity, performance, and reliability depend on the total demand levels in the overall organization. Figure 11.3 is an example of service specification.

Specific type of service	Application service	Hours of use
Batch processing	*Payroll application*	*2 hr/week*
On-line data entry	*Payroll application*	*daily**

System availability

Monday – Friday   Payroll input data entry will be available during normal working hours Monday through Thursday and Friday 8:00 through 24:00.

Saturday   Payroll will run around 03:00 Saturday morning and will be available to personnel 08:00 Saturday.

Sunday   Emergency service only.

System performance

* Four special terminals in the personnel area will be used for this work:, and the general agreement to subsecond response time is part of our understanding.

FIGURE 11.3   **SLA Service Specification**

Figure 11.3 illustrates the service to be delivered to the personnel function for payroll processing. It specifies data entry during the week with extra hours available on the last work day of the pay period. This occurs from terminals physically located in the personnel area for security reasons. Subsecond response times at the terminals and availability of the payroll output is documented.

Schedule and Availability

The schedule describes the period during which the system and the application are required to be operational. For on-line applications, the schedule should describe the availability throughout the day. It must also outline the conditions for weekend and holiday availability. The IT manager must allow for scheduled maintenance or make provision for alternate service during maintenance. The client manager should focus on peak demand periods, which may include weekends at accounting period close and other significant events. Both parties should consider reduced availability or reduced performance during off hours, if these considerations yield cost savings. Likewise, during especially critical periods for the client it may be attractive to negotiate exceptionally high service levels. A clear understanding on the part of both parties regarding requirements and capabilities will lead to the most effective agreement.

Timing

For batch production runs, the job turnaround time is usually a key measure. (Turnaround time is the elapsed time between job initiation and availability of output to the customer.) Many batch production runs are scheduled through a batch management system. This system accounts for the many interrelations between the applications and the databases. The data used as input to one job may have been created as output from several other jobs, hence the timing of the jobs becomes crucial. Clients should understand these dependencies. In most cases, they will develop a good understanding of scheduling constraints through the negotiation process. Frequently, large production runs occur overnight. In many instances, output availability from all overnight jobs at the beginning of the work day is the critical measurement. The output may consist of reports delivered to the client offices or, more likely, on-line data sets available for the user's review through personal workstations.

The most critical parameter for on-line activities is response time. (Response time is the elapsed time from pressing the program function key or the enter key to an indication on the display screen that the function has been performed.) Response time is highly dependent on the type of function being performed. It should therefore be specified by type of service. For example, trivial transactions should have subsecond response; i.e., the terminal operator should be unconstrained by the system when performing simple transactions. Trivial transactions require so little processing that the response appears to be instantaneous, on the order of several milliseconds or less. If subsecond response time cannot be achieved for trivial transactions, productivity drops substantially and the system becomes a source of annoyance to the user.

Many transactions are distinctly nontrivial. For instance, an on-line system for solving differential equations may provide subsecond response time when users enter parameters, but the numerical methods utilized for obtaining the solution may require the execution of many millions of computer instructions. The client usually understands this type of condition and appreciates the response time required for solution. Many other situations, particularly applications that query several databases to search for specific data, are decidedly nontrivial even though the query stated by the customer appears simple enough. Everyone concerned needs to understand these situations thoroughly prior to entering into agreements. User education through the information center can achieve the necessary level of understanding.

During the negotiation process, the IT management team must provide service availability and reliability information. An example of an availability statement is: "The system will be available 98 percent of the time from 7 A.M. to 7 P.M. five days per week." An example of a reliability statement is: "The mean time between failures will be no less than 30 hours, and the mean time to repair will be no more than 15 minutes." The availability and reliability measures must be unambiguous and easily measured by users and IT personnel. The remaining items in the SLA with personnel are shown in Figure 11.4.

Reporting process Reports will be made to personnel on a quarterly basis documenting the service levels delivered on the payroll application.

Financial considerations Terminals will be charged at the rate of $5 per hour of use. Batch processing will be at the rate of $35 per CPU minute. We estimate this will cost approximately $90:,000 during this 12 month period. Corporate overhead will absorb these charges.

Additional items: none

Signatures

by:_____ by:_____
                  IT organization   *Application owner*

FIGURE 11.4   **SLA Final Items**

Workload Forecasts

Workload forecasts and required IT resources are necessary ingredients of all service-level agreements. Production capacity installed by the operations department must be sufficient to contain the processing load generated by all applications in total.[9] The capacity must be sufficient to deliver the service promised to all the clients. IT workload should be defined for the period of the service agreement in terms that can be understood by client managers. These workload statements must be translated by IT managers into meaningful and measurable units for IT capacity planning. The forecast should cover batch workload volumes, volumes of on-line transactions, amount of printed output, and other resource

---

demands required by the client and provided by IT. In many cases, average workloads will suffice. But the negotiations must cover changes in workload if significant daily, weekly, or monthly departures from the average are expected.

On-line transaction volumes can vary by an order of magnitude during the day. They frequently display a morning peak, a lull in activity around noon, and a peak again during the afternoon. In some cases, the peak loads generated by one application occur during the valley in another application's load. Firms operating nationally experience workload waves throughout the day. Departments on the east coast open for business three hours before departments on the west coast. The air traffic control system in the United States illustrates this phenomenon, as flights begin to depart from east coast airports around 7:00 A.M. local time, hours before airport activity picks up on the west coast. Workload follows the sun in international operations.

The coincidence of workload peaks from many applications is a more critical issue for IT managers. Month-end closing of the financial statements is an example of an activity that usually generates this form of peak activity. In some cases, a number of activities conspire to form a huge bubble of workload for the IT organization. One instance of this phenomenon usually occurs at year end. Not only are the December closings in process, but other year-end activities are taking place as well. Year end is also a convenient time to implement new applications to supply the customer with new and different functions for the coming year. Many times these new applications are mandated by annual changes in the tax laws or by other government regulations. If the fiscal year differs from the calendar year, the workload pattern will have yet another form. Satisfactory service-level agreements are established with all these events in clear focus.

Unanticipated or unusual increases in workload are a frequent cause of missed service levels. Generally, increases in load do not come without warning. An alert organization will reopen the discussion on service levels when such increases are first anticipated. Everyone benefits from a good job of workload planning and generally all organizations suffer from one organization's poor planning. Sometimes the workload is difficult to predict because of changes in the organization's needs or changes in the organization itself, via acquisition or consolidation, for example. In these cases a reanalysis of the load and a new forecast should be prepared for the client organization.

Reasonably accurate forecasts are essential for providing satisfactory service, although the process for obtaining workload forecasts is frequently tedious and time-consuming. History is a good guide for most forecasts.

IT should provide clients with current workload volumes and trend information, particularly if IT has a cost accounting or charging mechanism installed. IT charging or cost recovery mechanisms are valuable in the SLA process because they help focus on cost-effective service levels. (IT accounting processes will be discussed in Chapter 15.) It is mandatory to IT and users that the load analysis and load forecasting proceed successfully.

Measurements of Satisfaction

Key service parameters contained in the service-level agreements must be routinely and continuously measured and reported. Specific quantifiable parameters are preferred to more ambiguous measures. In general, if the IT organization reports these critical measurements ambiguously, or if the reported data contains errors of fact, the client organizations will develop their own tracking mechanisms. Uncoordinated measurements and reports usually lead to conflict, result in accusations and finger-pointing, and needlessly consume large amounts of energy. The most effective approach is to provide unambiguous, credible reporting techniques at the outset.

An additional consideration must be addressed. Although it appears obvious, IT must measure and report service as it is seen by the user. It doesn't help to report job completion time if the output isn't available to the client organization until later! It only confuses matters to measure on-line transaction response times at the CPU. What the user at the terminal sees is all that counts.

User response times can be obtained easily if the computer operations department has a personal computer programmed to execute a representative sample of interactions with the CPU. The PC can log transaction data and report statistics on response times. It can be switched to various control units on the I/O channels, and it will provide valuable data on response time as seen by the user. This device can be the benchmark for establishing and monitoring on-line response times.[10]

It is essential that the measurement system employed by IT have high credibility with client organizations. This credibility leads to mutual trust and confidence and results in objective discussions when events don't transpire as planned. For any number of reasons, small excursions from the norms occur. These deviations are corrected most expeditiously in an atmosphere of open communication filled with mutual respect. Precision measurements of delivered service build trust and confidence between the IT organization and its users. Objective and credible reporting on service-level achievements paves the way for improvements in other activities.

Periodically the IT organization should solicit informal customer opinion on service-level satisfaction. Such surveys provide valuable insight into the customer's perceptions of IT performance.[11] Measured performance against the service-level targets should be tempered with user perceptions of service for maximum utility. It is not enough for service to meet stated criteria; the perception of service must also be satisfactory. IT managers must know the degree of user satisfaction, and the best approach is to ask the users directly.

Several avenues can be taken to obtain this information.[12] One method is to conduct an opinion poll of on-line users periodically through a brief questionnaire on their terminal screens. The questionnaire appears at the conclusion of the terminal session and asks for the user's perception of service delivered during the preceding period. To be effective, the questionnaire must be optional, anonymous, and well designed. A second approach is to survey a wider audience over a longer period and include a broader scope of questions. The results of these surveys are very useful in detecting problem applications. The results can uncover areas where service is poor or where perceptions of service are poor. They are useful for detecting incorrectly established service agreements, and for discovering users who are not sufficiently educated in the service-level process.

Client surveys should be used to improve internal service measures and to relate internal measures to perceived and actual customer service. The goal is to achieve a high correlation between internal and external service measures. Attaining this goal results in resource optimization, efficiency improvements, increased customer satisfaction, and reduced costs. IT organizations have an abundance of tools to perform this task.

Client managers and IT managers collectively should take action to correct unsatisfactory situations detected by the surveys or measurements. Managers can use the survey and measurement data to establish more effective relationships between users and providers of service. Their actions may take the form of better communication, better service, or more education. Both organizations benefit from such actions.

ADDITIONAL CONSIDERATIONS

In a complex environment, there are usually some additional operational items that must be included in the service agreement. During infrequent critical periods, some departments may need dedicated processors, large amounts of auxiliary storage, or other unusual services. Other departments may require special handling for data entry, or they may require specific analyses to be performed on a one-time basis. Prudent IT managers recognize the importance of responding enthusiastically to these vital client requirements, even though the service provided is beyond the service committed. Satisfying customers must be a priority goal for every IT manager.

The service-level agreement process detailed here is not necessarily endorsed by everyone. In some firms the process is frustrated by political considerations. It may be contrary to the corporate culture. Strong-willed managers in some organizations prefer to demand service without regard to cost. These managers believe their function requires high service levels purely because they perform important functions for the firm. They may believe that other organizations in the firm exist to serve their function, and that IT must do so also, no questions asked. Some managers abhor the scrutiny the service-level process brings to their activities. They prefer to conduct their business in an atmosphere of near secrecy. This attitude can seriously jeopardize the successful functioning of computer operations.

In some firms, the corporate culture is not conducive to the detailed planning and commitment process inherent in the service-level agreement discipline. Discipline is required; some firms just don't have it. These firms tend to behave in an *ad hoc* manner, reacting to situations as they occur. IT managers and other managers in these firms are continuously subjected to a myriad of conflicting forces that make a smooth operation nearly impossible. The service-level agreement process requires cooperation among many of the firm's managers. In this sense, the responsibility for successful production operations goes beyond the IT organization.

CONGRUENCE OF EXPECTATIONS AND PERFORMANCE

The goals of the SLA management process are to attain mutually acceptable levels of expectation, to develop an atmosphere of joint commitment, and to foster a spirit of trust and confidence between organizations. In this atmosphere, essential business information flows freely and openly.

It is not always easy to achieve these goals. For example, how can the firm handle the dilemma that occurs when users demand better service than IT can deliver?[13] Should the SLA describe service that can't be achieved, or should the SLA describe achievable service that is unsatisfactory to the user? One way to resolve this problem is to consider affordable costs. Through management processes described earlier such as strategizing, planning, budgeting, plan review processes, and steering committee actions, affordable IT costs should have been developed. If the firm can afford more capacity, it should be procured and the SLA developed accordingly. If costs limit capacity, then users' expectations and the SLA must reflect this consideration.

In some cases congruence of expectations and performance is achieved only after one or two plan cycles have elapsed. For firms that are just starting to systematize operational activities, the first iteration may not be perfectly satisfactory to everyone. For most firms the service planning process is not new.[14] It is an integral part of their planning cycle. For these firms, leveling expectations and performance is a structured and ongoing process.

Successful implementation of the service-level process is required for effective and efficient functioning of computer operations. SLAs are the cornerstone of the disciplined process leading to success in computer operations. Success in this area is critical for IT managers.

SUMMARY

Service-level agreements are the foundation on which the management systems of production operations rely. Service-level agreements are essential and valuable ingredients of the IT management system. But they are important for other reasons as well. The process of negotiating the agreements also develops and enhances understanding and mutual respect between the users and the service providers. The IT organization will develop a much clearer understanding of the business and what is required by the users to accomplish their missions for the firm. The client organizations will develop a better understanding of the opportunities and constraints inherent in the IT organization. A high level of respect and increased trust and confidence mutually benefits all organizations and the firm overall.

Review Questions

1. What are the elements of the disciplined approach to managing production operations?

2. How do SLAs deal with the issue of expectations in production operations?

3. What is the relationship of the SLA discipline to the other disciplines?

4. What is the purpose of the SLA? Who participates in establishing it?

5. What are the ingredients of a complete service-level agreement?

6. What is the distinction between turnaround time and response time?

7. Why can the concept of a service-level agreement be considered normal business practice?

8. What considerations should be used to break ties in the event of a deadlock during the SLA negotiation process?

9. Why are workload forecasts necessary for a satisfactory SLA process?

10. What difficulties arise in selecting a unit of measure for workload forecasts?

11. Why is it essential for the IT organization to measure and publish its service-level performance in a highly credible manner?

12. What are the benefits to the IT organization of conducting user-satisfaction surveys?

13. What are some of the social or political benefits that accrue to the firm from the SLA process?

14. What modifications to the SLA process may be valuable for strategic information systems?

Discussion Questions

1. Many of the activities of production operations are operational or tactical. Some are strategic or long-range in nature. What are some examples of the strategic activities?

2. Discuss the financial considerations that must be addressed as part of the process of establishing the SLA.

3. The business vignette in Chapter 3 implies that an important step in gaining line management involvement in MIS is solidifying the process of MIS/line management cooperation. How do the disciplines, particularly the SLA, help accomplish this goal?

4. Customers of information technology organizations frequently state: "I can't predict my demands on IT more than three months in advance." How would you handle this situation?

5. Discuss the relationship of the service-level process to the concepts of tactical and operational planning discussed in Chapter 4.

Assignments

1. Design a screen that appears at log-off time to obtain customer feedback regarding the on-line service provided by the IT organization.

2. Contact a service business firm in your area and investigate its service-level process. Does it have the equivalent of the SLA? Does it measure and report service levels? If customer satisfaction surveys are conducted, how are they handled?

ENDNOTES

[1] Byron Belitsos, "A Measure of Success," *Computer Decisions*, January 1989, 48.

[2] Security Pacific employs more than 40,000 and earned $655 million in 1990.

[3] Edward A. Van Schaik, *A Management System for the Information Business: Organizational Analysis* (Englewood Cliffs, NJ: Prentice-Hall, Inc., 1985), 162.

[4] This nomenclature stems from the author's experiences with these management processes.

[5] David R. Vincent, "Service Level Management," *EDP Performance Review* (April 1988): 3. Well-executed service-level agreements improve technology integration and assist users in finding more IT opportunities.

[6] See the appendix to this chapter for a sample service-level agreement.

[7] C. N. Witzel, "Service-Level Agreements: A Management Tool for Technical Staff," *Journal of Capacity Management* (June 1983): 344.

[8] John P. Singleton, Ephraim R. McLean, and Edward N. Altman, "Measuring Information Systems Performance: Experience with the Management by Results System at Security Pacific Bank," *MIS Quarterly* (June 1988): 325.

[9] Edward M. Ziska, "Customer Satisfaction and Capacity Planning," *EDP Performance Review* (November 1987): 1.

[10] Response times for trivial transactions at the user's terminal can vary significantly from those measured at the CPU. Response time at the CPU is mostly a function of application complexity, system loading, or contention for secondary storage devices. Response time at the terminal includes delays due to processing within the terminal, contention at the terminal control unit, loading of the telecommunications links, and delays in the communications controller or processor.

[11] Sophisticated firms conduct statistically significant customer satisfaction surveys daily and compare the results to internal service measures. The internal service measures are refined and improved based on survey data. These firms are able to improve service and to reduce costs significantly through improved resource utilization.

[12] Robert E. Marsh, "Measuring and Reporting User Impact of Substandard Service," *EDP Performance Review* (March 1989): 1. Marsh develops a service-level indicator that measures responsiveness, availability, and reliability. The article describes its construction and use.

[13] John Vacca, "Service-Level Agreement User Experience," *EDP Performance Review* (October 1987): 1. This article describes experience with service-level agreements at TRW.

[14] Kathryn Hayley, "CIO Challenges in the Changing MIS Environment," *Journal of Information Systems Management* (Summer 1989): 8. This article states that departments having service-level standards range from about 88 percent in the insurance industry to about 54 percent in the energy industry.

REFERENCES AND READINGS

Singleton, John P., Ephraim R. McLean, and Edward N. Altman. "Measuring Information Systems Performance: Experience with the Management by Results System at Security Pacific Bank." *MIS Quarterly* (June 1988): 325.

Van Schaik, Edward A. *A Management System for the Information Business: Organizational Analysis*. Englewood Cliffs, NJ: Prentice-Hall, Inc., 1985.

APPENDIX   Sample Service-Level Agreement

SERVICE LEVEL AGREEMENT

The purpose of this agreement is to document our understanding of service levels provided by _____ to _____ .
This agreement also indicates the charges to be expected for the service described herein and the source and amounts of funds reserved for these services. This is not a fixed price agreement. Charges will be made for services delivered at rates established by the controller.

The date of this agreement is _____ .

Unless renegotiated, this agreement is in effect for 12 months beginning _____ .

Specific type of service	Application service	Hours of use

System reliability _____

Reporting process _____

Financial considerations _____

Additional items _____

Signatures

by: _____   by: _____
        IT organization   *Application owner*

12 Problem, Change, and Recovery Management

A Business Vignette
Introduction
Problem Definition
What Is Problem Management?
  The Scope of Problem Management
  The Process of Problem Management
  The Tools of Problem Management
  Problem Management Implementation
Problem Management Reports
What Is Change Management?
  The Scope of Change Management
  The Change Management Process
Reporting Change Management Results
What Is Recovery Management?
A Business Vignette
  Emergency Planning
Contingency Plans
  Crucial Applications
  Environment
  Strategies
Recovery Plans
Summary
A Business Vignette
Review Questions
Discussion Questions
Assignments
References and Readings
Appendix A
Appendix B
Appendix C

HINSDALE, ILL. It began on a Sunday afternoon as a small electrical fire in an unattended Illinois Bell central telephone office and switching station in this Chicago suburb. But the fire, which blazed out of control for an hour May 8, 1988—and was not put out for another five—started a chain reaction of telephone and computer breakdowns that crippled telecommunications throughout northern Illinois.

Once the smoke had cleared, salvage workers discovered that 35,000 local telephone lines were out of service and 118,000 long-distance fiber circuits had been destroyed. Another 13,000 special circuits, most of them carrying computer-to-computer connections, were also destroyed.

As phone lines went dead, so did terminals tied to computers in business sections throughout the western Chicago suburbs. Hundreds of automated teller machines in Chicago, served by suburban DP centers, went blank, as did scores of Illinois Lottery terminals. O'Hare Airport suffered a temporary loss of flight information sent from a west suburban FAA tracking station over 28 data circuits. "We had to slow things down," FAA spokesman Mort Edelstein said. "We had to separate all planes by 20 miles, instead of three miles, just for safety." The FAA switched to 16 backup lines within hours.

"One of our vendors called us from her car phone," said Tim O'Neill, a computer operations supervisor at Unocal Corp.'s Refining and Marketing Division in Schaumburg, Ill. "She could call us, but we couldn't call her." Unocal was able to switch faulty phone lines to the company's national private network, which is powered by microwave links.

Sears, Roebuck & Co.'s telemarketing division in Downers Grove, Ill., sent dozens of phone sales personnel home on indefinite leave, according to Sears spokesman Gordon Jones. Some firms were paralyzed by the phone shutdown. "Right now, we're totally out of business," said Dennis Blondell, assistant director of operations at the Floral Network, Inc. a subsidiary of the Floral Transworld Delivery Association of Florists, late Thursday evening. Blondell said the outage, which hit on Mother's Day, stopped all phone traffic and halted the computer system that connects 12,000 of the association's 23,000 florists.

Blondell, among others, wondered how a single fire was able to knock out so much of Illinois Bell's network. The answer, according to Illinois Bell

Vice-President of Operations James Eibel, is that the system has only 25% redundancy at gateways like Hinsdale. Chicago hubs, in contrast, have 50% redundancy.

The Hinsdale switching center was unattended when fire broke out at about 4:00 P.M. As a matter of policy, the alarm did not ring through to the local fire department. According to Illinois Bell spokesman Tim Bannon, the company feared that without guidance from Illinois Bell technicians, local firemen would damage sensitive phone equipment with sprayed water.

A task force of 150, including disaster specialists from Northern Telecom, Inc., AT&T and Bell Communications Research Corp (Bellcore), worked to remove humidity in the building that was combining with soot to form corrosive acid. "Replacing the components is a trivial problem," scoffed one Bellcore adviser on the scene. "But replacing the 40,000 circuit connections is a long and tedious chore. What we had here was a complete disaster."

After a failed effort to link trunk lines to temporary microwave dishes atop the Hinsdale center, the company abandoned the idea of salvaging the old equipment. Instead, they installed a fully digital switch during the next two to four weeks. Long distance service was expected to resume a week after the fire, with the replacement of 60 Northern Telecom bays of multiplexing equipment that route long-distance toll calls over fiber-optic lines.

Meanwhile, Floral Network said it plans to erect a satellite dish to bypass the local switching office in the event of another phone line emergency.

INTRODUCTION

The preceding chapter laid the foundation for a disciplined and systematic approach toward managing production operations, a critical success factor for the IT manager. This chapter develops the tools, techniques, and processes of the next three disciplines. Discussion of the production operations management system continues by considering the essentials of problem management, change management, and recovery management. The relationship of these processes to the complete management system for production operations is displayed in Figure 12.1.

The processes of problem, change, and recovery management are grounded in the discipline of service-level agreements. They form a source of information for the discipline of management reporting. Service level agreements are the foundation, and management reporting is the capstone, of the disciplined approach. Problem, change, and recovery management are some of the processes forming the management system to ensure the attainment of service levels. The remaining processes are described in subsequent chapters.

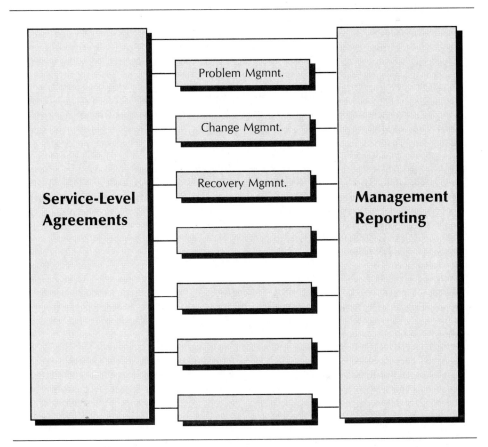

Service-Level
Agreements

Problem Mgmnt.

Change Mgmnt.

Recovery Mgmnt.

Management
Reporting

FIGURE 12.1   The Disciplines of Problem, Change, and Recovery Management

The focus of problem, change, and recovery management is on departures and potential departures from acceptable operation. The purpose of these management processes is to correct deviations from the norm and to prevent future deviations. When a problem occurs, management takes action to correct the problem and to remove the source of the problem. The problem-management discipline guides managers in taking corrective action to prevent future problems and to ensure that the same or similar problems do not recur. Frequently, system changes are introduced to increase system capability, to take corrective action on problems, or for other reasons. The introduction of change is itself a rich source of potential problems. Change management is intended to concentrate on the risk of system alterations and to provide management control for change activities. Uncontrollable events may occur that cause major problems to computer operations regardless of management diligence. Recovery management deals with these eventualities.

These disciplines are applicable not only to IT; they must be part of the management system for users. Users are vitally dependent on IT operations, and they must be intimately involved in the operational IT processes. In addition, as user departments acquire computing hardware and application software, they begin to assume responsibilities for managing problems, for controlling system changes, and for developing recovery and contingency plans. Users who own and operate equipment and applications must implement management disciplines or suffer degraded performance from poorly controlled systems. For user-owned systems, IT assumes a consultancy role. The disciplined approach to operations described in this chapter applies to IT systems and also to user-owned and user-operated systems.

PROBLEM DEFINITION

The word *problem* means different things to different people. In the business environment problems are in the mind of the beholder. They are not necessarily perceived in the same way by all observers. For the purposes of this discussion and to prevent ambiguity, problems are defined as incidents, events, or failures, however small, that have a negative impact on the ability of the production operations department to deliver service as committed in the service-level agreements. Problems include actual failures that result in service degradation or potential failure mechanisms that could lead to degraded service.

Many other kinds of problems faced by the manager of computer operations are excluded from discussion. For example, the salary budget for next year may be less than the requested amount and managers must revise their financial plans. Parking for employees may be disrupted due to temporary construction activity, thus creating problems, especially during periods of inclement weather. Management may need to resolve such issues but, unless they lead to missed service levels, these problems are excluded from the area of problem management. Many important issues face managers, but the thrust of the disciplined management processes is to provide committed customer service.

WHAT IS PROBLEM MANAGEMENT?

Problem management is a disciplined process for detecting, reporting, and correcting problems impacting the attainment of service-level objectives by the production operations department. The problem sources include hardware, software, network, human, procedural, and environmental failure that cause incidents or potential incidents of service-level disruption. Problems are mostly generated within the firm, but they may originate from external sources as well. Problems may originate in the user organization or in the IT organization. The problem management system must deal with problems from all sources that affect the delivery of satisfactory service.

Several objectives can be achieved by the production operations department through the use of the problem management system. These objectives are listed in Table 12.1.

TABLE 12.1   **Problem Management Objectives**

Reduce failure to acceptable levels.

Achieve committed service levels.

Reduce the cost of failures.

Reduce the total number of failures.

Implementation of an effective problem management system will reduce the number of unplanned incidents or events to an acceptable level and will reduce the cost of these defects. Successful implementation of the problem management system will result in achieving the service levels committed by IT in service-level agreements. The process will minimize the total number of problems requiring attention. Having to deal with fewer problems conserves organizational energy and reduces expense. In addition, the problem management system provides managers with tools for understanding the root causes of the relatively infrequent failures that do occur.

Problem management is a required process for achieving success in the management of production operations. It is one of the first steps in the management system. Problem management is a necessary condition for success, but is not necessarily sufficient for success.

The Scope of Problem Management

Problem management broadly encompasses many sources of potential and actual impacts to service levels. The scope of problem management is outlined in Table 12.2.

Failures in hardware systems are indicated by unscheduled system restarts, unanticipated or intermittent operations, or other abnormal operating conditions. These difficulties may arise from failures in I/O channels or failures in peripheral devices such as tape drives or disk storage devices. Failures can originate within the CPU as well. System software is also a problem source. Parts of the operating system may malfunction or some important function may not be available to the applications. Intermittent or unanticipated results may also originate within the system software.

TABLE 12.2   Scope of Problem Management

Hardware systems

Software or operating systems

Network components or systems

Human or procedural activities

Application programs

Environmental conditions

Other systems or activities

Most major systems are connected to networks, which are an occasional source of difficulty. Networks may function intermittently or may fail totally. The failure may originate in network hardware, network software, or operating system communication programs. Failures may result from incorrect human interactions with the network or improper procedures governing the human interactions.

System failures may occur at the person-machine boundary or may derive from improper manual procedures. The procedures themselves may lead to failure if they are incomplete, incorrect, subject to misinterpretation, or if they do not cover all situations. Operator failures may also result from not following established procedures or from incomplete execution of procedures.

The application portfolio is a frequent source of failure, particularly if it is undergoing major enhancement or maintenance activity. Chapter 7 pointed out that failures are especially probable if the applications are old and have had frequent maintenance actions in the past. The applications may yield unintended or unexpected results, or they may display unusual or abnormal program termination. The expected function may be missing or incorrect. These situations usually result in unsatisfactory performance and frequently lead to missed service levels.

The environment is another source of difficulty. Power disruptions, air conditioning difficulties, or failures in the heating system may all lead to service-level disruptions. On a larger scale, earthquakes, hurricanes, or other natural disasters may be an infrequent source of serious difficulty. In addition to the causes outlined above, if service levels are missed for any other reasons, these defects are also included in the scope of problem management. In general, if a situation impacts service or has the potential to impact service, that situation must be addressed within the scope of problem management.

The Process of Problem Management

The process of problem management includes tools and management techniques designed to detect, report, correct, and communicate the particulars of problems and their resolution. It embraces an informal organization and includes responsibilities for the members of this organization. It includes protocols and regimens for solving problems and for analyzing and reporting the results to management. Individuals and groups who are highly motivated to reduce incidents to an absolute minimum rely on the process to maintain service at the highest attainable level.

The Tools of Problem Management

The essential tools and processes of the problem management discipline are displayed in Table 12.3. Complete and effective implementation of these steps comprises the primary elements of the problem management system.

TABLE 12.3   Tools and Processes of Problem Management

Problem reports

Problem logs

Problem determination

Resolution procedures

Status review meetings

Status reporting

Problem reports record the status of all incidents. Each report begins with problem detection and records corrective actions until the issue is satisfactorily resolved. The report is initiated by the person who discovers the problem. It is filed with the individual responsible for the problem management process. The problem report includes the items listed in Table 12.4.[2]

TABLE 12.4   Problem Report Contents

Problem control number

Name of problem reporter

Time and duration of incident

Description of problem or symptom

Problem category (hardware, network, etc.)

Problem severity code

Additional supporting documentation

Individual responsible for solution

Estimated repair date

Action taken to recover

Actual repair date

Final resolution action

The problem report receives updated information during the time the problem is open or unresolved. When the incident is closed, the report is posted with the problem resolution data and filed. A record of all problems is maintained in a problem log. The problem log is a method for recording incidents, assigning actions, and tracking reported problems. The log contains essential information from the problem reports for all resolved and unresolved problems during some fixed period.[3] A convenient and useful period is the rolling last twelve months. The problem log and associated reports provide data for analysis by management. Careful analyss of prior problems is very useful for securing reductions in problem activity.

Problem Management Implementation

Problem status meetings are an essential part of the management process. Status meetings should be held regularly to discuss and document unresolved problems, to assign priorities and responsibilities for resolution, and to establish target dates for corrective action. A regularly scheduled, daily problem meeting is a reasonable approach. Severe problems require additional reviews. Some questions for these extraordinary reviews are: "How could this problem have been prevented?" and "How could the impact of this problem have been minimized?" Status meetings and reviews should focus on recurring incidents, problem repetition, and careful analysis of unfavorable trends. Areas experiencing high problem levels should be given special attention. Trend analysis will give clues to weak or vulnerable areas. Special consideration should be given to rediscovered problems because this may indicate incomplete or ineffective problem resolution.

Representatives of each business function having open problems and key individuals representing IT should participate in the status meetings. The IT representatives are members of systems programming, computer operations, and the applications programming groups. The person responsible for liaison with the hardware service organization should also be present. The problem management team leader should chair the meetings. When the process is operating effectively, it is not necessary for managers to be present. Managers are responsible for providing support to the effort and for ensuring participation and results, but generally the meeting will be more effective if participation is limited to nonmanagers. Managers tend to make assessments of people, but what is required at the problem meeting is the assessment of problems. People assessments should be reserved for another forum.

Problem determination and resolution are aided by classifying the incidents by categories such as hardware, software, and networks, and by establishing problem severity levels. The categories and severity levels should be determined by the team members. Individual problem resolution assignments should be based on documented individual responsibilities. The documentation will help clarify each individual's role and will ensure that proper skill levels are applied to problems. For example, if the problem is found to reside in the operating system utility programs, the systems programmer with responsibility for these utilities should be assigned this problem.

Problem resolution procedures include action plans and dates for resolution from the individual to whom the problem was assigned. Resolution status is reported to the team leader and to the affected areas or users. Implementation of corrective action is scheduled after review with all affected parties and with concurrence of the team leader. If the problem involves system changes, the change management process may be invoked. In any event, when it can be demonstrated that the problem has been resolved to the team's satisfaction, the problem report is closed and the log so posted. Figure 12.2 illustrates the process flow for problem management.

Automated tools can assist with the creation of problem reports and the maintenance of the problem log. Computer-based tools support interactive databases and make the task of generating reports for the team members and for management easy and efficient.[4]

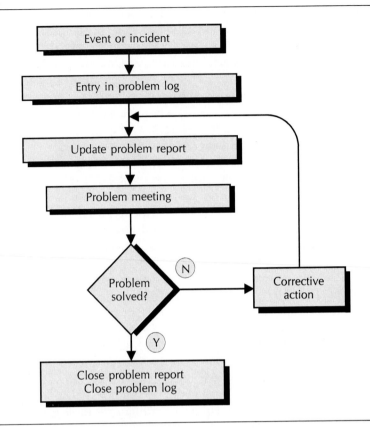

FIGURE 12.2   Problem Management Process Flow

Management's attitude toward the people involved in problem management is vital for success of the discipline. The process is designed to solve problems and to help prevent incidents. It must not be used by managers as a vehicle for individual performance assessment. IT managers must establish an atmosphere of free and open communication among all the participants. The environment should exclude finger-pointing and assigning blame. A productive environment encourages a spirit of mutual cooperation between service providers and system users. Jointly and separately the team members should feel responsible for resolving incidents promptly and for preventing incident recurrence. Taking responsibility for the resolution of an incident is not equivalent to accepting responsibility for causing the incident in the first place. Some of the most effective status meetings occur when the environment is supportive. When the environment is favorable, the presence of managers is not necessary.

PROBLEM MANAGEMENT REPORTS

IT managers must receive summary reports documenting the effectiveness of the problem management process. The summary data should reveal trends and indicate problem areas impacting the delivery of satisfactory service levels. These summary reports are useful in establishing future service levels. IT plans should account for problem activity when establishing resource and expense levels.

Problem reports may be produced weekly, monthly, or on request. The reports highlight numbers and causes of incidents by category and severity. The duration of service outages, length of time for problem determination, and time for resolution must also be reported. An aged-problem report is useful in determining the organization's responsiveness to incidents. An important derived result from the report is an analysis of duplicate or rediscovered problems. The rediscovery rate is important because it measures problem resolution effectiveness. Another additional important derived result stems from an analysis of problems caused by changes applied to the system. Problems caused by system changes can measure the effectiveness of the change management process.

Well-conceived reports enhance the process by providing concrete evidence of its effectiveness. Problem reports demonstrate IT responsiveness to IT customers. They lend credibility to service-level commitments and reinforce the alliance between the service recipient and the service provider. Problem summary information is especially useful when renegotiating SLAs. The entire process is essential to production operations success.

WHAT IS CHANGE MANAGEMENT?

Change management is a management system for planning, coordinating, and reporting system changes that have the potential to impact negatively service provided by the production operations department. It is a disciplined approach to planning, coordinating, and reporting system changes. Changes are a focus item because they have the potential to influence service delivered by the production operations department. Changes can involve modifications to hardware, software, or networks; they can originate from human processes, manual procedures, and the environment. The objective of the change management process is to ensure that system changes are implemented with minimum or acceptable levels of risk and that service levels are not jeopardized by change activity. The process foundations are service-level agreements. The production operations department may deal with many changes, but only those having the potential to impact service delivery are included in the change management discipline.

The introduction of modifications into a complex human-machine system raises significant, often unrecognized risks of disruption to system processing and reduction in service levels. Changes tend to generate problems. Frequently, well-intentioned individuals are motivated to introduce small changes in areas that may appear remote from the operation. Their intent is to improve system operation in some way. Ill-considered changes to sophisticated environments frequently lead to difficulties. A disciplined approach is required.

Rapid-paced technological advances, coupled with information technology growth within the firm, make change a way of life for the IT organization. IT and user organizations must cope with many issues when introducing change. A disciplined management approach to change implementation will reduce problem incidents and yield significant benefits to the firm.

Managing problems is tightly linked to managing changes. Problem resolution frequently requires the introduction of changes. Change management and problem management must act in concert to ensure that the solution to one problem does not result in another incident. In an undisciplined environment, it is common, on average, for the solutions to two problems to yield one additional problem requiring action. Some organizations report that 80 percent of corrective actions result in additional problems. An effective, disciplined approach has the potential for reducing the ratio from 2:1 to 20:1.[5] The resulting benefits to the firm are greatly improved effectiveness and significantly reduced costs. There are other less obvious but important advantages as well. For example, the image of the IT organization is enhanced and employee satisfaction in the user and IT organizations is improved when problems are minimized and changes are controlled.

The Scope of Change Management

Change management addresses many activities that could negatively impact the delivery of committed service levels. The major sources of changes that have a bearing on service levels are listed in Table 12.5.

Most changes to computer-based information systems or to their environment bring risk of service disruption. All changes that induce risk must be controlled with the change management process. Hardware changes of all types are candidates for inclusion. Any alterations to operating systems, network software, utility programs, or other support programs must be considered. Application programs undergoing maintenance or enhancement are prime candidates for the change management discipline. Earlier chapters noted that the application-enhancement process is error prone; it frequently leads to operational failures. Changes to application programs must be monitored carefully.

TABLE 12.5   Change Management Scope

Hardware changes of all kinds

Changes involving software

Production program changes

Changes in the environment

Procedure alterations

Equipment relocation

Problem management–induced changes

Other changes affecting service

The Change Management Process

Change management includes tools, processes, and management procedures applied to the analysis and implementation of system changes. The ingredients of the change management process are itemized in Table 12.6.

TABLE 12.6   Change Management Ingredients

Change request

Change analysis

Prioritization and risk assessment

Planning for the change

Management authorization

Changes intended for the production environment are entered into the change management process via a change request.[6] The change request document is a permanent historical record detailing the change actions from initiation through implementation. It records the type and description of the requested change and identifies prerequisite actions. This document is a record of the risk assessment, the test and recovery procedures, and the implementation plan. It is also an approval vehicle. The individual responsible for introducing the change is recorded, and management authorization is included in the document. Table 12.7 identifies the major items in the change request.

TABLE 12.7   Change Request Document

Description of the change and its log number

Problem log number if resulting from an incident

Changes that may need to precede this one

Type of change

Priority of change and risk assessment

Test and recovery procedures

Project plan for major changes

Requested implementation date

Individual responsible for managing the change

Individual requesting change if different from above

Management authorization

The change request document is developed as the change is analyzed, planned, and implemented. Upon completion of the change, the change request is filed for future reference and analysis. The change request must also be recorded in the change log.

Changes are approved for implementation by the change management team. The team is composed of individuals representing computer operators, systems engineers, application programmers, and individuals responsible for liaison with hardware and software vendors. The facility manager must be represented because changes in the physical environment (heating, power, space, etc.) are critical to the operation. The change management team is headed by one of the team members representing the operations group. It is usual and desirable to have some overlap in membership between the problem team and the change team. The change team meets regularly and frequently for short periods to analyze, review, and approve the plans for changes.

The focus of the change management team is on risk analysis. It is essential to categorize changes by potential risk to the operation. Change analysis should focus on factors that quantify the magnitude and significance of alteration to ongoing operations. For example, if the change significantly impacts system performance or capacity, or if it has a large impact on the user community, it probably carries major risk. If this change interacts with other pending changes or if it mandates future changes, it must be reviewed critically. Changes that affect more than one functional area or that require individual training are riskier than those without these conditions. A sound test and implementation plan assists greatly in risk reduction.[7] Successful teams ensure low or acceptable risk to the ongoing stream of changes.

Major or extraordinary changes require a thorough project plan that identifies people, responsibilities, target dates, and implementation reviews. Change team attention to numerous details is required for major changes. The change team acts as technical auditors in the process and, on special occasions, may request additional reviews by nonpartisan technical experts. The change will be approved only when the change team has high confidence that it can be implemented with low risk of failure. The team must also be confident that, if the implementation is not successful, the recovery plan will be effective.

The goal of the change management process is that system alterations occur with low or acceptable risk. This goal applies to all changes whether they are major and substantial or relatively minor in scope. The replacement of a large mainframe may require change planning and management over the period of a year or more. Installing the latest version of the operating system on the mainframe may take months of planning, but adding another direct access storage device may be completed in a week or so. Altering the format of a batch program may require only a day or two. In any case, all changes large and small should be undertaken only after all risks to the firm have been considered and evaluated.

Changes must be capable of being tested, and the team must evaluate the test plan for thoroughness and completeness. The test cases and expected test results are documented in advance and compared with actual results upon implementation. Variances between expected and actual results require analysis and rationalization. Variances may reveal incomplete preparation or ineffective implementation. Significant variances may indicate more problems or additional changes in the future.

REPORTING CHANGE MANAGEMENT RESULTS

Successful change implementation must be communicated to all affected parties, and the details must be entered into the change log. The log should be reviewed periodically to ascertain trends and assess process effectiveness. The assessment focus is on problem areas, expectations versus results, required emergency actions, and areas requesting or requiring numerous changes. One measure of process effectiveness is the percentage of changes implemented in a problem-free manner. This number can exceed 95 percent in a well-disciplined environment.

Periodically the production operations department must analyze trend data and summary information obtained from the change-log review. IT production managers are responsible for ensuring effective operation of the change management process. They need information from the change log to make their assessments. Like problem management, change management does not need continuous, detailed management involvement. The process needs management direction and support but not continuous intervention.

WHAT IS RECOVERY MANAGEMENT?

Information technology managers recognize that the resources they manage are critical to the firm's success. They are keenly aware that unavailability of major on-line applications will have immediate and serious consequences and that degraded operation of customer support applications jeopardizes client relations. The consequences may be loss or deferral of sales and revenue or possible damage to the firm's reputation.

An outstanding example of failure occurred on May 12, 1989, when the American Airlines Sabre reservation system crashed and was down for approximately 13 hours. The crash was caused by the failure of a utility program formatting large direct-access storage devices. The program went astray, destroying the volume serial numbers on 1080 volumes of data and making access to the data impossible. The entire system, based on eight IBM 3090-200E mainframes, was inoperable from 23:38 on May 11 until 10:29 on May 12. It gradually came to full operation in the next 90 minutes. This system failure disabled American's load management process and permitted travel agencies and corporations to book deep-discount tickets in an unlimited fashion and to rebook higher fare tickets at discounts.[8] The extent of the losses to American from these activities has not been published.

In other situations the firm may incur degraded internal performance and loss of efficiency. Vital information assets can be damaged or rendered inoperable by numerous natural or manmade actions. In the face of this reality, prudent IT managers will engage in contingency planning designed to cope with the potential loss or unavailability of information resources. Recovery management deals with the possibility that IT resources may be lost or damaged, or that they may become unavailable or unusable for any reason.

Many managers experience difficulties dealing with this topic. They view the risk of loss of the computer center from a tornado, for example, as extremely low. Or they consider the solution to this situation to be extremely complicated and difficult to manage. For these reasons and others, some IT managers rationalize that this type of planning has a low priority—low enough that little or no effort is devoted to it.

On the other hand, events such as an emergency power outage or a head crash on a direct-access storage device have a higher probability of occurrence. Management of these contigencies is easier to visualize and implement, and managers usually appreciate the need to prepare contingency plans for these events.

IT managers in financial institutions are required to have a recovery plan. Federal regulations require that banks and other financial institutions maintain disaster recovery plans on file for inspection by federal auditors.[9] The Office of Management and Budget has directed federal agency attention to this subject since 1978 through Circular A-71, "Security of Federal Automated Information Systems." Updates to this circular have been issued directing the Commerce Department to develop standards and guidelines and to assist other federal agencies in developing their plans.

IT managers are dealing with a continuum of possibilities involving likelihood of occurrence versus seriousness of the consequences. The recovery management process deals with these uncertainties and articulates actions in those cases where management believes the risk involved is worth the effort expended.

Recovery plans deal with emergency situations. They provide back-up resources and plan restoration of service from the effects of major disasters. These are plans everyone hopes never to need—they are like insurance policies. One is willing to pay the premium but hopes never to need to collect on the policy. Prudent IT managers implement recovery plans because they know that complex and vital resources can suffer from many types of loss. Recovery from minor emergencies is more than a once-in-a-lifetime experience. Recovery from major disasters is something no manager wants to experience but something for which every manager must plan.

Recovery management is one of the disciplines of production operations. It involves people, processes, tools, and techniques. Recovery management originates from the conviction that the IT organization has a commitment to defined service levels and that service levels can be effectively insured through contingency planning. Like the previous two processes, recovery management involves managers and nonmanagers representing the IT organization and the client organizations working together. Recovery management requires that managers cooperate to achieve mutual goals.

Because recovery management is a critical part of every firm's plans, IT managers must be especially attentive to recovery planning. The importance of recovery management is illustrated in the next Business Vignette.

In 1989, Hurricane Hugo and the California earthquake reminded IS professionals of an important principle: Good contingency planning can counteract the effects of natural disasters. Providing protection against disasters can also be very costly, however. It can be said that IS can survive almost any contingency, providing the plans, the equipment, the backups, and the money are put in place before the disaster. It can also be said that firms which elect not to prepare for disastrous events may need to rethink their positions.

A medium-size credit union in Charleston, SC decided on August 15, 1989, to sign a hot-site disaster recovery contract. A month later the hurricane blew Charleston apart, along with its power supplies. The credit union had a mainframe computer upon which all its operations, including automatic teller machines, depended. Victims of the hurricane needed cash.

To the rescue: the disaster recovery contract with Sun Data, of Norcross, GA. Under terms of this contract, the credit union transferred its data processing to a hot site in Atlanta, and with the help of local communication links it was able to provide service to its customers. Credit union depositors needed money to get back on their feet after the departure of Hugo, and federal officials wanted to know when funds would be available. Using the backup facility, installing some manual processes, and by cooperating with other financial institutions in the area, the credit union provided critical service to their 52,000 account holders.

Reciprocal arrangements with other similar institutions do not always work. This credit union might have set up a disaster plan with another Charleston bank, for instance—and both of them would then have been blown out of business by Hugo. One of the most common traits of major disasters is electrical power outages extending for prolonged periods over fairly large areas. Facility sharing with other nearby firms, while effective for local difficulties such as burst water pipes, is ineffective for dealing with regional disasters. Power problems usually accompany other problems, and firms are increasingly considering uninterruptible power sources which can bring down the mainframe and large data stores gracefully prior to switching to the hot site. These UPS devices, though expensive, also protect against power surges.

Sound disaster recovery planning and effective implementation of recovery plans protect the firm and its reputation, provide customer service sometimes when most needed, and indicate responsible IT management. No firm can afford to be without them.

Emergency Planning

The emergency planning process addresses situations resulting from natural disasters such as floods or wind storms and from events such as riots, fires, or explosions. Emergency plans must contain the steps needed to limit the damage and deal with the problems. Typically these events have a low probability of occurrence.

The most effective mechanisms for dealing with these emergencies involve early detection and containment procedures to limit the damage. Detection mechanisms include fire alarms and detectors for smoke, heat, and motion. In some cases liaison with the civil defense authorities is appropriate. Additionally, the plans for handling emergencies usually include procedures for evacuation, shelter, containment, and suppression.

The plan must identify evacuation conditions, establish means to communicate evacuation plans, and contain procedures to ensure that evacuation has been successfully completed. Shelter plans include protection from rain, leaking pipes, and water from other sources in addition to protection from wind damage and flying debris. The plans for containment usually involve storage rooms for fuel and other hazardous materials and for vital information such as tape volumes and critical documents. These storage facilities are constructed with fire- and water-retardant walls, floors, and ceilings. Suppression plans describe the procedures to extinguish fires, stop water flow, clear the facility of smoke, and maintain security in the process.

Emergency plans must be clearly and succinctly documented and communicated to everyone having potential involvement with emergency situations. Responsibilities must be established and directed and individuals must be trained to respond effectively. Emergency plans must be tested periodically to ensure that individuals are completely familiar with their responsibilities and that employees and managers know how to respond effectively. IT managers are responsibile for ensuring effective emergency planning.

Contingency planning ensures successful performance of critical jobs when services or resources are lost or unavailable. Contingency planning responsibility for critical applications resides with the application owner. For example, the production control manager who owns the materials requirements planning system is responsible for managing production in the event of a system hardware failure. However, the responsibility for correcting the system hardware failure resides with the IT manager, the provider of service. In this case as in most others, the owner of the application has more options and more flexibility for dealing with the problem than the service provider.

The split in responsibilities outlined above calls for more rather than less interaction between application owners and service providers. The application owner should obtain critical planning information such as probable frequency of occurrence of outages, likely duration of outage, and anticipated extent of service loss from the IT organization. Additionally, IT may recommend helpful action plans to the application owner. As with the disciplines discussed earlier, all affected parties must work together in a disciplined manner to serve the firm's best interests.

Crucial Applications

IT managers, in consultation with application owners, should identify the most critical applications in the firm's portfolio. This is not an easy task, but it is very important. It is not easy because the critical nature of an application can only be seen in context with other applications. That is, an application may or may not be critical depending on whether other applications are available. For example, order processing for the manufacturing plant is an important function, but on-line order entry may be more critical. Failure in order entry means loss of business, but failure in order processing may only delay start of production. Interactions among applications are important in determining which applications are critical.

Timing of the emergency is another factor to consider. The hardware and software supporting on-line applications such as air traffic control, oil refineries, and reservation systems, for example, are always critical. The payroll application is generally critical weekly, but the ledger may be critical more frequently. Finally, the expected duration of the outage is another complicating factor. Most applications become critical if the service outage is prolonged.

There are two reasons for performing the analysis in spite of the obvious difficulties and ambiguities. The first is to know which applications to recover first. In case of widespread failure, a predetermined priority sequence organizes the recovery work and helps prevent chaos. The second reason is to understand the possible trade-offs. If hardware or telecommunications capacity is the problem, then it is very useful to know what load can be shed in favor of higher priority applications. Definition and analysis of these trade-offs must include system owners and users. A well-organized approach directs users' attention to the most critical tasks.

Environment

Each critical application requires a specific environment for successful operation. The important environmental factors are identified in Table 12.8.

TABLE 12.8 Application Environmental Factors

Hardware system components

Operating system and utility software

Communications resources

Databases

Sources of new input data

Knowledgeable users

Critical support personnel

Plans must be developed for each important application that include all the critical environmental parameters. The interaction between the critical applications and all other related applications or systems must be developed as part of the plan. This planning effort may be quite large. Prior to undertaking this work managers must understand the trade-offs between planning costs and the cost of loss or operational failure.

Strategies

Many options are available to contingency planners. Among others, these options include manual operations, backup systems, or data servicers.[11] Backup systems may be located at the same firm or at cooperating firms. The firm may also utilize the services of so-called "hot site" providers. These enterprises maintain operating hardware and software

---

configurations for use in emergencies.[12] The credit union in the previous vignette survived Hurricane Hugo by using a remote hot site.

Disaster recovery services businesses are large and are expected to experience growth rates of 20 percent per year through 1995 in response to customer demand. According to a 1989 report prepared by the Ledgeway Group in Lexington, MA, the disaster recovery market will exceed $1 billion by 1995. Services consist of hot site providers, contingency planning and consulting services, and software systems for disaster recovery planning.[13] Many vendors are willing to work with IT managers on disaster planning and recovery.

Resorting to manual procedures is usually the least effective response to disaster. If the system is relatively simple and the outage of short duration, manual processing may be effective. But in many instances manual processes are impossible to implement and completely ineffective. For example, manual processing will not sustain an airline in event of a reservation system failure nor will it help engineers when their computer-aided design system fails. However, with relatively uncomplicated systems and for short outages, some form of manual fallback is usually present. For example, with a low-volume, on-line order-entry system, clerks can manually complete order forms while the system is recovering from a brief outage. When the system returns to normal operation, the data on the forms is entered into the on-line system. Short-term manual methods may suffice for simple, noncritical systems.

For severe outages, some form of backup is required for all systems. A multiple system with distributed architecture may be effective if the firm has geographically separated processing centers linked with broadband communications lines. The distributed architecture can be particularly effective if there is a reasonable degree of hardware and operating system compatibility. Compatibility is very important for backup or recovery purposes and may be desirable or necessary for other reasons also.

Telecommunications networks are potentially effective in transmitting programs and data between sites for backup purposes and for communicating results during recovery. IT managers, sensitive to recovery management issues, strive for this architecture.

Arrangements can sometimes be negotiated with cooperating firms to mutual advantage. Each firm needs to provide for contingencies. If technical details receive sufficient attention, cooperation may be mutually advantageous. Documents of understanding outlining the terms and conditions of the cooperative agreement must be prepared and signed by representatives of the firms. Effective agreements outline procedures for testing the backup and recovery procedures and for implementation in event of need.

Some firms provide data processing centers for others' use in the event of a severe outage. These "insurance installations" write a policy that provides fee-based system capacity for use in an emergency. Technical issues of hardware and software compatibility and program and data logistics must be addressed. Network availability and capacity must also be factored into the decision to use these services. In addition, some major firms provide disaster planning tools; others provide total disaster planning services for organizations unprepared to undertake this activity for themselves.

Service bureau organizations are another alternative in the recovery-planning process. These data processing organizations are frequently well equipped to handle additional processing loads. Your firm may already employ service bureaus for peak-load processing or for special applications, and it would be natural to develop an emergency backup arrangement with them. In this instance, the logistics of backup and recovery may have been partially developed, thus simplifying the situation considerably.

Telecommunications systems require special contingency planning. Special attention is required because these systems are usually highly critical to the firm and because they offer unique planning opportunities for critical applications.[14] Firms with multiple locations linked together via telecommunications networks must consider two issues. The first issue is how the firm maintains intersite communications in the event of network disruption; the second is how the firm utilizes the intersite network to assist in backup and recovery for its individual locations.

There are several important considerations in maintaining intersite network systems. Network redundancy, alternative routing, and alternative termination facilities are usually considered first. Modern networks usually provide ample opportunity to exploit these alternatives. In an emergency, it may be possible to exchange voice and data facilities or to utilize the traditional dial network for data. Some firms employ value-added network firms for backup purposes. As an absolute last resort, manual processes may be considered. Manual processes for network operations are very unattractive even for emergency purposes.

Using the network for major application hardware and software systems backup and recovery must be addressed as part of the firm's system architecture. If recovery management is an architectural consideration in network design, considerable advantage can be obtained in recovery planning, and usually significant side benefits are realized as well. For example, networks that exhibit recovery advantages usually permit efficient load sharing with attendant cost reductions. Networks pose special problems and offer special opportunities for the recovery management planner. Because of their critical role, networks require special consideration in the recovery planning process.

RECOVERY PLANS

The goal of the recovery management process is to develop, document, and test action plans covering the contingencies facing IT and the firm. The actions must include all portions of the organization that engage in information technology activities.[15] The firm must be prepared to cope with adversity in a logical, reasonable, and rational manner. The firm must protect its major information technology investments located throughout its constituent parts.

Individuals in the IT organization are a critical and perhaps indispensable resource. Because of their unique knowledge, skills, and abilities, IT personnel are especially vital in times of emergency. Most recovery plans assume that people will survive the disaster and will be able to carry out recovery activities. But sources of outside help should be considered in the planning, since this assumption may not always be valid. The personnel strategy must address the availability of required skill levels from within the firm and from external sources. Recovery plans must include provisions for notifying key people and for managing work assignments during the recovery process.

Disaster planning must include equipment and space necessary to conduct essential operations. Since major data processing equipment requires power, air conditioning, and raised floor, space within the owned facility should be earmarked for emergency use. Additional space not currently owned or leased by the firm should be considered for planning purposes. Sources and availability of this space will change constantly, thus the options should be reviewed periodically to ensure currency.

The recovery plan must include documented emergency processes outlining actions to be taken by the recovery teams. Departures from normal operating procedures are expected during emergencies. For example, in the event of an emergency, systems programmers may be assigned to all local and remote data centers to implement network recovery procedures. Since networks are critical, key technical people will ensure their operation as a first priority.

The emergency personnel roster and recovery team assignments and responsibilities are an important part of the documented processes. This information must be distributed to all employees and managers involved in the recovery operation. To avoid destruction, the plans must not remain at the work site but must be retained elsewhere by key individuals.

Routine disaster-plan testing is necessary to ensure successful implementation. It is as important to test recovery plans periodically as it is to perform fire drills occasionally. Testing will maintain employee awareness of their roles and responsibilities and will focus attention on this easy-to-defer activity. Telecommunications links must be exercised and data retrieved and restored to test the completeness and validity of the recovery process. Not all processes can be tested, but sufficient testing should be accomplished to ensure recovery readiness. IT managers must be confident that recovery plans are thorough, effective, and capable of implementation when necessary.

The importance of recovery management and the extent to which it applies in the daily operation of firms is illustrated in the next business vignette.

SUMMARY

The disciplines of problem, change, and recovery management are central to the effective operation of a computing center. These disciplines are built on service-level agreements; they support service commitments. They provide some of the necessary conditions for success. Effective operation of these disciplined processes along with the implementation of disciplines to be discussed subsequently ensure successful production operations. Successful production operations is a critical success factor for IT managers.

The disciplined approach lends itself well to daily operations in a computing center. This approach also meshes well with the tactical and operational planning essential to production operations. The disciplines discussed thus far are based on the premise that thoughtful and conscientious individuals can work together to minimize the effects of the inevitable difficulties that arise in a complex environment. Both users and IT benefit from the structured approach because it maximizes the results to everyone.

Detection of problems that adversely affect service levels and subsequent problem correction and prevention is the vital first step. This chapter discussed processes such as problem meetings and described procedures such as problem logs and analyses, which comprise the elements of problem management. Problem management reduces problems, lowers costs, and increases customer satisfaction.

Changes resulting from problem correction or required for other reasons must be carefully controlled and implemented. The change management discipline identifies a system for dealing with the risk of change. Risk assessment, testing and backup procedures, and implementation planning form the backbone of change management. Change is a continuous

Risk Management Succeeds

The 1989 San Francisco earthquake brought death and destruction to the Bay Area. Roads, bridges, buildings, and utilities were damaged or destroyed but many computer installations survived virtually intact—in many cases with less damage than the buildings in which they were housed. Computer system survival in an area known for its computer design and development activities was the direct result of disaster planning and advance preparation.

Many of Silicon Valley's computer technology firms are critically dependent on computer systems for their operation—indeed their survival—and they employ extensive risk management techniques.

One of the predictable risks of operating in the valley is loss of electrical power. Protection against momentary or prolonged power outage requires these companies to install uninterruptable power supplies that take over when normal power fails. These UPS systems offer great protection against a common threat, but they are quite expensive.

At one firm in the valley, mainframe computers continued to operate during the quake though personal computer equipment, which was less well protected, sustained damage. Mainframe operations were supported by a two-tiered backup power supply—a huge battery system that kicks in immediately followed later by diesel-powered electrical generators. The company closed one facility to check for damage and to make minor repairs but re-opened the following day.

Another large company in the valley kept most of its facilities open immediately after the quake. In anticipation of quakes, the company had reinforced its buildings at a cost of about $1 million. This precautionary investment reduced the extent of the damage.

These elaborate and expensive undertakings are considered so essential and so vital that one Silicon Valley firm has appointed a director of corporate disaster recovery to oversee these operations.

process in a modern computer center: Change management brings order and discipline to the process and permits it to proceed under control.

Occasionally activities go astray for reasons beyond managers' control; plans must be formulated to contend with these contingencies. Recovery management processes and contigency planning models were discussed to deal with natural and manmade disasters and other emergencies. IT managers must handle this issue well. For those who need assistance, service providers are available to provide help of many kinds.

As end-user computing evolves and becomes widespread and as decentralization increases, the disciplines of production operations become more widely applicable. Problems and changes are a fact of life for end users. Recovery management techniques are vital to owners of data handling facilities, whatever their organizational affiliation.

Review Questions

1. What is the definition of *problem* in the context of problem management? What is problem management?

2. Why are human and procedural issues included in the scope of problem management? How should these issues be treated differently from the others?

3. What sociological considerations surround problem management implementation?

4. What actions can management take to sponsor and promote effective problem management meetings?

5. What connections are there between the disciplines of problem management, change management, and service-level agreements?

6. What are the tools and processes of problem management?

7. What items are contained in the problem management report?

8. What are the ingredients of an effective problem resolution procedure?

9. Change management deals with understanding and controlling risk. What change management actions accomplish this?

10. How can managers ensure that they are sufficiently involved in the change management process?

11. What is recovery management? How does it differ from emergency planning?

12. What interactions take place between IT and the users during the development of contingency plans?

13. Why is it difficult to develop a list of critical applications?

14. What special problems and opportunities surface in developing recovery plans for a telecommunications network?

15. Why should disaster recovery strategies be developed as part of the usual IT strategy and planning process?

16. Why is it difficult to test recovery plans? What are some ways to deal with these difficulties?

Discussion Questions

1. What issues about recovery management are shown by the Illinois Bell fire?

2. What special considerations enter into the recovery management process when third-party service providers are utilized?

3. Compare and contrast the effects of the disaster at Hinsdale on the firms discussed in the first vignette. What options were available and what actions did each take?

4. Relate the subject of this chapter specifically to the notion of critical success factors introduced in Chapter 1.

5. For the topics presented in this chapter, discuss the balance between bureaucracy, effectiveness, and efficiency.

6. How might you quantify the economic benefits derived from effective operation of problem and change management?

7. Discuss some approaches a firm might use to evaluate contingency and emergency plans and test recovery plans.

8. If you were the chief information officer of a firm that relies on end-user computing extensively, how would you organize people to implement the disciplines discussed in this chapter?

9. What are the advantages and disadvantages of informal organizations to accomplish the goals of these disciplines? If you were the firm's CIO would you prefer formal or informal organizations for these tasks? Why?

10. Discuss some ways in which you could quantify the effectiveness of the processes discussed in this chapter.

11. Assume you are the manager of a department in which each of your employees has a workstation interconnected to all others in the department through a LAN. The LAN itself is connected to the firm's mainframe. Discuss the important factors involved in the recovery management process for your department.

12. During which stage of growth is the subject of recovery management likely to arise? Why?

Assignments

1. Itemize the elements of information from the problem management process that you think are important to report and design the format for this management report. How frequently should it be published? To whom should it be sent? If you were the IT manager, what action would you expect to follow after the issuance of this report and what action might you take?

2. Visit your university computer center or the computer center of a local firm and review its problem management system. Prepare a critique and report your findings to the class.

3. Design a change log patterned after the problem log in Appendix B. Be sure the log makes clear the relationships between changes; i.e., the log must identify prerequisite changes, corequisite changes, and mandatory future changes.

ENDNOTES

[1] Copyright 1988 by CW Publishing, Inc., Framingham, MA, 01701. Reprinted from *Computerworld*.

[2] Appendix A to this chapter contains a sample problem report.

[3] See Appendix B to this chapter for a sample problem log.

[4] Most major computer vendors supply automated on-line tools to assist in problem management.

[5] These figures are based on the author's experience with problem and change management.

[6] A sample change request is contained in Appendix C to this chapter.

[7] John Kador, "Change Control and Configuration Management," *System Development* (May 1989): 1. This article discusses how to test and implement changes with low risk.

[8] David Coursey, "Sabre Rattles and Hums, Crashes," *MIS Week*, May 22, 1989, 6.

[9] Ron Levine, "Disaster Recovery in Banking Environments," *DEC Professional* (January 1988): 104.

[10] John Mahnke, "Planning Saves MIS from Disasters' Wrath," *MIS Week*, October 30, 1989, 1.

---

[11] Bill Zalud, "Here's How To Survive a Disaster," *Security* (May 1989): 48.

[12] Some representative hot site providers include: Comdisco Disaster Recovery Services, Inc., Rosemount, IL; Sungard Recovery Services, Wayne, PA; and IBM Corporation, White Plains, NY.

[13] "Disaster Recovery Expected to Boom," *MIS Week*, April 3, 1989, 12.

[14] Common carriers must have disaster recovery plans also. Within days of the Hinsdale fire, AT&T and MCI connected 31 major Illinois Bell customers to long-distance networks substituting for Illinois Bell connections.

[15] Phillip J. Rothstein, "Up and Running: How To Ensure Disaster Recovery," *Datamation*, October 15, 1988, 86.

REFERENCES AND READINGS

Rothstein, Phillip J. "Up and Running: How To Ensure Disaster Recover." *DATA-MATION*, October 15, 1988, 86.

Stamps, David. "Disaster Recovery: Who's Worried?" *DATAMATION*, February 1, 1987.

Toigo, Jon William. *Disaster Recovery Planning: Managing Risk & Catastrophe in Information Systems*. Englewood Cliffs, NJ: Prentice-Hall, 1989.

APPENDIX A   SAMPLE PROBLEM REPORT

Problem control number _____

Individual reporting the incident _____

Time and duration of incident _____

Description of the problem _____

_____

_____

_____

Problem category _____

Problem severity code *(circle one)*   1   2   3   4   5

Individual assigned to correct the problem _____

Estimated repair date _____

Actions taken to recover *(append additional pages if required)* _____

_____

_____

_____

_____

_____

_____

Actual repair date *(problem closed)* _____

Final resolution actions, if any _____
_____
_____
_____

APPENDIX B   SAMPLE PROBLEM LOG

Problem sequence number _____
Date of incident _____
Problem category _____
Brief problem description _____
_____
_____
_____

Severity code *(circle one)*   1   2   3   4   5
Duration of repair action _____

APPENDIX C   CHANGE REQUEST DOCUMENT

Change log number _____
Problem log number *(if applicable)* _____
Description of requested change _____
_____
_____
_____

Prerequisite changes *(indicate change numbers)* _____

Corequisite changes *(indicate change numbers)* _____

Category of change _____

Priority *(circle one)* low   medium   high
Risk assessment *(circle one)* low   medium   high
Requested implementation date _____
Individual requesting change _____
Change manager _____
Management authorization _____
Attach test and recover plan.
Attach project plan if applicable.

13 Managing Production Operations

A Business Vignette

Introduction

Managing Systems
  Batch Systems Management
  On-Line Systems Management

Performance Management
  Defining Performance
  Performance Planning
  Measuring Performance
  Analyzing Measurements
  Reporting Results
  System Tuning

Capacity Management
  Capacity Analysis
  Capacity Planning
  Additional Planning Factors

The Link to Service Levels

Management Information Reporting

Summary

Review Questions

Discussion Questions

Assignments

References and Readings

A Business Vignette

Profile: Jerry Lenders

Position: Director of Technology, Infomart

Mission: "To learn enough from mistakes to prevent them from happening again."

Lenders Maintains Information Flow[1]

Journalists are a demanding bunch, and nobody knows that better than Jerry Lenders, director of technology at Infomart in Toronto. He oversees what the company calls an "electronic printing press," an on-line database system that delivers news retrieval and library services to journalists at some of the top newspapers in North America.

"It has been said that we are the heart of the organization, and I think that is true," he notes. "If the system is not up, it causes aggravation for everyone, not just the people on-line at the time. Obviously, there is no revenue coming in, so it is crucial that we stay up because that revenue does not get replaced."

The database service, called Infomart Online, also serves as a gateway to Southam News newswire, Dow Jones News/Retrieval Service and Datatimes, three on-line systems offering full-text on-line retrieval of newspapers and other services. Lenders also manages an on-line system called Private File Service, a customized on-line database service used by 45 clients who need to store and manage text and numeric data.

The success of an electronic information business hinges on being able to offer subscribers ready, easy access to the information stored in its "electronic warehouses." "Newspapers don't want anything to get in their way," Lenders says. For Lenders, that means he must make certain that his system can provide more than 1,000 subscribers with instant access virtually 24 hours a day, seven days a week to information published in dozens of daily newspapers as well as from a variety of other sources.

At the core of the Infomart system is a cluster of four DEC VAX minicomputers. Four other VAXs are available for overflow or as backup in the event the first cluster crashes. A triple redundancy in communications links helps insure that lines stay open. Some 20G bytes of information, about half of it text from newspapers, is stored on the system's disk farm of 15 drives.

Managing the variety of information that must be constantly available at a subscriber's fingertips can be daunting.

"This business has a lack of historical data as to what users really want: archival information or current information," Lenders says. "My gut feeling is that they only want to access current information, but we're monitoring it to find out." Lenders estimates that the database is growing at the rate of about 11M bytes per day. "We recently bought four DEC RA90 hard disk drives capable of storing 1.4G bytes each," Lenders says. "Those will last us until May or June of next year."

Keeping this growth under control is a challenge, he says. "We need to decide what information must be put on-line, what is put on high-speed storage drives and what will not be kept at all," he says. The newspaper files are backed up every night. "We operate the system for newspaper customers seven days on 22 hours; for others it is seven days on 20 hours," Lenders explains.

Lenders, who majored in computer science at the University of Waterloo in Toronto, began working at Infomart as a systems engineering representative (a programmer) in 1981 and two years later, moved up to become a systems engineering manager. In June 1986, he become director of technology with responsibility for the computer operation and its 16 employees.

He says that he constantly wrestles with providing adequate service at the lowest possible cost to the company and its subscribers. "The bigger challenge is that the technology is changing so rapidly," Lenders says. "That makes it difficult to stay on top of what is cost-effective. I am very conscious of the bottom line," he adds. "How do we maximize the bottom line and service availability? That is a fine, tricky line."

Since computer failure can be disastrous if its cause cannot quickly be found and resolved, guarding against that prospect is one of his primary responsibilities, Lenders says. "We have an all-in-it-together attitude here, and we methodically resolve problems," he says.

"I don't mind when we make mistakes, but I always make sure that we learn enough from them to prevent them from happening again," he says. "Once is fine as long as we learn from them and plug the holes so that we can go on."

INTRODUCTION

Preceding chapters discussed service-level agreements, problem management, change management, and recovery management as the first elements in the disciplined approach toward managing production operations. Successful production operations are a critical success factor for IT managers; a well-organized process is essential for managing this activity. Jerry Lenders understands this very well. His mission is to learn from problems and to prevent their repetition. He manages change in a rapidly growing environment.

This chapter builds on the foundation elements of service-level agreements and the processes of problem, change, and recovery management. It develops tools, techniques, and processes with which the remaining disciplines can be implemented. Discussion of the disciplined approach to production operations continues with the essentials of batch systems, on-line systems, performance management, and capacity planning. Figure 13.1 displays the relationship among these processes and relates them to management reporting.

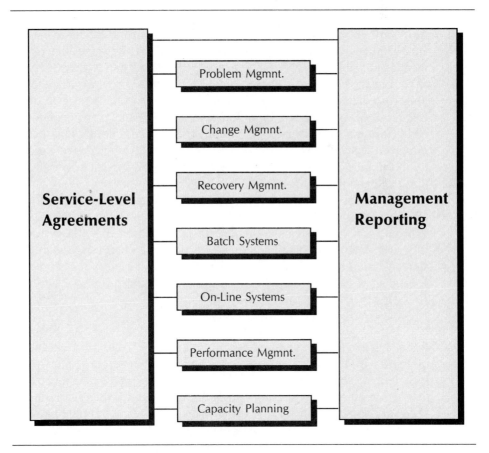

FIGURE 13.1   The Processes Continued

Problem, change, and recovery management processes, which are grounded in the discipline of service-level agreements, provide information for management reporting, the capstone of the disciplined approach. The underlying reason for developing and implementing batch management, on-line management, performance management, and capacity management is to ensure attainment of service-level objectives. Attaining these objectives is the key to achieving customer satisfaction.

MANAGING SYSTEMS

Careful system performance management helps ensure that service-level objectives are achieved through the effective and efficient use of systems resources. Users of system services and the providers of these services have expectations regarding service delivery. These expectations should be quantified through formal service-level agreements as discussed earlier. But whether they are specifically quantified or not, it is a basic management responsibility to utilize system assets effectively and efficiently. And, as the business vignette discloses, management must provide service availability according to customer expectations in spite of uncertainties. Jerry Lenders must provide continuous access to a large and growing database for customers who are uncertain about their specific needs. His success hinges on providing extraordinary system availability and adequate capacity in the face of growing but uncertain demand. Consistently high system performance is his number one objective.

The performance management process defines the management system for accomplishing measurably satisfactory levels of output from hardware and software systems. System performance observed by the customer is derived from both batch and on-line programs. Therefore, the discussion of batch and on-line system management processes must precede a detailed discussion of performance management.

Batch Systems Management

Batch systems management is the process for controlling the execution of regularly scheduled production application programs. This work includes receiving incoming transactions, processing batch data and transactions, storing or distributing output data, and scheduling the resources necessary to accomplish these tasks. Scheduled work must be planned and defined in service-level agreements. Reports of measured activities must be prepared for management review and analysis.

Major computing centers have many examples of batch systems. Some manufacturing plant examples are inventory reconciliation processing and production requirements generation. Daily accounts-payable processing or routine updating of the general ledger are examples from the firm's financial department. Many product development laboratories have large simulation programs that are processed as batch systems. Usually production jobs from personnel, sales, marketing, or service are regularly scheduled batch work also.

Scheduling and controlling these production applications are normally handled through an automated workload scheduling system—itself an application program. The scheduling program initiates jobs upon successful completion of prerequisite jobs. Thus it manages application dependencies. The system maintains an orderly flow of work through the computer center operation. The job scheduling process ensures that operational systems are utilized efficiently, that batch work proceeds in an orderly and organized manner, and that service-level agreements for batch operations are achieved. In complex mainframe environments, a computerized scheduling system is required to manage and control hundreds of batch operations in a well-disciplined manner.

Normal computer-center operating procedures include complete and unambiguous instructions for processing the scheduled work in the absence of unusual occurrences. This means that job initiation information is complete and that conditions necessary for successful job completion are known and can be met by the operations staff. The instructions and information are part of the input data for the scheduling program. In addition, instructions for handling potential difficulties such as unsuccessful completion of a prerequisite job or job failure resulting from insufficient secondary storage space must be part of these procedures. These instructions are developed as part of the recovery management process.

Developing the schedule for the computer center is part of operational planning, which consists of a long-range plan, a daily plan, and a current plan.[2] The long-range plan describes the applications scheduled for the next several weeks or so. This plan includes daily applications, weekly applications, month-end applications, and jobs scheduled on demand. A plan for each day's production is developed and is input to the daily scheduling system. The scheduling program provides current status of the batch production work to the system operators and managers.

Effective problem management requires recording of deviations from established procedures that impact service-level agreements or that have the potential to do so. Abnormal or unusual events such as unsuccessful job completion, unanticipated output, or abnormal operator intervention must be logged and entered into the problem management system where corrections must be made. Changes to production schedules or to resource levels or types of resources required for production work must be reviewed by the change management team. Batch operations are closely linked to problem and change management processes.

Production operations managers must ensure that transfer of responsibility between shifts proceeds effectively throughout the day. An operations transition meeting helps communicate the current status to the incoming operations crew. An operations log, maintained by each shift of operators, should be reviewed with the senior people from the entering shift so that an orderly transition occurs.

The batch management process must be assessed periodically by production managers and their senior staff members. The review focuses on process effectiveness. Thus it must evaluate process scope and methodology. The assessment must include job scheduling and resource management activities and it must examine the effectiveness of the linkages between batch management and problem and change management. The review should develop recommendations and directives for process improvements. The findings are important to individuals engaged in service-level agreement negotiation and preparation.

There must also be a strong linkage between individuals responsible for hardware performance and those controlling the batch management process. Alterations to the work schedule, variations in the volume of input transactions, or changing requirements for on-line storage frequently affect system performance. They may also precipitate changes in the hardware or software system configuration. Configuration changes to obtain optimum throughput may be required for customer satisfaction or to achieve service-level commitments. Prompt and effective communication of changing conditions will reduce the number of future problems and service-level disruptions.

IT managers and their key staff members are vitally concerned with production operations since it is a critical element contributing to their success. The process of batch management itself and the reports of process effectiveness serve to highlight trends in operational activity. Successful IT managers scrutinize the reviews and reports carefully for variances between planned, scheduled, and actual workload accomplished. Variances from normal operations that have been reported to the problem management system and potential or developing bottlenecks in the work flow are high-interest items for IT managers. Successful batch management achieves service levels and maintains satisfied customers.

On-Line Systems Management

On-line systems management is a disciplined process consisting of information, tools, and procedures required to coordinate and manage on-line application activities. On-line applications support users, both internal and external to the firm, in their use of remote devices. System software and hardware required for on-line applications are included as part of the application activity. The objective of on-line management is to ensure that on-line application service levels are achieved through the efficient and effective use of computer resources.

There are many examples of on-line systems. Some well-known programs such as the order-entry systems and airline reservation programs discussed earlier are on-line systems. Other examples include electronic mail systems, inventory update programs, computer-aided design systems, and service dispatch and reporting programs. Many application systems formerly operated as scheduled batch programs are now available as on-line programs. Typical examples of these applications include language compilers for application development, personnel database update and retrieval applications, and decision support systems.

The scope of on-line management includes coordinating the people, tools, techniques, and facilities to monitor and control on-line application systems services. This coordinating activity is best accomplished through common, well-defined interface points for users, vendors, and support groups. Specific groups to be addressed include operations personnel, technical support individuals, and applications development and maintenance programmers, among others. Normally a single interface point or control point should be established for each group. For example, the operating system support department might be designated the contact point for operations personnel and for hardware vendors. Complex on-line systems such as multiple processors in multiple locations may demand multiple coordinated interface points.

Several functional responsibilities must be assigned to specific persons at the assigned control point(s). These responsibilities consist of user interface functions and network system functions. The user interface function relates to on-line application users and to application development and maintenance programmers. On-line application programming groups may reside in the IT organization or in the application-using organization. For example, an on-line application specialist from the system support department may be assigned the responsibility to address technical application issues with application developers, or an experienced application programmer may be assigned to work with end users to provide technical guidance regarding on-line application development. In each case, application developers have a contact point through which their concerns can be addressed. Activities included in the user interface or responsibility are shown in Table 13.1.

TABLE 13.1   User Interface Functions

Receive user communication.

Respond to user queries.

Perform problem determination.

Obtain technical assistance when required.

Initiate problem management action.

Provide guidance to users and support groups.

The user interface control point receives, records, and reports all inquiries from on-line system users. Callers may describe current problems or may seek assistance or advice regarding on-line application usage. Users also frequently inquire about the status of problems and changes. The person at the user control point performs elementary problem determination and may obtain assistance from appropriate support functions for complex problem determination. Unresolved problems are recorded and reported to the problem management system for resolution. In addition, the user interface may provide guidance to users and to technical support personnel for on-line as well as batch applications.

TABLE 13.2   On-Line and Network System Functions

Initiate, monitor, and terminate applications.

Coordinate system terminal operator's activities.

Monitor volume and performance indicators.

Implement predefined network procedures.

Develop problem determination procedures.

Train personnel to use procedures.

Coordinate application problem determination.

Table 13.2 lists the activities of the on-line and network control point, individual or function: initiate key on-line applications, monitor their operation, and terminate the applications according to procedures. The control individual coordinates the activities of remote master terminal operators and system console operators. Volume and performance indicators for on-line applications are also maintained at the control point. Examples of these indicators include the number of individuals using the application, current response times experienced, and system resources devoted to the application. This data is required for service-level management and is an important ingredient of the management reporting process.

The on-line application control point individual or function must implement predefined system and network backup and recovery procedures. These procedures are usually developed by technical support specialists in application development or network management departments. Control point persons develop and test problem determination procedures and ensure that these procedures are customized for the users and their applications. They coordinate the resolution of problems associated with network devices and communication controllers. Their activity requires close cooperation with the problem and change management systems and with network management facilities.

The on-line management process must be reviewed periodically to assess its effectiveness. Findings and recommendations from each review must be documented for management's attention and action. Summary data documenting the effectiveness of the on-line management process should be directed to IT management and, when appropriate, to the user-managers of on-line systems. User-managers are very interested in service levels and appreciate information on process effectiveness. Effectiveness reports help IT management recognize trends in on-line services and service objective attainment. When this information is shared, trust and confidence between managers in both organizations is improved. Good relations are important to both groups.

PERFORMANCE MANAGEMENT

Performance management is a management system for defining, planning, measuring, analyzing, reporting, and improving the performance of hardware and operating systems, application programs and system services. The objective of performance management is to ensure that performance targets and performance service levels are achieved through the effective and efficient use of systems resources. Six processes form the performance management discipline. They are outlined in Table 13.3.

TABLE 13.3   The Processes of Performance Management

1. Defining performance	4. Analyzing measurements
2. Performance planning	5. Reporting results
3. Measuring performance	6. System tuning

Performance management processes are applied to systems hardware, systems software and operating systems, and application program resources. Services provided by production operations such as application scheduling are included in the management process as well. The coordination of resources utilized in combination with each other determines the overall performance of production operations. Therefore, the performance management discipline embraces these resources in total.

Defining Performance

Computer system performance is defined as the volume of work accomplished per unit of time. Performance of the central processing unit is measured by the rate at which it executes instructions. Comparison of the actual instruction rate to the maximum rate determines the effective CPU utilization. For example, if the CPU is capable of 20 MIPs but is in the wait state half the time (the hardware is not executing instructions but is waiting for work to do), the effective performance is 10 MIPs and the efficiency is 50 percent. System performance measurements encompass the CPU performance and the associated hardware and software performance. Planning, measuring, analyzing, and reporting hardware performance of large mainframes is centered about the concept of system performance. Tuning activity endeavors to balance the CPU hardware, I/O channels, I/O devices, and the software and applications. The objective of performance management is to keep all hardware components relatively busy while maintaining continuous, high-quality service to the end user.

From the users' viewpoint, performance is defined by the number of jobs completed per unit of time, the number of transactions processed per unit of time, or the job turnaround time. Turnaround time is the elapsed time from job submission to job completion. For all these activities the unit of work is not uniform. Therefore, the planning and measuring processes are statistical in nature. The reporting and tuning activity must account for the statistical nature of the work.

The goal of performance management is to ensure that performance objectives are achieved through efficient and effective use of system resources. Performance objectives are established through performance planning.

Performance Planning

Performance planning establishes goals and objectives for the throughput of computer systems. Performance planning also develops processes, techniques, and procedures to ensure that the system delivers the desired throughput. Performance planning is an integral part of the disciplines discussed previously; in particular, it is an essential step in service-level attainment. The performance plan must account for the amount and type of work to be performed as well as its distribution over time.[3] The plan is based on known or assumed capabilities of the combined hardware and software system and must include the performance characteristics of the applications themselves. For example, if rapid response is desired for on-line transactions, system loading must be carefully controlled. Response declines rapidly as system loading increases.

There are other examples. If an application is made more efficient through program enhancement activity, the total system will perform better. If the workload is more evenly distributed during the day, system throughput will improve. In other words, system performance and system capacity are tightly linked. Improving system performance is equivalent to increasing system capacity.[4] For this reason, performance management and all activities affecting performance must be analyzed and clearly understood prior to capacity planning. Capacity planning prior to or in the absence of performance management is wasted effort.

There are no generally established parameters that define system performance, but some measures are usually found to be valuable. These common measures are system response time, transaction processing rate, CPU time in the wait state versus in the system state, CPU time in the system state versus in the problem program state, and system component overlap measures. For instance, if the CPU spends a higher percentage of time in the problem program state than formerly, this may indicate increased application throughput. Throughput also increases when input/output and processing operations are highly overlapped. System programmers in the technical support department can identify many additional performance factors.

However, detailed performance measures depend on the type of processing. For example, system performance measures for an airline reservation system are different from those for a computer-aided design system. Response time to travel agents' requests in the reservation system is highly important, but instruction processing rates are more important for the complex mathematical calculations in design systems.

Installation performance planning is usually accomplished by the technical support staff using both general and installation-specific performance indicators. These indicators also are the basis of performance measurement.

Measuring Performance

Response time and throughput measurements under a variety of work-load conditions are the basis of performance management. Key performance data and measurements include the time interval or elapsed time for processing key categories of user work. Examples of these measurements are service time for a transaction; transaction rates or number of transactions per unit time; and response time for transaction initiation. Additionally, the amount of work or the number of transactions the system performs in total and for each user is an important performance measure.

The process for collecting measurements, establishing the range of acceptable performance levels, and tuning applications and services must be well documented. Many performance measurements will relate directly to service-level agreements and will be referenced in them. For example, the service level agreement may specify subsecond response times for trivial transactions or may indicate some number of transactions per minute for the on-line order-entry system. The primary responsibilities for performance management, such as collecting and analyzing performance data and enhancing systems performance via corrective actions, must be assigned to an IT manager.

Performance management is concerned with the utilization of key hardware components. These hardware components include direct-access storage space, input/output units, main memory utilization, and paging/swapping subsystems. Performance management also includes the control and management of system software components such as supervisor modules, program modules, and system input/output buffers. System programmers in the technical support department possess the knowledge and skills to adjust system parameters and configurations using performance monitoring tools.

There are many ways to measure or monitor computer system performance.[5] Hardware devices or monitors that can be attached to the computer system to accumulate statistics on system component utilization are widely available. These devices count events occurring in the system and report the events and their rates. For example, a hardware monitor can be attached to a computer input/output channel to measure the number of transactions processed and report the percentage of time the channel is busy. This information is useful to system programmers for balancing input/output activity among the channels on the system.

Computer system performance can also be measured with software monitors. Software monitors can be part of the operating system or part of the applications themselves. Software monitors measure CPU activity such as CPU busy versus CPU in the wait state, CPU in the system state versus in the problem-program state, channel busy time, channel overlap time, and input/output device busy time. Numerous other measurements such as memory allocation and use, operating system module utilization, and memory swapping or paging activity can also be obtained.

Application program monitors can be customized to provide specific application data. This data may include utilization of storage devices and frequency of module execution. From the application monitor the programmers may determine transaction processing times in the application and frequency distributions of terminal usage by terminal type or location. The amount and type of performance data available is limited mostly by the application programmer's imagination.

Hardware monitor devices are independent of the computer system operation and provide precise detailed measurements for later analysis. Their use requires specific hardware system knowledge. Hardware monitors are relatively expensive. Software monitors degrade system performance somewhat and measure only information available to the computer instruction set. But software monitors are easy to install, easy to use, and relatively inexpensive. For these reasons, software monitors are much more popular than hardware monitors.

Analyzing Measurements

Performance measurements must be obtained for appropriate periods including peak and off-peak workload times. Measurements are most useful if they provide information to identify trends in system performance. Performance data analysis, especially from peak workload periods, may reveal system bottlenecks. Identification of bottlenecks and subsequent corrective action improves performance and yields additional capacity at no additional cost. The analysis of performance parameters must include comparisons with the performance plan. Causes for departures from the plan must be understood so that service-level integrity can be maintained and the planning process improved. Intelligent use of performance measurements greatly assists in improving system performance through system tuning efforts. It is also invaluable in the capacity planning process.

Reporting Results

Performance measurements and comparisons to historical trends and to the plan must be reported. IT managers require the results to perform service-level risk assessments and to assess the integrity of the performance management discipline.

System Tuning

When established performance levels are not achieved, IT managers must develop plans to correct the unsatisfactory situation. Corrective actions for alleviating low performance levels include hardware configuration changes, input/output balancing, operating system performance improvements, system memory tuning, and direct access storage space reorganization. Additional activities such as tuning application programs that consume large amounts of critical resources and limiting the maximum number of concurrent users may also be required. Restricting the number of concurrent system tasks or other limiting actions may also be required to improve system performance.

The role of user satisfaction surveys was discussed earlier. User surveys should be developed to validate the credibility of performance measurements. Measurements must reflect actual user experience. Surveys are valuable in establishing user satisfaction and relating customer opinion to current levels of the key performance measures. Continual correlation of customer perceptions with internal measurements is the key to high levels of customer delight.

Periodically the performance management process itself should be reviewed to determine its effectiveness. The analysis of findings and the recommendations should be presented to the IT management team for review and possible action. Reports documenting performance levels achieved for all major application programs must be produced regularly for IT management and for application owners and users of IT service as well. Performance-level reports for each application and service are essential for capacity management and planning activity.

CAPACITY MANAGEMENT

Capacity management is the process of planning and controlling the quantity of each system resource required to satisfy users' current and future requirements. Capacity management also includes forecasting the quantities of computer room facilities (electrical power, air conditioning, chilled water, and raised floor) needed to install additional system resources. The disciplined processes discussed previously are prerequisites for effective capacity management and capacity planning. Computer center capacity can be determined only when managers deal effectively with problems and changes and when they control the performance of batch and on-line systems. Capacity management is an essential step toward meeting service levels.

The objective of the capacity management process is to identify additional system resources required to achieve service levels. Additional resources may result from workload changes, additional applications, or service-level improvements.[6] Capacity management processes are also useful for identifying surplus or excess capacity, or obsolete capacity that can be removed from the installation. Capacity management relates directly to IT's financial performance and indirectly to the firm's financial performance. Effective capacity management is absolutely essential to successful IT managers.

Capacity Analysis

The capacity management process includes an analysis of current system resource requirements. Based on present-day applications and services, this analysis establishes the benchmark for comparing proposed system configurations. Current system resource utilization is normally obtained from the system itself. The data is obtained from performance measurements. It can be analyzed for average workload and for peak periods such as peak hour, peak workload day, and monthly financial closing days, for example. Additional analyses should be conducted by service elements such as transactions processed, applications serviced, user-group workload, and department activity. These analyses are the basis for capacity planning.

As an example of capacity analysis, consider on-line secondary-storage capacity. Secondary storage is critical to production operations and is relatively easily measured. If the measurements reveal that dataset storage extensions during peak processing take most of the available storage capacity, then it can be predicted that jobs will fail for lack of storage as transaction volumes increase. In this instance, storage capacity measurements indicate that adjustments in system configuration or workload parameters are required to avert failure. Capacity analysis of storage devices should also detect datasets that are no longer used or so infrequently used that they should be moved to off-line storage. Similar measurements and analyses can be made on CPU instruction rates, channel-busy percentages, and input/output activity. This type of capacity analysis is required for all system resources on a continuous basis.

The results of capacity analysis permit systems managers to know the extent to which system components are being utilized throughout the day. The analysis reveals processing bottlenecks or potential bottlenecks and also indicates unused or underused system resources. Capacity analysis is the basis for capacity planning.

Capacity Planning

Capacity to satisfy future requirements is based on business volume growth, new application services, and improved service levels. This load information must be obtained through careful consultation with user managers and is normally obtained when service levels are established. The fundamental basis for establishing workload information is the business plan. Workload data must be analyzed for consistency with the business plan and with current service levels. Load forecasting begins with SLA negotiation; it forms the basis for system resource capacity forecasting.

The task of capacity planning involves the input parameters of volume changes in current applications, load increases stemming from new applications, and planned changes in service levels. Current capacity excesses or deficits from the previous analysis are additional inputs. The data are used to forecast future capacity requirements. For example, if present direct access storage is 90 percent utilized and volume changes and new applications are expected to increase storage requirements by 20 percent, then additional storage of at least 20 percent will be required. Conservative planning may indicate that storage should be increased by 30 percent, perhaps.

All other system resources must be analyzed in a similar manner. For instance, if response times to on-line transaction processing is deteriorating, capacity planners have several choices. They can tune the on-line system to improve its performance, or they can adjust the operating system configuration or priorities to improve response. They can adjust or reschedule other work running in the multiprogramming environment to improve response, or they can increase system resources. System resources can be increased and response improved by installing higher speed storage devices for the on-line program or its data. As a last resort, a faster CPU or an additional CPU can be acquired. System planners have many options for handling performance and capacity issues.

The final result of capacity management is an optimized configuration of hardware, system software, and application programs. This optimum future configuration satisfies the needs of the business with management-determined effectiveness and achieves the performance committed in the service-level agreements. Information derived from the capacity management process is vital input to the IT planning process.

The capacity management process must be documented for IT managers and senior user-managers. The responsibility for collecting user requirements and forecasting systems capacity must be assigned to an IT manager. Many techniques can be utilized in forecasting systems capacity. For example, if CPU capacity is critical to the operation, capacity management may hinge almost completely on CPU parameters. For most systems, however, capacity analysis involves a composite of many measurements. In these cases, system simulations may be required for complete analysis. The key to capacity management is selecting the simplest technique that provides a satisfactory degree of accuracy.

Most successful computer installations maintain a database containing previous workload projections and capacity requirements. These historical databases are essential for studying trends in workload and capacity growth and for improving forecasting techniques.

Capacity planning and management is the conclusion of all previous disciplines. This is true because all the disciplines impact capacity requirements. For example, the manner in which problems are handled directly relates to workload because good problem management reduces problems, lowers workload, and reduces capacity requirements. The same is true for change management. Efficiencies in the operation of batch and on-line systems directly translate into capacity reductions. In other words, poorly managed batch and on-line operations require capacity increases in comparison with well-managed operations. Finally, capacity planning in the absence of performance analysis and planning will be largely ineffective. Capacity cannot be well planned if significant performance uncertainties exist.[7]

Capacity analysis and planning, essential to any successful computer operation, requires antecedent processes for their effectiveness. Some IT organizations attempt to size capacity without the required prerequisites, and some publications address capacity without recognizing the inherent dependency on performance analysis and planning.

Additional Planning Factors

Senior IT managers must remain alert to conditions that can affect computer system capacity. They must carefully scrutinize additional business information that has a bearing on capacity forecasting. Some factors to be considered in the capacity planning process include the following items:

1. Changes or alterations in strategic direction destined to improve or increase IT services

2. Business volume changes in either direction

3. Organizational changes (always a potential impact on IT resources)

4. Changes in the number of people who use IT services[8]

5. Changing financial conditions within the firm[9]

---

6. Changes in service-level agreements or service-level objectives having a bearing on system performance requirements

7. Portfolio management actions such as new applications or changes to current applications that impact system throughput

8. System resources required for testing new applications or modifications to current applications

9. Application schedule changes initiated by operations or user-managers

10. Schedule alterations for system backup and vital records processing

11. System outage data and job rerun times from the problem management system

The first five items above result from changing business conditions within the firm and are usually reflected in business plans. Some important factors such as volume changes and changes in financial or budget changes can occur between plans, usually as the result of external conditions. IT managers must be particularly sensitive to these factors. For example, if the firm is experiencing an unplanned growth in sales and revenue, some user organizations may receive an unplanned increase in budget. The user organization may typically spend 10 percent of its budget on IT services, but its marginal spending for IT may be 50 percent. This occurs because, as revenue rises and functions receive additional discretionary resources, they increase their spending on automation to handle the increased volumes. The reverse is also true. The fluctuations in demand for IT services are likely to be wider than the fluctuations in the firm's business activity. This phenomenon is critical to senior IT managers.

Information about the remaining six items on the list is available from the disciplines of production management or from individuals within the IT organization. Individuals responsible for problem and change management, batch and on-line application specialists, information center personnel, and IT customer service representatives are responsible for gathering information that impacts workload projections. Users also must be coached to alert IT managers when anticipated changes in requirements are first detected. Effective capacity planning depends on a continuous stream of critical information. Information gathering is a continuous process, but it must be accompanied by validation procedures. As part of the validation process, the IT steering committee should examine and concur with unusual or unplanned workload projections.[10]

THE LINK TO SERVICE LEVELS

Equipment plans developed from the capacity management process specify the hardware components required to satisfy user service levels. The equipment plans translate into equipment costs and installation and setup costs, and they specify additional supporting equipment. After developing the optimal system configuration with the additional required components, the computer room facilities must be evaluated to ensure that the proposed configuration can be satisfactorily housed. Additional facilities, space, or equipment may be required. The capacity management discipline is highly important because it drives the budgeting and planning process and specifies resources needed by vital applications.

Periodically the capacity management process should be reviewed to assess its effectiveness. Close agreement between previous capacity forecasts and actual capacity requirements is one effectiveness criterion. The predicted and the actual results of workload, capacity, and service levels should be compared for each user application. These comparisons should be undertaken even if hardware or software changes have not been made. The results of the capacity review, including findings and recommendations, should be documented for further scrutiny and analysis and should be retained as future reference material. Reports documenting the capacity required for each application should be produced for IT managers and for user managers. Subsequent service-level negotiations require this information.

MANAGEMENT INFORMATION REPORTING

Service-level agreements are the foundation of the management system for production operations, and management reporting is the capstone. Throughout the discussion of the disciplined processes within this management system, reporting played an essential role. Each process informs managers of its results so that sound decisions can be taken and process improvements can be made. This management system for production operations is effective. It enables managers to succeed in a vital and critical area. The free flow of important and introspective information is essential to process success. Each management process must be examined for effectiveness. This evaluation must be used to improve the process further. Thus, for several reasons, management reporting is essential.

Management reports are not exclusively for use by the IT organization: They are essential to IT customers as well.[11] Mature, successful IT managers take every opportunity to share information because they know that sharing important data improves performance. As Jerry Lenders pointed out, "We have an all-in-it-together attitude here, and we methodically resolve problems." The attitude displayed by Lenders and the others at Infomart is extremely important to managers in all firms. Most managers in most firms are problem solvers, and, given a chance, they are reasonably adept at it. The system must give them a chance.

Management reporting and all the other informal communications are highly important to efficient operations. Reporting problems, failures, and successes exposes the individual and the organization to criticism and to praise. More importantly, open communication leads to increases in trust and confidence. Building trust and confidence is important to managers, even if it may mean vulnerability in the short run. Good managers understand the value of good communication. Effective managers insist that reporting processes be honest and complete.

SUMMARY

At Infomart, Jerry Lenders balances requirements for customer service, system availability, and recovery protection by providing double redundancy in CPUs and triple redundancy in communications links. Part of the redundant capacity stems from uncertainty in customer demand and part from the need for high availability. All capacity costs money, and his essential trade-off is between system costs and system benefits to Infomart. In addition, Lenders must understand technology advances and relate these to the needs of the business. Lenders exemplifies the role played by many successful IT managers in firms today.

This chapter presented tools and techniques for managing production systems—batch systems and on-line systems. Operational planning data is fed into batch scheduling systems that control job initiation and provide instructions and status information to system operators and managers. Departures from normal operations lead to problem-management action items. Defined control points and responsibilities are required for effective on-line operations. The control points ensure that user questions are resolved, that problems and changes are properly coordinated, and that on-line systems are coordinated and controlled in the computer center. Because users of on-line systems are frequently widely dispersed, the control point functions as a communications hub among system developers, operations personnel, and users.

Batch and on-line applications depend critically on computer capacity and on the performance of system resources. Tools and techniques for performance management and capacity analysis and planning are presented in this chapter. There is a strong link between system performance and system capacity because improvements in performance through system tuning actions effectively increase system capacity. In fact, this may be the lowest cost capacity available. This chapter demonstrated that capacity planning can be effective only when all other disciplines are operating well.

Reports from all the disciplined processes are presented to management so that they are kept informed on the performance of the production operation. The reports also permit managers to assess the effectiveness of the processes themselves.

The management processes described in this and the previous chapters form a roadmap to success in production operations. Production operations is a critical success factor for IT managers. The management system described above ensures success if implemented completely. Implementation may seem arduous and time consuming, but shortcuts or circumventions lead to inefficiencies and to failure. Failures in production operations lead to failures elsewhere. They always cause extreme distress for IT managers. Successful IT managers embrace the management system described above; they thrive on its benefits and they enjoy its rewards.

Review Questions

1. What is the relationship of service-level agreements and management reporting to the disciplines discussed in this chapter?

2. What are the essential reasons for developing and implementing the management processes described in this chapter?

3. What is the relationship between system performance and management expectations at Infomart?

4. What is meant by batch management?

5. What are some examples of batch systems? Do you think on-line systems are more difficult to manage? Why?

6. Why are batch systems difficult to schedule? What functions can a computer program perform in this task?

7. What are some of the inputs computer operators may provide to the problem management system?

8. Production operations frequently operates around the clock. What special problems does this pose, and how are they solved?

9. What are the linkages between operations personnel and the processes of performance management, capacity management, and problem management?

10. What is on-line management? What kinds of communications difficulties are raised by on-line systems? How can they be solved?

11. What is the scope of on-line management?

12. What are the differences between responsibilities for the user interface function and the network system function?

13. The on-line and network system function initiates, monitors, and terminates applications. Where should this function reside organizationally?

14. With whom do the network control point personnel interface?

15. What is performance management? What are the processes of performance management?

16. How is the performance of a CPU usually measured?

17. How do users measure performance?

18. What does performance planning accomplish?

19. What tools are available to measure system performance?

20. What is the connection between system bottlenecks and system tuning?

21. What is capacity management? What are its objectives?

22. How are capacity analysis and capacity planning related?

23. How does capacity management depend on the previous disciplines?

24. What is the link between capacity planning and service levels?

25. Why is management reporting called the capstone of the management processes discussed in this and the previous chapter?

Discussion Questions

1. Discuss the essential difficulties faced by Jerry Lenders in his job at Infomart.

2. Why are each of the disciplined processes essential to service-level attainment?

3. Discuss the connections between batch and on-line systems and the disciplines of problem management and change management.

4. Current trends indicate that batch programs are being converted to on-line systems. What are the implications of these trends for performance management and capacity planning?

5. Throughout this chapter and previous chapters the text states that the management processes must be examined periodically for effectiveness. Why do you think this is important? How might you go about doing this?

6. Reporting the results of the performance management process is important to several groups. Discuss the importance of these reports to system operators, system programmers, application programmers, users, and managers.

7. System performance means different things to system programmers, for example, than it does to application users. Why are these differences present? Why must a common ground be found?

8. Discuss the advantages and disadvantages of hardware monitor devices and software monitoring tools. Why might a combination of devices and tools be most effective?

9. Analyzing performance measurements, reporting results, and tuning systems are iterative processes. Discuss the role of IT managers in these processes.

10. Discuss why capacity management must be preceded by the processes of problem management through performance management. What essential difficulties are present if capacity analysis is not preceded by the recovery management discipline?

11. Discuss the trade-offs among the processes of IT planning, service-level agreements, recovery management, and capacity planning. What risks are always present regardless of the manner in which trade-offs are made?

12. Discuss why organizational changes are always a reason for reviewing system capacity factors.

13. It was stated earlier that information technology managers are change agents. How does this statement relate to the answer to the previous question?

14. What issues discussed in the first chapter are attacked through the process of management reporting?

Assignments

1. List the reports generated by the management system outlined in this and the two previous chapters. In your opinion, who should prepares these reports and who should review them? Prepare a schedule for these reporting and reviewing processes and describe the resulting management process.

2. The processes described in the disciplined approach to production operations seem most suitable to mainframe installations. How can the principles contained in these processes be applied to minicomputer installations? What ideas are relevant to end-user computing? How would you propose to implement them?

ENDNOTES

[1] Michael Alexander, "Lenders Maintains Information Flow," *Computerworld*, December 12, 1988, 87. Copyright 1988 by CW Publishing Inc., Framingham, MA 01701. Reprinted from *Computerworld*.

[2] Israel Borovits, *Management of Computer Operations* (Englewood Cliffs, NJ: Prentice-Hall, 1984), 191.

[3] H. Pat Artis and Alan Sherkow, "Boosting Performance with Capacity Planning," *Business Software Review* (May 1988): 67. This article states that workload characterization is the cornerstone of all performance and capacity planning programs.

[4] Richard K. Wheeler, "Capacity/Performance Management: A Success Story at Chase Lincoln Bank," *EDP Performance Review* (March 1988): 1.

[5] Various types of performance measurement techniques are discussed in Borovits, *Management of Computer Operations*, 153-179.

[6] A quantitative approach to hardware selection based on capacity measures is given by Borovits, 55-64. The process includes system software and hardware. It must account for the performance and changes in performance of application programs. Most installation managers prefer to simplify the problem by treating each of these elements separately.

[7] John Kador, "Capacity Planning for Competitive Advantage, *Datacenter Manager* (January-February 1989): 34.

[8] Douglas J. Howe, "A Basic Approach to a Capacity Planning Methodology: Second Thoughts," *EDP Performance Review* (October 1988): 1. End-user computing increases the complexity of capacity planning and extends the planning horizon out to 10 or more years according to this article.

[9] Leilani E. Allen, "When the CEO Says 'No Upgrade,'" *EDP Performance Review* (November 1988): 1. This article describes how to handle the need for short-term reductions in workload.

[10] The steering committee may not agree with the projected additional load. The committee can help the firm balance additional load and therefore additional capacity with business financial objectives.

[11] Richard K. Wheeler, "Capacity/Performance Management," 1. This article stresses the need to involve senior managers in the capacity/performance process.

REFERENCES AND READINGS

Alexander, Michael. "Lenders Maintains Information Flow." *Computerworld*, December 12, 1988, 87.

Borovits, Israel. *Management of Computer Operations.* Englewood Cliffs, NJ: Prentice-Hall, 1984.

14 *Network Management*

A Business Vignette

Introduction

The Importance of Network Management
  Networks Are Strategic Systems
  Networks Are International
  Networks Facilitate Restructuring

The Scope of Network Management

Management Expectations of Networks

The Disciplines Revisited
  Network Service Levels
  Configuration Management
  Network Problems and Changes
  Network Recovery Management
  Network Performance Assessment
  Capacity Assessment and Planning
  Management Reporting

Network Management Systems

International Considerations

The Network Manager

Summary

Review Questions

Discussion Questions

Assignments

References and Readings

Greenville, S.C. — Get a remote Token-Ring. It pays. At least that is the experience of Metropolitan Life Insurance Co.[2] The organization is currently reaping benefits from a first-time ever adaptation of IBM's Token-Ring network that is configured to speed up the insurance carrier's remote data-entry operations.

The unique implementation provides the key to a significant reduction in network operating costs, higher productivity and faster response times by eliminating equipment and providing a more direct link to an IBM 3090 host. Thomas Waltz, manager of the firm's Greenville computer center, which supports the heaviest part of the company's online network, states: "Today, 50% of all transactions are done under one second, whereas prior to the Token-Ring implementation, about 8% were done under a second. That is equivalent to those units being directly attached locally to the host," he said.

These achievements are mostly chalked up to a little entrepreneurship and to a willingness to fix what wasn't broken. "We tend to encourage people with the ability to use a lot of entrepreneurship to try things," said Daniel Cavanagh, a senior vice-president responsible for all of Met Life's data processing, and telecommunications facilities and services.

There was nothing particularly wrong with the previous state of the company's remote computing operations. Claims approvers were eking out an average response time of 2.5 seconds for more than two million transactions per day. "We are constantly looking for ways to improve response time," said David Zimmerman, a vice-president and staff controller for the firm's remote sites.

The company's data-entry operations have experienced immediate jumps in productivity. "We saw more transactions being processed the day after conversion than there were the day before with the same amount of people," Cavanagh said. "You hear people in the remote offices talking about the fast machines. It's actually the same physical terminal as before, but once hooked up to the Token-Ring, it operates a lot faster," Zimmerman said.

Also very real are the cost savings, which so far amount to more than $16,000 a month in leased-line costs alone. Some of these savings have been used to offset the cost of the new equipment. What Met Life did was draw up a plan, targeting the 10 heaviest volume producers among its 50 remote data-entry sites. "We estimated the cost of the new hardware for the 10 offices at $1.7 million, with a payback period of 31 months," Waltz said. All 10 offices will be converted by the end of 1989. The projected savings are $900,000 over a five-year period.

Met Life's remote data-entry centers operate nonstop from 6:00 AM Monday to 4:00 PM Saturday. That leaves a mere 38 hours to execute a complete system changeover—with no room for error. Waltz credits cooperative planning with virtually pain-free rollovers to a remote Token-Ring system.

"We go out to the offices and get them prepped long before we ever get any equipment delivered," Waltz said. During the preconversion visit, an installation team determines where new equipment and circuit terminations will be located and what changes will be needed to facilitate the conversion. To prevent any unpleasant surprises, a test run is made by setting up a control model in the computer data center. Once everything is powered up, the computer center remotely dials into the IBM 3720 at the remote site, configures it and downloads an internally built Network Control Program into the box. The NCP software controls the 3720 and connected devices.

On the weekend an installation team, made up of computer center personnel and home-office data communications staff, disconnects the coaxial cable from all old control units, attaching them to replacement computers. On Sunday, the computer center brings up the new system, activates the remote devices, and tests every application. On that first Monday, the team remains on-site. From the user's perspective, nothing has changed. "They sign on as usual, and their screens remain the same," Waltz said. The only visible difference, he noted, is a much faster response time.

Last year, the company's revenue totaled $15.2 billion, yielding a net income of $494 million. Met Life has about 24,000 terminals installed for its 38,100 employees.

INTRODUCTION

The previous three chapters emphasized the management of information systems operations by discussing the role of customer expectations and by developing the disciplines of problem, change, and recovery management. Performance analysis, capacity planning, and management reporting completed the management processes for production operations. Production operations in today's firm are tightly linked and highly dependent on data communications networks because they transport much of the data entering or leaving computer systems. Many of the firm's important systems are on-line, processing data as it arrives. Because networks play an extremely important role in modern information systems, their management deserves special attention.

Networks are part of the information processing infrastructure in firms of all sizes. LANs connecting personal workstations are the most rapidly growing segment of the telecommunications market, indicating that small firms and departments in large firms are rapidly adopting the technology. Major telecommunications projects designed to link operations electronically are underway at Ciba-Geigy, H. J. Heinz Company, G. D. Searle, and many other firms. Merrill Lynch, for example, is planning to spend more than $200 million on networks over the next five years. The networks will link their New York and New Jersey headquarters facilities with 500 domestic branch offices and international branches in Tokyo, London, Singapore, Sydney, Bern, and Toronto. Merrill Lynch expects to spend a significant portion of its telecommunications budget on network management.[3]

Networks greatly improve the firm's ability to process information, but they add complexity to the firm's information processing infrastructure. However, many of the management practices and systems described in preceding chapters for traditional information systems serve very well as a basis for network management. For example, strategy and plan development, the management of expectations, and operational disciplines are all especially important to network managers. As we learned in the business vignette, performance management and capacity planning played a major role in Met Life's network architecture and management. Met Life intends to improve system and network performance and to reduce costs.

Network performance and many network operational and management issues are addressed in this chapter. The chapter's purpose is to develop a foundation of management systems and principles upon which network managers can rely for success.

THE IMPORTANCE OF NETWORK MANAGEMENT

Networks Are Strategic Systems

Network management is becoming increasingly important to businesses and to their IT managers. There are many reasons for the intense interest in network management, but the growing strategic and operational value of networks is the most important. As we learned earlier, most strategic systems rely heavily on telecommunications technology. Reservation systems, brokerage products, and many other information systems are telecommunications-based.

Not only Met Life, but most other insurance companies are active users of telecommunications services too. Security-General, for example, expects independent agents to sell its policies rather than competitors' because of superior service. Their system provides E-mail communications to headquarters, a library of policy forms, and the ability to generate policies on-line. Security-General expects to recover its costs by charging the agents $3000 to $4000 per workstation.[4]

Reducing or eliminating time and distance barriers through telecommunications is a valuable strategic thrust. Reductions in time accelerate important business processes and reduce information float. Reducing the effects of distance can help firms capture economies of scale by linking small units together; it permits management control over far-flung business operations. Well-conceived and well-implemented telecommunications systems allow firms to establish excellence in sales and service and to penetrate new markets.[5] Rapid advances in telecommunications technology promise additional opportunities for alert managers.

Networks Are International

International trade and commerce and the globalization of business enterprises are facilitated by telecommunications technology. For example, significant capacity expansion is being planned for the Pacific basin, not only to carry additional traffic, but to provide backup in case of network failure. Fiber-optic cables connect the U.S. with Japan and Hawaii. Additional connections will link Hong Kong, Korea, China, and the Philippines,[6] and are under construction in the Atlantic, the Caribbean, and the Mediterranean. Connections through the U.S. enable communication between the Far East and Europe. Soon high-speed, broad-bandwidth telecommunications links will encircle the globe.

The European Economic Community agreements, which establish a united economic European nation in 1992, together with the current political alterations in the Eastern European bloc of nations have generated intense interest among Western companies. The interest is broad but is especially intense in the field of information systems and telecommunications. Companies like IBM, which already have a foothold on the continent, are likely to compete head-to-head with well-established firms in Germany and other EEC countries for a share of the Eastern European information processing market.

The interest is not just one way, however. The giant British telecommunications firm, British Telecom, agreed to purchase Tymnet, a large value-added network, from McDonnell Douglas Corp. for $355 million. Earlier BT bought a 22 percent stake in McCaw Cellular Communications Inc. for $1.5 billion. Tymnet serves the top 10 industrial companies in the U.S. and one-third of the top 100 U.S. corporations.[7] These actions by British Telecom indicate a strong interest in the American market and an increased stake in the international value-added network business.

Leading-edge firms around the world recognize the opportunities and are beginning to capitalize on them. The business vignette in Chapter 6 illustrated how the Volvo Corporation uses international networks to link manufacturing and sales operations in Europe with those in the U.S. The advantages to Volvo are obvious. It cannot operate its business without these linkages. Eastman-Kodak links tens of thousands of employees with its voice-messaging system and connects computers internationally through its communications facility. DEC manages its far-flung business operations through extensive use of international communications facilities. Messaging through global networks of PROFS systems enables IBM employees to communicate internationally on product development, manufacturing, or sales issues. High-speed data pathways permit firms to develop hardware and software products on an international basis.

The global infrastructure for international communications is rapidly falling into place. Large segments of our economy already take advantage of the system; the travel industry is one example and financial institutions are another. It is estimated that more than one trillion dollars move through international telecommunications systems daily. Executives in many industries in all countries are intensely interested in capitalizing on the potential of international networks. Telecommunications promises to be one of the great opportunities of the 1990s.

Networks Facilitate Restructuring

Decentralization of business activities and of IT operations; alliances, partnerships, and joint ventures; and information coupling between firms and their suppliers and customers are important current trends shaping future corporate networks. Networks are strategically important because they enable decentralization and corporate restructuring: They support the information flow demanded by dispersed or remote units. Networks are critical assets of decentralized organizations. They provide efficient and effective communication of knowledge; they enable vital control information to flow to and from dispersed organizations.

For example, five companies in five countries are collaborating to build a new generation jet engine. The companies, located in the U.S., Great Britain, Japan, Germany, and Italy, will jointly build jet engines worth approximately $4.3 million each for 2000 new aircraft over the next 20 years. This cooperative effort is possible only because a 24-hour communications network transfers data on design control, bills of material, parts catalogs, tool design, and many other items among the firms.[8] Connectivity of systems, data, and applications is the critical component of decentralized businesses. Organizational restructuring and trends toward decentralization are facilitated by network technology. Thus network management is critically important to corporate strategies.

There are other reasons for the growth in size and importance of telecommunications systems, such as the rapid growth of personal computers. It has been estimated that about 50 million PCs will be installed by 1992. About half of all business PCs will be networked through 2 million LANs.[9] But PCs stimulate only part of the growth in networks. Much of the rest comes from electronic mail, facsimile, telex, telephone, voice mail, audio conferencing, computer conferencing, electronic document transfer, and video conferencing.[10] These technologies are being used in combination with each other; for example, there may be as many as 500,000 facsimile boards installed in PCs by the end of 1991. An all-purpose workstation handling personal audio, video, text, and image communication is in the near future.

Network management is also important because networks are becoming more complex and difficult to manage. They are becoming more complex because the number of users is increasing rapidly and their communication paths are longer and more convoluted. There are many more network products, such as concentrators, multiplexers, and modems. Customer equipment is proliferating too. There are many suppliers of network and customer equipment; their products are not always completely compatible. In spite of significant progress in standardization, incorporating multisupplier products into networks poses a management challenge.

IT managers and their firms desire network integration, but this objective strains present network management systems. Most firms are planning significant increases in expenditures for network management systems, but some firms are limited by the availability of technical and management skills. Because of the large installed base of networks and network devices and because of their rapid growth, there is a large built-up demand for network managers and network management systems.

Business and government executives are intensely interested in networks, their capabilities, and the opportunities they afford their organizations. This interest is reflected in the key issues reported in Chapter 1. Information architecture, partly a networking issue, ranked at the top of the 1989 list, and telecommunications systems ranked tenth. Global systems now appears as one of the issues on the list. In addition, industry deregulation in the U.S. and elsewhere has focused critical executive attention on telecommunications systems for the firm.

THE SCOPE OF NETWORK MANAGEMENT

Network management consists of systems, processes, and procedures to achieve efficiency, effectiveness, and customer satisfaction in network operations. These management practices are applied to hardware, system software, and application software, which provide the transfer of voice, data, or image information throughout the firm. Network management focuses on the assets employed for communication. The assets themselves may not all be owned by the firm; some, like bandwidth, may not be tangible or physical. Some communication assets will be found in the firm's portfolio of application programs. Thus, the scope of network management is broad and diffuse, rather than narrow and concentrated.

Physical assets subject to network management include customer terminals, local cabling, concentrators, modems, multiplexers, and lines or links. Private branch exchanges and computer processors frequently form part of the physical system as well. Communications software, some application programs, and databases supporting the hardware and software are also included within the scope of network management. The traditional boundary between telecommunications management and information system management is rapidly disappearing because the technologies themselves are indistinguishable. The separate management subjects have been replaced by one: information technology management.

Network management and the discipline of on-line management are also merging. Firms with extensive traditional data processing operations use the management systems discussed earlier to manage these processes. Organizations with extensive telecommunications systems have implemented the management systems described in this chapter. Firms with both must blend network management into their management systems throughout.

IT managers must not make arbitrary distinctions that may cause discomfort to system users. For example, if application users experience problems, they want solutions. They don't care whether the problem originates in RBOC equipment, in local cabling, or in their application. Effective customer support can be achieved only if portfolio management, production operations management, and network management are well coordinated and thoroughly integrated by IT. The IT management system must provide seamless, transparent telecommunications services to users. Users expect high-quality service; their expectations are entirely reasonable. Successful IT managers meet this challenge.

MANAGEMENT EXPECTATIONS OF NETWORKS

Users' requirements for applications incorporating telecommunications capability are driving integration of voice, data, and image communications. For instance, automatic number identification is very attractive in customer service applications and telemarketing systems. New kinds of applications are being formulated by software developers as they tie computer screens into the telephone system. The firm's PBX is playing an expanded communication role in the organization. Users are demanding access to databases through the telephone system to provide services ranging from health care to metal reclamation to loan application. The integration is occuring at the application level and at the data transport level.

The growth of network applications is accompanied by increased requirements for more and better network management systems. Appropriate tools, processes, and organizations must be available so that managers can operate and control networks efficiently. Corporations are willing to make sizable investments in large, sophisticated networks, but they insist that the implementation be manageable when completed. Users expect that vital network capability will be reliable, available, and serviceable. Users also expect networks to be cost effective. In short, investments in network hardware, software, and applications demand management tools, techniques, and processes that ensure successful network operation.

Executives expect telecommunications systems to link customers and suppliers to the firm through sophisticated electronic data interchange (EDI) applications. These linkages are vital to the firm: They provide competitive advantage. In some cases, these linkages alter the firm's structure, change the industry in which the firm operates, or support business expansion. Leading-edge information systems blend communications technology with information processing capabilities. Corporate executives expect information technology to provide effective linkages with others outside the firm here and abroad. They expect to employ information technology for competitive advantage.

As we learned in previous chapters, unfulfilled expectations are an important source of difficulty for IT managers. The emergence of valuable telecommunications capability and the integration of this capability into the firm's mainstream operations serve to increase the attention that must be devoted to managing expectations. Successful IT managers are well served by management systems designed to cope with the expectations of executives and users throughout the firm.

High expectations and an abundance of impressive technology give rise to many network management problems for which there are no easy solutions. There are many network component vendors and a large number of options for managers to consider. Even with the growing attention to standards, most systems are vendor-specific or proprietary; there are no complete network management solutions.[11] Integrated and multivendor network systems increase the difficulties of network management. Better network management systems are urgently needed.

THE DISCIPLINES REVISITED

Customer expectations must be carefully managed. Expectations of networks include availability, reliability, and responsiveness at affordable and reasonable costs. Network managers require specific management tools, techniques, processes, and organizational structures to maintain network components that are highly responsive and highly available. Communication network management is a system of people using management information and tools to maintain physical and logical control over network operations. Operational control involves solving physical and logical network problems and monitoring performance to ensure attainment of contracted service levels. Like production operations, network management benefits substantially from a disciplined approach.

> Merrill Lynch selected MCI and IBM to design and operate an advanced network management and control system for the firm's worldwide telecom network. Merrill Lynch noted that this new contract is valued at $50 million over a 5-year period. It is in addition to a second 5-year contract, valued at $150 million, that was awarded to MCI for worldwide voice and data services. MCI is the prime contractor, with responsibility for network design, capacity planning, and disaster-recovery planning. IBM will be charged with providing network-management services, including performance and problem management, as well as change and configuration management.[12]

MCI and IBM utilize the disciplined approach in managing the Merrill Lynch network just as firms throughout the world use these disciplines for computer-center management. Service-level agreements that incorporate customer expectations regarding reliability, responsiveness, and availability are the basis of the network management system. Management reporting is the capstone of the system. Network managers must be concerned with problem, change, and recovery management.

Performance planning and analysis, capacity planning, and configuration management are crucial to the management system for network managers.

Some references define the activities or disciplines of network management somewhat differently than this text. For example, the International Standards Organization (ISO) defines the management-system functions as fault management, configuration management, accounting management, security management, performance management, and adjacent areas such as planning. This text considers change management and recovery management as essential disciplines; accounting and security are also important but are discussed in a broader context in subsequent chapters. The disciplines of network management are shown in Figure 14.1.[13]

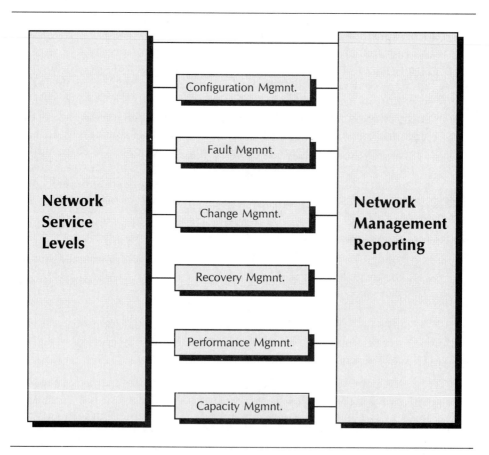

FIGURE 14.1   The Disciplines of Network Management

Network Service Levels

Network service levels can be established in much the same way as service-level agreements for computer center operations. The agreement establishes mutually defined levels of service to be delivered by the network manager to client organizations. Costs and benefits of the network service must be established and must be balanced to provide a cost-effective way for IT to serve clients. Usually, network service agreements are negotiated at fairly high levels in the firm; individual user managers are frequently not involved in these negotiations. For example, the service agreement for the network supporting the firm's office system will probably be established between the network manager and the administrative manager responsible for office systems. Likewise, service levels for order-entry systems will be resolved between the order-processing manager and the network manager. A well-functioning network management system should be transparent to most individual managers and to all users.

Network service-level agreements describe the type of service provided and contain measures by which network service is monitored. In general, this means that network and user managers must document network availability, reliability, and responsiveness. The latter usually involves both response times and workload or volumes, and is frequently referred to as network performance. Network managers must ensure that user managers understand the inverse relationship between volume and response time for most applications.

For most on-line applications involving simple transactions, subsecond response time is a reasonable expectation. Response time for on-line transactions is highly visible to users: Poor response degrades both productivity and morale. When Met Life installed network improvements, response times improved and productivity immediately increased. Since total costs also declined, unit effectiveness improved substantially. Network managers have considerable leverage to improve business operations.

Network service-level agreements are documented in much the same manner as the SLA for production operations. The main ingredients are shown in Table 14.1.

The network SLA begins by describing the types of network services provided to the client organization. The service could be as simple as a local area net for the immediate office area or as complex as a broadband international net employing satellites for transoceanic communication. The services go beyond the physical implementation. They must include types of transmission (voice, data, or image) and transaction volumes. This part of the agreement is intended to capture the user's expectations of types and volumes of communication and expected geographic coverage.

TABLE 14.1   Network Service-Level Agreement

Date of agreement

Duration of agreement

Network service provided

Availability of service

Service reliability

Response times committed

Network cost/benefits statement

Service reporting

Signatures

Users have expectations concerning the availability, reliability, serviceability, and cost of network service. The network SLA is where these expectations are described and documented. For some networks, availability is limited to normal working hours. For others, the network must be available continuously. For example, the office system LAN is operational during usual working hours but the service may be extended beyond normal hours upon request. International operations may require long-haul network capability nearly continuously. Response time requirements will also vary widely. LAN users expect subsecond or near instantaneous response for their transactions; international applications may be less demanding. The service agreement describes the level of network service committed to the users by the network manager.

The network SLA must account for the cost/benefits factors of network service. Network managers and applications managers must establish cost-effective service levels for the firm. As the Met Life example shows, advances in telecommunications technology can sometimes be employed to reduce costs and to improve service or response time.

Service levels attained versus service goals must be carefully reported. The reported results must have a high degree of credibility with users. Obvious verification mechanisms must be available to all interested parties. As with production operations, user satisfaction surveys and opinion polls can be very effective in maintaining client satisfaction.

Configuration Management

Configuration management is the management system for controlling network topology, physical connectivity, and network equipment and for maintaining supporting data. It also includes allocation of transmission bandwidth to various applications and consolidation of low-speed traffic onto higher-speed circuits for more economical transmission. It is an essential step in the disciplined process of network management. Table 14.2 indicates the scope of configuration management.

TABLE 14.2   Configuration Management Scope

Physical connectivity

Logical network topology

Bandwidth allocation

Equipment inventory

Equipment specifications

User information

Vendor data

Configuration management is concerned with the physical network elements such as customer terminals, cluster controllers, modems, multiplexers, and links and their physical and logical connections and topology. The discipline requires an accurate inventory of equipment including physical location, technical capability, and intended customer application. Not all the equipment may be owned by the firm; some may be leased from equipment suppliers, and some communication capabilities may consist of leased bandwidth.

Configuration management controls the physical connections and interrelations of the telecommunications equipment. It manages the logical topology of the network through routing tables. It assigns and controls bandwidth allocation between applications. Configuration management is a critical discipline because it allocates and controls network assets. The configuration databases must describe the network configuration at any moment in time. They must be maintained for use by subsequent disciplines.

In addition to inventory data, other databases are required for efficient network management. Physical device addresses, application requirements, and other technical information must be readily available to reconfigure the network. Reconfiguration may be necessary because of defective operation or changing bandwidth requirements. Information about vendors must also be maintained to track service calls, install additional equipment, and manage equipment enhancements and upgrades. Configuration management databases are essential for managing problems and changes.

Network Problems and Changes

Network problem mangement (frequently called fault management) is similar in many respects to problem management for computer-based application systems. Faults or problems are indicated by network users, network operators, or by automatic alarming devices. The fault is documented in a problem report and tracked in a problem-management log.

The problem-resolution process involves network technicians, vendor equipment service personnel, network users when appropriate, and application specialists. During the problem resolution process, information contained in configuration databases is available for analysis and diagnosis. The process of problem management consists of locating the fault, isolating the faulty component through reconfiguration so the network can continue operation, replacing the faulty component to restore the network to its original status, and repairing the faulty component and placing it in inventory.

When the problem has been satisfactorily resolved, the configuration databases are updated with any new information. For example, if the problem originated in a cluster controller and the device was replaced with a new unit, the field inventory records would be updated and vendor maintenance activity would be initiated. If a link became inoperative, routing table information would be changed after proper analysis, traffic would take alternate paths, and link repair would begin.

The trouble report is closed when the problem has been corrected. Then the problem log is updated to reflect the corrective action.

Network problems arise from a variety of sources; physical problems are most common, however. The relationship between problem frequency and protocol layers is presented in Table 14.3.[14]

TABLE 14.3 Network Problem Sources

Protocol Layer	Percentage of Problems
Application	3
Session	7
Presentation	8
Transport	10
Network	12
Data link	25
Physical	35

There are many tools and many different pieces of test equipment to help network technicians isolate faults. Some rather simple testers can discover faults in the physical cabling, connectors, and switches. These devices cause many of the problems with networks. Incorrect addressing or unplanned configuration changes also lead to problems. User intervention with network components should not be permitted because it usually causes more problems than it solves.

Networks frequently are designed to be highly flexible systems. Their physical and logical configurations are subject to constant change from fault correction, bandwidth allocation adjustment, application or user mobility, and network growth. Changes to the network's physical or logical configuration are a prolific source of faults if not managed carefully.

In some network systems, bandwidth changes and some configuration changes occur routinely and automatically. The network system may include dynamic bandwidth allocation, for example, or it may perform automatic routing changes. Some changes result from problem correction and some from the need to improve performance. Other changes are designed to expand the network to serve more users, to add equipment, to expand services, or to reduce costs. Such changes originate from thoughtful planning and occur over longer periods. The change management process described in Chapter 12 is a good model for managing the nondynamic network changes. Network changes involve risk. The disciplined process of change management manages and controls risk.

Network Recovery Management

It is not possible to eliminate all risk of failure in network operations because some events are beyond the network manager's control. Indeed, in many firms vital network resources are owned and operated by third parties. Natural events such as hurricanes or earthquakes and disasters such as the Hinsdale fire underscore the need for network managers to have disaster recovery plans.

Network managers must have plans to recover from local disasters that affect LANs and their connections to local mainframes. The firm's PBX is a vital communication asset. Recovery plans for it and for its connections to common carrier facilities or to centralized computer systems must be in place.

There are some high risks with networks that rely on common carrier services. The rapid deployment of fiber-optic links is being concentrated along rights-of-way owned by communications companies, railroads, pipeline companies, and utility companies. Because rights-of-way are expensive, common carriers install multiple fibers in one trench. To economize further, several carriers may use the same trench. These networks are dangerously vulnerable because an accidental breach of the trench may sever all the links of several carriers. Network managers cannot be sure that leasing links from several companies offers backup; they must examine the routing as well.[15]

Recovery management is an important and critical task for network managers. But the many alternative network configurations available and the need to provide network solutions for computer system recovery planning make the task manageable.

Network Performance Assessment

Most network applications require that the network perform at or above certain minimum performance levels. For example, voice transmission requires a minimum bandwidth of 4000 Hertz in order to be up to our normal expectations. Likewise, high fidelity sound transmission requires significantly higher bandwidth. Network performance must be maintained at levels that satisfy application and user expectations as recorded in service agreements.

Network performance must be monitored and controlled. Monitoring allows managers to understand network performance; controlling permits managers to adjust capacity and change performance. Performance tools measure traffic and enable managers to compare throughput to capacity. Managers must understand changes in throughput, they must discover bottlenecks, and they must have measures of user response time. In addition, network managers must have measures of mean time between failures and mean time to repair. These measures are required to comply with availability, reliability, and serviceability commitments in the service agreement.

In general, network availability expressed in percentage is calculated by dividing the mean time between failures (MTBF) by the sum of the mean time between failures and the mean time to repair (MTTR). This can be written as:

$$\text{Availability } (\%) = \frac{\text{MTBF}}{\text{MTBF} + \text{MTTR}} \times 100$$

For example, if a network has a mean time between failures of 200 hours and has a mean time to repair of one hour, then the availability is about 99.5 percent.

Availability increases as the mean time between failures increases, so it is important to have reliable network components. However, most network managers are strongly motivated to achieve availability by reducing mean time to repair to a very low number or by eliminating it entirely. This can be accomplished by installing network redundancy, providing alternative capacity, or automatic rerouting. This strategy provides network reliability to the customer as well. It is a favorite way to achieve service levels.

Transmission accuracy is usually not a concern with today's networks because of built-in error-detection and error-correction mechanisms. However, error rates must be monitored because retransmissions add to network loading. They may also be a leading indicator of future failures or faults.

Capacity Assessment and Planning

Network capacity must be assessed periodically, and capacity planning is almost a continuous process. The rapid pace of technology introduction and adoption and the rapid growth of applications mean that network managers must constantly monitor throughput, user-experienced response times, and network utilization. This data is required to determine the current load in detail. When analyzed over time, it also gives managers the trend information needed to project future network bottlenecks and constraints.

Load projections of current applications and load forecasts from planned future applications are the base for accurate projections of future load. The growth in current application volumes may result from increased number of users (end-user computing may be expanding its customer set) or from increasing transaction volumes with current systems (E-mail usage may be increasing). New applications, such as linking the distribution system to the firm's customers through EDI, may place new, increased demands on the network. Business plans, functional plans, and service agreements may reveal additional demands or new network uses.

A new capacity plan must be developed to handle the incremental load. It may specify additional bandwidth, more user-oriented devices, additional controllers or concentrators, or different, more effective links. The objective of the capacity plan is to install cost-effective network solutions that achieve user expectations specified in service-level agreements. Because of the dynamic nature of telecommunications systems and applications, network managers should update the capacity plan quarterly. It should be reviewed periodically through monthly status reports.[16]

Network planners should use the strategic business plan as the starting point for their capacity plan; they should develop tactical plans to implement it. Network architectures must be flexible to accomodate changing business conditions. If possible, the network should be kept simple and easy to understand. Network installation requires skilled people and takes significant amounts of time for testing and reconfiguration.[17] The change-management discipline is very helpful in implementing capacity enhancements.

Management Reporting

Many of the performance assessment tools provide management data that can be shared with network users. Users must receive reports relating to their service-level agreements. They should be informed of service availability, reliability, cost, and system responsiveness. In turn, they should respond to IT user-satisfaction surveys.

Network managers and senior IT managers need additional information. Throughput, or the transaction-processing rate, network utilization, problem analysis, change-management results, and capacity plans are essential. The results of performance analysis and capacity planning are important for tactical planning purposes.

NETWORK MANAGEMENT SYSTEMS

Network management systems consist of computer hardware, software, and a variety of application systems. They are designed to provide operational control, to collect data about network operations, and to monitor and report network usage and performance. Data collection, storage, and retrieval capabilities are useful for managing problems, monitoring and managing performance, and planning network capability and growth. Systems to perform these tasks are becoming more sophisticated and more integrated. Managers of networks require systems that support multivendor hardware configurations and provide automated operational controls. Suppliers of network management systems are driving toward these objectives.

Until recently network management systems were designed and operated to support specific network elements. For example, a LAN would receive management support from software installed in the master system on the net. A large mainframe system would obtain support from system software in the network control program. Support for each network element would be separate from the others, and it would be vendor-unique. Integrated network management systems are designed to coordinate and integrate network management support across network elements and across vendors.

The goals for an integrated network management facility have not yet been attained but progress is evident. Major vendors like IBM with

Netview, AT&T with Unified Network Management Architecture/Accu-Master, and DEC with Enterprise Management Architecture are assembling the essential elements. However, these systems are still largely vendor-unique. Upwards of 50 vendors supply network management systems of all types. It is estimated that large U.S. companies will spend nearly $500,000 each on network management products in 1991, up 75 percent from the previous year.

Most network management tools are workstation-based and consist of specific hardware and software. Most contain alarm monitoring, traffic monitoring, network status reporting, and remote facility testing capability. Some support capacity planning, resource utilization analysis, and configuration management. Others provide chargeback and billing functions. The most capable systems cost about $250,000. The choices are wide. The selection is highly dependent on the firm's current network configuration and its future plans.

The Open System Interconnect, or OSI, standards offer hope for resolving some of the network management difficulties. International standards are being proposed for network-management systems based on the OSI model. The implication is that, as network product vendors move toward OSI network standards, system-management vendors will provide tools to operate in the standard environment. Future network management systems will go beyond providing tools; they will contain expert systems to automate further some of the tasks previously reserved for network specialists.

Expert systems will be valuable for problem determination, for recovery operations, and for some operational-control tasks. They may help design new network configurations and advise network managers during change management. Sophisticated expert systems may also analyze performance data in near real time and advise management about impending system overloads and capacity constraints. Because network monitoring systems gather enormous amounts of data and because much of the data has high short-term value, expert systems have the potential to act quickly on the information to affect system changes. They have the potential to optimize the use of system resources and to reduce system costs.

INTERNATIONAL CONSIDERATIONS

Earlier, telecommunications was labeled the enabling technology. Telecommunications enables firms to capture strategic advantage through use of voice, image, and data communications. Software applications, utilizing these forms of information exchange, enable the firm to relate to its customers and suppliers more effectively and to attain internal efficiencies. The firm increases the effectiveness of its value chain through improvements in time-to-market, for example. Telecommunications improves effectiveness, increases efficiency, and improves old relationships or creates new ones.

Global networks enable the firm to attain these advantages on a worldwide scale. "The globalization of commerce and communications is changing the way we work, the way we do business, the way we learn, and the way our nation competes in the global marketplace. Why global? We might as well ask: Why breathe?" states William G. McGowan, chairman of MCI Communications Corporation.[18]

International links based on ISDN are being established with both Europe and the Far East. Nippon Telegraph and Telephone and Illinois Bell Telephone Co. joined forces to offer an ISDN link between Tokyo and Chicago. The link utilizes AT&T's Switched Digital International services and the Japanese carrier KDD's international ISDN service to join the countries. Within Japan, NTT supplies the domestic ISDN service; in the Midwest, Illinois Bell provides ISDN service through its Integrated Network Service. The first customer for the service is Anderson Consulting, which will connect its Tokyo and Chicago offices for facsimile and videoconferencing applications.

France Telecom will interconnect its ISDN system, called Numeris, with ISDN in the U.S. This connection will give its customers access to applications in the States and provide U.S. customers with connections to applications in France.[19] France Telecom widely advertises its service to and within Europe. Satellite service to the U.S. from Europe for a 64 kbps link is $7911 per month, while the U.S. to Europe and intra-European network connection is advertised at $40,198 per month.

The field of telecommunications in Europe is mostly the exclusive territory of the postal telegraph and telephone (PTT) branches of the European Community of nations. However, the European Economic Commission will allow private companies to offer value-added services which, in effect, will start deregulation. Resale of leased lines, already in effect in Britain, will eventually take place on the continent. Restrictions on the marketing of terminal equipment designed to connect to the public network are also forthcoming.[20] It is expected that these developments will lead to significant growth in telecommunications in Europe and will foster growth of firms using these services.

Recognizing these trends, the Netherlands privatized its PTT in January, 1989, not only to meet its domestic needs, but to capture a share of the world's telecommunications business. The Dutch are preparing to be at the center of the open market which will exist in Europe after

1992. They are rapidly installing ISDN and fiber-optic cables, and they expect to invest $1.8 billion in 1989 alone.[21] The new Dutch company gives the Netherlands a strong competitive edge in the global telecommunications market.

The dynamic nature of telecommunications in all countries places many responsibilities on international network managers. They must be knowledgeable about a broad range of topics such as the technical characteristics of long-haul links including satellites; services provided by international carriers including value-added networks; tariff schedules from many suppliers in many countries; and many facets of national and international rules and regulations. Transnational data flow is subject to a large number of regulations, guidelines, and laws. Managers of international networks must be aware of these rules and must have a means for staying current on changing policy and regulatory matters in this country and abroad. The telecommunications manager in a company with international operations has a challenging position.

THE NETWORK MANAGER

Networks and network management are rapidly growing businesses fueled by technological, economic, and policy considerations. Rapid advances in telecommunications technology are coupled with an explosion of products, services, and bandwidths driven by increasingly favorable economics. Policy and political changes are cooperating to accelerate system adoption and implementation on a global scale. Adoption of the technology mandates structural changes in organizations. Structural changes usually require advanced systems, so the cycle reinforces itself.

Management tools, techniques, and processes lag technology adoption by a considerable margin. Thus IT managers are not completely prepared to orchestrate the communication technology revolution. But orchestrate they must. Because network assets cross organizational, political, and geographic boundaries, their management demands the skills of a generalist. Corporate networks probably place higher demands on the IT manager's general management skills than the traditional information systems do.

What are the functions of the senior network managers? What responsibilities do they have?

Corporate network managers must develop the network strategy for the firm in support of the firm's business strategy. They must evaluate technical and regulatory developments in the industry on a worldwide basis and provide input to the firm's strategic plan in order for the firm to capitalize on the new trends. They must shape the firm's strategic direction with their technical and business insights. In short, the corporate network manager must be a superior business strategist.

Corporate network managers must also be outstanding tacticians. They must clearly understand the costs and benefits of the current network system; they must evaluate emerging technologies and new products; they must understand changes in the legal, regulatory, and business environment; and they must use this knowledge to develop and implement near-term plans. The near-term plans must move the IT organization and the firm toward the goals spelled out in strategies and strategic plans.

Operational considerations must also have the attention of corporate network managers. They must review the reports from the management disciplines and must ensure that the disciplines are effective. Operational control is one of their important responsibilities.

Corporate network managers have both line and staff responsibility; they must be consultants to the firm's executives, and they must provide advice and counsel to functional managers throughout the organization. They have important jobs with considerable responsibility.

Network management has been a subject largely learned on the job by individuals whose training and experience has been in computer science, information systems, or telecommunications. Many telecommunications managers were trained before deregulation. They worked with the local telephone company to arrange for local and long-distance service; corporate executives considered their service to be a utility similar to water services or electrical power. Today the job is much more complex because there are many more products, services, and vendors; telecommunications is no longer just a utility. It generates revenue and creates strategic advantage. In most firms, the telecommunications function has merged with the information systems function as the technologies themselves have merged.

Until recently there were scant opportunities for students to learn network management in colleges or universities. Carnegie-Mellon University, in a joint project with Bell Communications Research Corporation, is bringing together courses in telecommunications, computer science, and management to award a master's degree in network management.[22] Other institutions are developing programs too; academic interest in network management is on the rise.

SUMMARY

Networks are an important part of the information processing infrastructure of most firms. They are important because they are an integral part of many strategic systems; they help provide competitive advantage. Telecommunications systems shrink time and distance. They provide leverage to firms through improvements in operational efficiency and effectiveness, and they can make these improvements on a global scale. Networks are also important because they facilitate organizational change. They can provide communication and control mechanisms to firms as they restructure or as they move toward decentralized operations.

Network management includes processes and procedures to ensure efficiency, effectiveness, and customer satisfaction from networks. Managers have expectations of networks expressed in service-level agreements; the subsequent disciplines of configuration management, problem and change management, recovery management, performance assessment, and capacity planning are designed to achieve planned service levels. These disciplines are the management framework for success in network operations.

There are many tools or network management systems to help monitor and control networks. Some of the major tools provide great assistance to network managers but none are capable of integrated network management at this time.

Corporate network managers need many skills to carry out such tasks as strategizing, strategic planning, tactical management, and operational control. If they manage or use global networks, many international considerations must be a part of their thinking as well. The network manager has an important and challenging job, one that is exciting and can be very rewarding.

Review Questions

1. What network management issues surfaced in the Volvo vignette in Chapter 6? What issues are unique to an international network?

2. Why is network management important to most firms today?

3. What is the connection between the disciplined approach to production management discussed in the previous chapters and network management?

4. What are some of the capabilities offered by networks which are important to firms strategically?

5. What technologies are spurring the growth of networks? How are some of these technologies working together to increase the growth of networks even faster?

---

6. Review the key issues found in Chapter 1 and itemize those relating to telecommunications systems and opportunities.

7. Why are networks important to corporate organization and corporate restructuring?

8. Why is network management a difficult topic?

9. What is the scope of network management?

10. What do users expect of networks? What do executive managers expect?

11. What are the disciplines of network management? What are the differences between network service-level agreements and applications service-level agreements?

12. What are the essential ingredients of the configuration management discipline?

13. What databases are maintained in the configuration management system?

14. What processes are involved in network problem management?

15. According to the text, most faults originate in the physical portion of the network. Why do you think this is so?

16. What network changes occur automatically in some networks? What types of changes are planned through change management?

17. What risks are associated with using common-carrier networks? How can network managers reduce these risks?

18. How can network managers maintain high system availability?

19. What information is needed to develop a new capacity plan? From what sources does this information come?

20. Why is it difficult to develop automated management tools for today's networks?

21. What special knowledge must managers of international telecommunications systems have?

22. What responsibilities do corporate network managers have? What responsibilities do they have regarding the disciplines?

Discussion Questions

1. Discuss the importance of telecommunications systems to each of the strategic systems discussed in Chapter 2.

2. International networks are expanding rapidly. Discuss why this is happening by considering technical, economic, regulatory, sociological, and political factors.

3. Discuss how a firm might use a telecommunications system as part of its implementation strategy for decentralization. What role might telecommunications systems play in the strategy of a firm that wants to grow by acquisition?

4. Itemize the reasons why network management is a difficult task. Which reason is most important, in your opinion? Why?

5. Define the boundaries of network management. Where is the boundary with respect to other internal operations of the firm? How far does the boundary extend outside the firm?

6. What are the sources of the expectations that executives may develop regarding telecommunications systems? What role should IT managers play in developing executive expectations?

7. How do you think the issues of information infrastructure and organizational learning are related to telecommunications systems in an organization?

8. Compare and contrast the service-level agreement for a traditional information system with that for a network system.

9. Why is the configuration management discipline so important to network managers?

10. Why must configuration management be a real-time operation for managers of sophisticated networks? There may also be a modest configuration management system for the mainframe computer center. Discuss the similarities and differences between configuration management for the mainframe and configuration management for networks.

11. Discuss the evolution of the disciplined management systems as a firm migrates from a traditional, mainframe data processor to an adopter of end-user computing. What changes might take place as the firm adopts EDI?

12. Referring to the formula for availability in this chapter, discuss the relationship between availability, reliability, and serviceability. What actions can managers take to improve availability?

13. Discuss the capabilities available with today's network management systems.

14. Discuss the potential applications of expert systems to the management of networks.

15. Discuss the responsibilities of a network manager in an international firm by writing his/her job description.

16. Discuss the implications to application portfolio management of a firm's strategic direction to link its dispersed operations to its headquarters through networks.

Assignments

1. Using Chapter 12 as a guide, design a fault or problem report and a problem logging system appropriate for network managers. Develop a rationale for combining network and data processing problem meetings or for holding them as separate events. What are the most important factors in this decision?

2. Study one of the major network management systems such as Netview from IBM or AccuMaster from AT&T. Develop a list of its functional characteristics and itemize its strengths and weaknesses.

3. In cooperation with one of your classmates who studied a different system, compare and contrast the two systems.

[1] Excerpted with permission from Patricia Keefe, "Met Rings In Remote Sites," *Computerworld*, December 12, 1988, 1.

[2] Metropolitan Life spends about $225 million annually on information technology.

[3] Mark A. Kellner, "Merrill Lynch Dials MCI for $150 Million Pact," *MIS Week*, June 12, 1989, 15.

[4] John Mahnke, "Insurance Carrier Hooks Agents on Network," *MIS Week*, August 28, 1989, 13.

[5] Michael Hammer and Glenn E. Mangurian, "The Changing Value of Communications Technology," *Sloan Management Review* (Winter 1987): 65.

[6] Dennis W. Elliott, "North Pacific Cable Meets Asian Connectivity Needs," *Telecommunications*, February 1990, 39.

[7] Emily Leinfuss and Andrew Collier, "The Brits Are Coming for Tymnet," *MIS Week*, August 7, 1989, 1. Tymnet is the second largest U.S. provider of VANs with an annual revenue of $250 million and is an employer of 1500 people. The flow of funds is not one way, however. AT&T agreed to purchase Istel, a British firm specializing in computer services, in late 1989 for $288 million. This activity follows two cross-investment agreements by AT&T, one with Italtel of Italy and another with Phillips of the Netherlands, earlier in 1989.

[8] Wayne Ryerson and John Pitts, "A Five-Nation Network for Aircraft Manufacturing," *Telecommunications*, October 1989, 45.

[9] Mickey Williamson and Tammi Harbert, "The End of Anarchy," *CIO*, November 1989, 39.

[10] Cornelius H. Sullivan and John R. Smart, "Planning for Information Networks," *Sloan Management Review* (Winter 1987): 39.

[11] Jeanne Iida, "Integrated Network Management: Progress, But Still No Solutions," *MIS Week*, May 22, 1989, 14.

[12] *Trends in Communication Policy* (Boston, MA: Economics and Technology, Inc., September 1989).

[13] Kornel Terplan, *Communications Network Management*. Englewood Cliffs, NJ: Prentice Hall, Inc., 1987, 2. Terplan approaches this topic through initial success factors which lead to the activities of operational control, administration, analysis and tuning, and capacity planning.

[14] Rolf Lang, "Diagnostic Tools Decipher LAN Ills, Reduce Downtime," *Federal Computer Week*, June 26, 1989, 48.

[15] James B. Pruitt and David Kull, "Tenuous Connections: Public Network Flunks Key Test," *Computer Decisions*, April 1989, 28.

[16] Kornel Terplan, "Network Capacity Planning," *EDP Performance Review*, August 1988, 1.

[17] Michael Puttre, "How to Avoid the Classic Blunders of Networking," *MIS Week*, October 2, 1989, 21.

[18] Willian G. McGowan, "Why Global?" *Telecommunications*, April 1989, S-6.

[19] Lorie Teeter and Jeanne Iida, "First International ISDN Links Established," *MIS Week*, December 4, 1989, 12. NTT's net, called INS-NET, serves 600 subscribers with 3321 lines. Numeris will serve the entire country of France by the end of 1990. Currently Numeris has about 2000 subscribers but expects 100,000 by 1992 and 1 million by the end of the century. It was estimated that ISDN lines in the United States numbered about 60,000 at year end 1989.

[20] James Fallon, "Users To Have Choice in Europe-Wide Data Nets," *MIS Week*, September 4, 1989, 10. The European Community's telecommunications market is estimated to be 25 percent of the $381 billion worldwide market. It is expected to grow from 3 percent to 7 percent of the EC's gross domestic product by the year 2000.

[21] Peter Lablans, "Privatization of the Dutch PTT: New Telecommunications Opportunities," *Telecommunications*, November 1989, 61.

[22] David A. Ludlam, "Managing Global Networks," *Computerworld*, October 2, 1989, 86. About 35 colleges offer degrees in telecommunications but only 3 grant the Ph.D. degree. A degree in information systems is offered by about 360 colleges and universities in the U.S. and 60 of these offer the Ph.D.

REFERENCES AND READINGS

FitzGerald, Jerry. *Business Data Communications*, 3rd ed. New York: John Wiley & Sons, 1990.

Gantz, John. "How To Succeed." *Telecommunications Products + Technology*, October 1987, 17.

Housel, Thomas J., and William E. Darden III. *Introduction to Telecommunications: The Business Perspective*. Cincinnati, OH: South-Western Publishing Co., 1988.

Karp, John, and World Communication Works, Inc. "Network Management: What Every Executive Needs to Know." *Fortune*, Advertisement Supplement, January 30, 1989, 15.

Keefe, Patricia. "Met Rings In Remote Sites." *Computerworld*, December 12, 1988, 1.

Orazine, Ron. "Why MIS Managers Are Becoming Network Experts." *Telecommunications*, January 1988, 57.

Stallings, William. *Business Data Communications*. New York: Macmillan Publishing Company, 1990.

Terplan, Kornel. *Communication Network Management*. Englewood Cliffs, NJ: Prentice-Hall, Inc., 1987.

Part
Five

15 Accounting for Information Technology Resources

16 Information Technology Controls and Asset Protection

Controlling
The Information
Resource

Business control is a basic management responsibility. Controls are required by law in many instances and are increasingly important with advances in automation. Managers must understand financial and operational control issues surrounding information technology and must maintain satisfactory control as technology penetrates the organization and business practices are reengineered. This Part's chapters include Accounting for Information Technology Resources and Information Technology Controls and Asset Protection.

Technology advances require managers to have an increased awareness of business control considerations. Managers must develop and use refined tools, techniques, and processes to discharge their control responsibilities effectively.

15

Accounting for Information Technology Resources

A Business Vignette

Introduction

Why Account for IT Resources?

The Objectives of Resource Accountability

Should IT Recover Costs from Users?
  The Goals of a Chargeback System

Alternative Methodologies
  The Profit Center Method
  The Cost Center Method

*Some Additional Considerations in
Cost Recovery*
  *Funding Application Development and
  Maintenance*
  Cost Recovery in Production Operations
  Network Accounting and Cost Recovery

Relationship to Client Behavior

Some Compromises To Consider

Expectations

Summary

Review Questions

Discussion Questions

Assignments

References and Readings

As the computer industry struggles with its current slump, it is important to remember what caused the slump in the first place. Many reasons are advanced, but the most plausible centers on measurement—or lack of measurement.

Senior management closed the floodgates on the information technology buying binge of the early 1980s when confronted with the fact that MIS directors could not measure the return on the investment in new equipment. The concern was not so much that the investment return was inadequate, but that MIS did not know what the return was. No measurement system was in place. In the drive to achieve computer literacy, however it was defined, we were flying by the seat of our pants.

How does one measure the return on investments in information technology? The answer is in terms of the impact of the investment on business operations at the organizational level. The impact may be directly financial—dollars saved or costs avoided—and measurement is relatively easy. More frequently—especially with today's investments—the impact is broader, perhaps leading to a better positioning in the marketplace or an improved distribution operation or better customer service. Such gains are far more difficult to measure, but they are equally valid.

Advanced office systems have proven to be the most difficult investments to cost-justify. The lessons learned to date in that arena are instructive. Once a knowledge worker begins using a machine, say a personal computer, and integrates the use of the machine into regular work processes, it is not possible to differentiate the contribution of the machine from the contribution of the person. They become a single entity.

Further, once a machine is tied to a network, its contribution is no longer associated with a single individual, but with the total organization. The conclusion that mature users have arrived at, after struggling with various measurement approaches over the years, is that investments in advanced office systems (and in fact in any network-based system) can only be measured in terms of improvements in organizational effectiveness. The beneficiary may be the individual, but the gain, to be cost-effective, must be measureable at the organizational level.

A good CIO should have a clear understanding of the mission of the organization for which he or she is responsible, agreement with superiors as to what constitutes success and how success is to be measured, and a prioritization of problem areas. Information systems should be measured in terms of how they contribute to success and help solve problems, as both are defined by management. Granted, these are not objective measures such as profit contribution, actual versus plan, market share, factory throughput, and other quantifiable yardsticks that might enter into the equation. Ultimately, it is senior management's subjective decision as to what constitutes success that really counts, and information systems performance measures must reflect that fact.

Unfortunately, not every company has in place a good system for measuring organizational effectiveness. Far too often, they tend to fall back on the numbers coming out of the accounting system, and measurement is limited to budget versus actual. In such cases, measurement of information systems performance will be equally limited, and the tendency will be to revert to approaches that concentrate on head-count reduction and time saved—both of which are virtually meaningless statistics in measuring operating performance.

In this context one begins to see the value of the chief information officer concept. As a member of senior management, with a special knowledge of information systems and their impact, a CIO can take a broad view of company requirements and raise questions about the adequacy of the company's systems for measuring organizational effectiveness—questions originally triggered by difficulties in cost-justifying equipment investments. The all-important change is that, while MIS directors are concerned with the management of technology, CIOs are involved in the management of the enterprise. In fact, a thoughtful CIO will begin to challenge other practices that inhibit the ability to measure performance in any meaningful way. The need to value investments in machines and people on some common ground is one area that CIOs should address in seeking better ways to measure performance.

Experience shows that the knowledge worker and his or her machine must be viewed as a single entity. The performance of one is inextricably entwined with the performance of the other. Yet the accounting system does not permit this view. Machines are considered a capital investment to be amortized over time. People are an expense to be written off immediately. One is a balance-sheet item, the other is not. Management invests in machines to enhance the investment in people, but the accounting systems do not look at people as investments, so there is no way of tracking investment performance through the accounting system.

One lesson we have learned from using office systems over the past 10 years is that they are an investment in people. In the final analysis, the ability to compete more effectively depends on the talents and intellectual prowess of knowledge workers. The ultimate "mission-critical" systems are those that expand brainpower and help knowledge workers throughout the enterprise become more comfortable with technology as an operating tool. Once that comfort level is established and operating personnel can communicate with information-technology specialists on a common ground, ideas as to how technology can be used as a competitive tool will flow with ease.

What office systems have done is bring to the forefront the need to find a more rational way of measuring investments in people. Investments in office systems can only be measured properly in terms of their contribution to the maximization of human capital—the extent to which they help knowledge workers perform more effectively.

Another measurement anomaly of importance to CIOs is the accounting treatment of systems and software investments. It is not just the computers that are the asset; it is also the systems and software that run on them. Yet just as there exists a dichotomy in the valuing of people and machines, investments in equipment are capitalized while investments in systems and software are expensed. As a result, the accounting system is totally inadequate as a yardstick for telling us what our total investment in information technology is or how that investment is performing.

The measurement problem must be solved. We cannot trivialize it either by collecting and analyzing minutiae or by ignoring the problem and hoping it will go away.

We invest in information technology today to help achieve corporate goals—numeric and otherwise—as those goals are defined by senior management. The measurement systems must reflect that objective. The challenge facing today's CIO is to convey that message to colleagues in the senior ranks of management and to help design measurement systems that evaluate information systems in terms of the goals of the enterprise.

INTRODUCTION

Information technology resources comprise a large and growing portion of most firms' budgets. Because of cost pressures and increasing competition, many organizations today are intensely scrutinizing these budget items. Many firms are spending 2 to 5 percent of revenue on IT activities; IT managers are expected to explain these expenditures and account for them to the firm's officers.[2] As corporations become more information intensive, these expenses grow rapidly and become increasingly visible. The consolidation of telecommunications departments with information processing departments; the growing importance of telecommunications to the firm; the widespread adoption of end-user computing; office and factory automation; and the increasing complexity of the application environment all heighten executives' interest in IT expenses.

Accounting for IT resources is a complex and difficult task because the expenditure of the resources occurs in a variety of ways. Some of the expense mechanisms are tied directly to the internal workings of complex hardware and software systems, some are incurred in the operation of sophisticated networks, and some are related to labor. For example, accounting for the detailed operations of a multiprocessing operating system requires computer resources. It is difficult to explain to users. It seems to be understood only by highly skilled computer specialists. Accounting for these activities can sometimes lead to endless discussions over the methodology.

Accounting for IT resources is difficult because the resources themselves have many forms and are generally widely scattered throughout the organization. In addition, as the business vignette emphased, the accounting system is ill-suited to measure many of the important aspects of information technology.

Accounting for IT resources and charging for the use of these resources alter incentives and impact organizational behavior. High costs discourage usage. Low costs or free services encourage high and perhaps unwarranted consumption. Consequently, accounting and charging processes alter cost and expense flows, influence the economics of organizations within the firm, and change the behavior of individuals and organizations. In addition, the administration of the IT cost-accounting system may be a time-consuming task, but it is an important task. The IT organization and structure will be affected by the cost-accounting system too.

This chapter introduces IT resource accounting and discusses some common chargeback methods. Several alternatives for handling application development, program maintenance, and production operations are discussed. These variations illustrate the subtleties of charging mechanisms and provide insight into some of the financial motivations experienced by IT users.

WHY ACCOUNT FOR IT RESOURCES?

Most IT resources are in critically short supply. Skilled people, sophisticated networks, complex and strategically important systems, and modern hardware and software systems are valuable. In some cases they take a long time to acquire and represent a large investment. IT investment decisions and the allocation of future IT resources are critical tasks for the firm's managers. To make these investment decisions intelligently, managers need reasonably accurate knowledge of past expenditures and plausible projections of future expenditures. A relatively accurate cost-accounting system is essential to begin to quantify the costs and benefits of IT activities. The accounting system also quantifies and calibrates the costs of delivering IT services to user organizations.

There is no meaningful alternative to accounting for IT expenses. Some firms treat IT expenses as overhead and pool similar expenses together. For example, all labor expenses for IT are collected and summarized, hardware expenses are pooled, and supplies and other costs are also summarized. In some firms these pooled expenses are treated as corporate overhead affecting the firm's profit at a high level. In other firms, expense catagories for IT are granular and are passed through to the ultimate benefactors of the service. There are many alternatives between these extremes, but there is no alternative that avoids accounting for expenditures. The choice of accounting methodology, however, has many important consequences.

Accounting is a process for collecting, analyzing, and reporting financial information about an organization. Accounting can have two forms: financial accounting and managerial accounting. Financial accounting provides information about the organization to outside individuals or institutions such as banks, stockholders, government agencies, or the public. Because the information is used by the public, financial accounting is subject to carefully crafted accounting rules. Financial accounting is oriented toward providing a balance sheet and an income statement for the firm.

On the other hand, managerial accounting is concerned with information that is useful in the internal management of the organization. The organization can establish its own rules and can tailor them to its specific needs. The goal of managerial accounting is to provide the firm's managers with information that will enable them to optimize the firm's performance. Accounting within the IT function provides the IT management team with information to optimize and control IT activities. IT accounting information is a valuable and critical resource.

Innovative managerial accounting can overcome some of the difficulties presented in the vignette. For example, application development costs can be capitalized and amortized over the anticipated life of the application. Traditional financial accounting methods do not reflect the importance of intangible assets in the service and information industries. Price

Waterhouse is pushing the FASB to consider allowing companies to capitalize costs associated with developing their own software. "Companies spend enormous amounts of money on internal computer systems for production, purchasing decisions, sales analysis," says Arthur Siegel, vice chairman for accounting and auditing at Price Waterhouse, "but general practice is to expense those costs as incurred. We need to rethink this."[3] A firm does not need to wait for the FASB to act in order to implement some of these changes internally.

Within the firm purchased system software and purchased applications can be treated like balance sheet assets and depreciation can be taken to reflect technological and business obsolescence. Accounting for human or social capital is a problem that remains to be solved. These and many other aspects of accounting for IT resources are the main themes of this chapter.

Managers are concerned about the financial aspects of information technology. IT costs and expenses are increasingly being dispersed throughout the firm as distributed data processing and end-user computing blossom. And, as computer resources become widely dispersed, the budgetary responsibility for these expenses also becomes dispersed. In addition, the ingredients of the expenses are changing due to the declining cost of hardware, the increased use of personal workstations, the rapid growth of networks of all kinds, and the trend toward purchased software. For all of these reasons, it is becoming more difficult to understand the total cost of information processing in the firm.

As the dispersion of information processing continues and departments throughout the firm step up their spending for this activity, the central information systems organization's budget is not declining.[4] Trends in the central IS organization are to reduce spending on mainframe computers but to increase the purchase of application programs and network hardware and software. In addition, the costs of personnel have been rising at a steady rate. As a result, in most firms the total expenses for information processing have been rising.

As the technology becomes increasingly pervasive in the firm, costs and expenses are more difficult to identify and quantify. Training costs associated with the introduction of new information technology are a good example of this difficulty. When training in information systems activity was largely confined to the central IS organization, accounting for this activity was relatively straightforward. As office systems and end-user computing flourish, training and other startup expense are increasing and are less easily identified. The dispersion of all the training and startup efforts, particularly those of an informal nature, make accurate accounting for the activity extremely difficult. For instance, the introduction of office automation may cost as much as $12,000 per workstation. About half of this cost is for additional support staff, startup expenses, and individual training. Many of these expenses are not accounted for through the usual accounting system.[5]

In spite of the difficulties, accounting for IT expenditures is not optional. What is largely optional, however, is the methodology to accomplish the accounting process. Current rules and regulations and commonly accepted accounting practices leave considerable room for customization. The degree of customization and the manner in which IT expenses are allocated to users of IT services are important issues for the firm and for the IT organization. IT resource accountability is a critical issue.

THE OBJECTIVES OF RESOURCE ACCOUNTABILITY

Accounting for IT resources is a logical continuation of the information technology planning process described in Chapter 4. The management system for developing strategies and plans leads directly to the topic of resource accountability. The IT planning model discussed earlier includes applications, production operations, resources, and technology items. In addition, the planning process includes feedback mechanisms based on measurement and control activities.

The objectives of resource accountability are to help measure the progress of activities found in the operational and tactical plan and to form a basis for management control actions. Control is a fundamental management responsibility. It is a critical success factor for IT managers. Specifically, IT managers must deliver services of all kinds, on schedule, and within planned costs. IT resource accountability deals directly with these measurement and control processes.

The IT planning process and the firm's planning process culminate in a budget for the firm and a budget for IT resources in support of the firm's objectives. The firm's budget describes the resource expenditures required to meet the objectives in the firm's plan throughout the plan period. Likewise, the budget for the IT organization includes the resources necessary for IT to meet its organizational goals and objectives. The conclusion of a successful planning process is a budget for the firm that incorporates the separate budgets for each organization within the firm. The budgets are all tightly coupled and support the firm's goals and objectives.

Control and planning activities are entwined and merge with one another. Control is the process through which management assures itself that members of the firm act in accordance with the policies and plans of the firm. In other words, policy setting and planning activity precede control actions. Generally the same people are involved in both planning and controlling the firm's operations. IT accounting systems are a way for the people within the firm to communicate with each other regarding IT plans and actions. IT accounting information can also be used for motivating individuals and for appraising individual and unit performance.

For example, consider a firm that embarks on a strategy to improve its product development and manufacturing effectiveness through the introduction of computer-aided design and computer-aided manufacturing systems (CAD/CAM systems). The expression of this long-term goal will be found in the firm's strategy. The details will also be found in the functional strategy statements for the product development, product manufacturing, and IT organizations. The firm's plans describe how each of these organizations will expend resources to accomplish this goal. Each organization's plans will describe how resources will be applied to accomplish its part of the overall task. The IT organization will develop strategies and plans to support the CAD/CAM installation, and the IT budget will also contain resources to accomplish the CAD/CAM installation. To complete the process and to control the installation activity, the IT organization needs a management system to measure and control the expenditure of budgeted resources.

Budgetary control is the process of relating actual expenses to planned or budgeted expenses and resolving the variances, if any. All firms have a process for accomplishing this task. The IT organization will be included in this process; the IT manager will receive information periodically from the controller on its financial position relative to the budget. The important task of controlling spending must be monitored by IT managers. They generally receive assistance from the firm's financial staff.

The tasks of planning, controlling, communicating, and budgeting are all forms of management decision making. Accounting information from the IT organization is very useful in making decisions regarding IT activities. These activities are also valuable because they cement relationships between IT and other parts of the firm. The budget plays an especially important role in this process. IT resource accounting processes accomplish several important objectives. These are listed in Table 15.1.

Is it sufficient to control IT activities with budgetary processes alone? Are the best interests of the firm served if using organizations are not financially involved in IT expenses? Since the IT organization provides valuable services to the firm, perhaps it is best if the costs of these services are recovered from the users in some way. Many firms believe so; chargeback systems are receiving increased attention.[6] Most firms charge users for IT services in one form or another.

TABLE 15.1   What IT Resource Accountability Accomplishes

Provides continuity between planning and implementation.

Establishes a mechanism to measure implementation progress.

Forms a basis for management control actions.

Links IT actions to the goals of the firm.

Forms a communication vehicle for plans and accomplishments.

Provides information on which to appraise performance.

SHOULD IT RECOVER COSTS FROM USERS?

Many benefits derive from an IT cost recovery process; most firms consider the advantages to outweigh the disadvantages. The most important advantage of IT cost accounting and cost recovery is the basis it provides for clarifying the costs and the benefits of IT services. A well-structured and well-operating IT cost-accounting system attacks an important IT issue—the role and contribution of IT. An important secondary advantage is the communication linkages it strengthens between IT and user organizations. Table 15.2 lists some benefits of IT cost recovery.

TABLE 15.2   Benefits of IT Cost Recovery

Provides a basis for clarifying costs/benefits of IT services.

Strengthens communication between IT and user organizations.

Permits IT to operate as a business within a business.

Increases employees' sensitivity to costs and benefits.

Spotlights potential unnecessary expenses.

Encourages effective use of resources.

Improves the cost-effectiveness of IT.

IT cost-accounting and cost-recovery methods focus managers' attention on services that have low or marginal value. For instance, if users are charged for the operation of a regularly scheduled application, the output of which has low value, they will balance costs and benefits and may terminate the application. If production operations is considered a free utility, processing of marginal application programs will continue, perhaps indefinitely. Cost accounting provides tools to detect unnecessary or cost-inefficient services. Inefficient services can be altered or eliminated, thus reducing expenses and improving effectiveness.

Users who receive a bill for IT services have information they can use to obtain maximum gain for the costs incurred. They may search for more effective processing schedules or increase the use of services having high marginal value. For example, the operation of an on-line application may be expanded when users have tangible evidence that the expansion in services is financially attractive. Likewise, if users are charged for application program enhancement, they will be motivated to request financially viable enhancements. The charging process encourages the effective use of scarce resources.

A well-designed charging mechanism also enhances the IT organization's effectiveness. IT costs are scrutinized more carefully when they are well known and have sufficient granularity. The need to explain IT charges to users directs IT managers' attention to the cost elements. Attention to the cost elements motivates IT to improve the cost effectiveness of its customer services. Insufficient attention to the cost elements or poor granularity in the cost structure promote ineffective resource utilization. For instance, if all telecommunications costs are grouped together, degrees of effectiveness among the various services may go undetected and the organization will incur the resulting inefficiencies. IT effectiveness will suffer also.

Pricing IT services and producing customer billings instill a heightened awareness of IT costs and benefits among the firm's employees and managers. IT personnel become sensitized to the money the firm is spending to support their activities; they become more cost conscious. Likewise, users who receive bills for the services they are procuring from IT become more value conscious. They appreciate what the firm is spending on their activities through IT; they tend to make better use of IT services. Chargeback processes increase the cost and value consciousness of employees and managers.

The primary disadvantage of IT cost recovery is its administrative overhead. The cost-recovery process is not free and must itself be cost effective. Cost recovery may not be justified for organizations that are not information intensive. Also, cost recovery may not be acceptable within the cultural norms of some organizations, which prefer not to involve users in the costs of IT services.

Not all IT organizations favor user chargebacks. Some IT managers prefer to keep their finances clouded in secrecy rather than expose their operations to whatever criticism may result from charging. Some firms operate with relative insensitivity to costs once the budget has been approved. In some firms, employees are relatively insensitive to expense; in these firms the motivations in chargeback methods are ineffective.

Not all firms favor IT chargeout, and not all firms are prepared for effective operation of a chargeout system. James Wetherbe concludes that four necessary conditions permit chargeouts to operate effectively:

1. Expenditures are discretionary.
2. Customers who most need service have the necessary resources.
3. Customers understand how the chargeout system works.
4. Costs and benefits can be uniquely identified by customers.

He argues that these conditions are not fulfilled in many cases.[7] One alternative to chargeouts is to distribute IT costs through a committee or to allocate them during the planning process. Although these are simpler approaches, they suffer from many disadvantages. The allocation or distribution methods are frequently political in character, they are inaccurate, they are relatively inflexible to changing conditions, and they reduce line management accountability and responsibility. Another alternative is to bury IT costs and expenses in general corporate overhead keeping them relatively invisible to IT and user employees. This approach completely eliminates the advantages of chargeout systems and is not favored by many firms.

Given the relative advantage of cost-recovery processes to most firms, what goals and objectives should the firm strive for in the process? What can the process gain for the firm and how must it be established for maximum effectiveness?

The Goals of a Chargeback System

Chargeback mechanisms for the IT organization should serve the firm and its constituent organizations by improving the firm's effectiveness and efficiency. The IT organization should benefit, but the firm as a whole should benefit as well. Table 15.3 shows some goals of an effective chargeback system.

Budgetary controls are important, but they are subject to the same criterion of effectiveness as other activities within the IT organization and the firm. The chargeback process adopted by IT must be easy to administer and must be easily understood by IT clients. The administrative cost must be small in proportion to the resources being administered. For example, if the cost to account for an IT service and to bill an IT customer is $5, the total billed over the accounting period should be considerably more than $5 on the average. In other words, the granularity of chargeable services and the chargeback process itself must be cost effective.

TABLE 15.3   Goals of a Chargeback System

Is easily administered.

Is easily understood by IT customers.

Performs cost distribution effectively.

Promotes effective use of IT resources.

Provides incentives to change behavior.

The algorithm for generating or computing the customer charge must be easy to explain and easy to justify for the chargeback mechanism to be accepted by IT customers and managers. A complex billing system based on obscure and hard-to-understand parameters will alienate customers and create planning and budgeting difficulties. If the algorithm is constructed from simple parameters and is easily understood, the charging process is also more easily administered.

The chargeback mechanism must be designed to perform IT cost accounting in what users perceive as a fair and equitable manner. For example, the accounting for programmers' time may be based on hours worked using two rates depending on the programmer's skill level. Since there are many more than two different levels of productivity among programmers, this algorthim is not completely accurate, but most user organizations will consider it fair. It is also relatively routine to administer and, therefore, is cost effective.

IT chargeback mechanisms alter the financial incentives for organizations to use IT services. Usage patterns and relationships between the IT organization and user organizations are influenced by the choice of chargeback method and by the manner in which it is employed. Therefore, another goal of the chargeback process must be to promote cost-effective use of IT resources. For instance, if there is no charge for the use of on-line direct access storage devices (DASD), then there is no financial incentive to use the storage effectively. Users are not encouraged by the charging process to delete obsolete data sets, for example. On the other hand, if direct access storage is priced unreasonably high, users may resort to inefficient methods of data handling to reduce costs. For example, they may employ tape storage when DASD is more effective for the organization.

If users are sensitive to charges, charging mechanisms can be used to steer the organization toward certain technologies and away from others.[8] New technologies can be attractively priced to encourage usage, and older systems intended to be replaced by new technology can be priced unattractively. The price may not reflect the actual costs: The new equipment may be more expensive than the older, fully depreciated hardware. In this instance the pricing algorithm is designed to provide incentives, not to recover actual costs.

Chargeback systems that are easy to understand and administer, distribute costs effectively, and promote effective use of IT resources are valuable additions to the IT management system. Successful IT managers place extensive reliance on them. Successful IT managers also carefully consider the advantages and disadvantages of the alternative chargeback methods before selecting one.

ALTERNATIVE METHODOLOGIES

There are two major accounting alternatives that the IT organization can employ to handle cost recovery: the cost center method and the profit center approach. Each method distributes IT costs to using organizations, but the methods vary considerably in other respects. The profit center method is designed to generate revenue from the users in excess of the costs and expenses incurred by IT. The excess corresponds to profit and may be used by the IT organization for its approved purposes. Of course, if costs and expenses exceed revenue, then the profit center loses money. These losses must ultimately be recovered in some way.

The cost center method, on the other hand, seeks to break even financially. The amounts received for services are expected to match costs closely. Methods to make costs and revenues match exactly are frequently employed. Details of these financial arrangements are discussed in the next sections.

The Profit Center Method

IT organizations established as profit centers operate as a business within a business. All costs and expenses incurred by the IT organization must be recovered through charges for services rendered to customers. The normal relationship between revenue and expenses prevails, and the organization may operate at a profit or a loss. The profit center method is easy to understand and to explain. It has other advantages, which are listed in Table 15.4.

TABLE 15.4   Advantages of the IT Profit Center

Is easy to understand and to explain.

Promotes business management.

Provides for outside comparisons.

Establishes financial rigor.

Enables outside sale of services.

IT managers who operate a profit center develop important skills as business managers. They use many of the disciplines used by independent business people. They learn business management in a way not usually possible in other environments. IT profit center managers appreciate the sometimes invisible expenses of corporate overhead, employee benefits, and equipment depreciation, and they learn about pricing strategies in a fundamental manner. Business management experience extends to IT customers as well. They may gain exposure to the financial consequences of issues of which they were not previously aware. The profit center approach is favored by firms for this reason, among others.

Profit center prices reflect the real cost of doing business within the firm. They can easily be compared with the price of similar services from outside suppliers. These comparisons provide a mechanism to measure IT effectiveness, and they promote IT efficiencies. If the firm has a policy permitting users to purchase competitive services, then IT is usually highly motivated to remain competitive. If the IT organization cannot remain competitive with external suppliers, then the firm may decide to purchase services externally. This process may also highlight inefficiencies within the firm because it directs attention to general overhead expenses and corporate accounting policies. The profit center approach has benefits that extend beyond IT and its clients.

Another advantage of the profit center approach is that it provides a high degree of rigor and discipline to the financial relationship between IT and its client organizations. Costs are more carefully established, expenses more thoroughly evaluated, and prices and charges are determined with more insight. The discipline in this method assists in other processes as well. For example, accurate costs to achieve service levels are available, thus negotiations on service-level agreements proceed more rigorously. Application portfolio managment is improved and make/buy decisions are made with heightened confidence. Financial management in general is improved.

Establishing the IT organization as a profit center enables IT to sell services outside the firm with confidence that the activity is profitable for the firm. Selling IT services may be advantageous to the IT organization and to the firm. The profit center methodology is a prerequisite to becoming an external revenue-generating service bureau. The IT profit center may be established as an independent business with its own financial accounting system. In this case, the profit center accounting system will link to and closely support the financial accounting system for the organization. The accounting rigor of a profit center enables the firm to capitalize on its IT resources as a source of revenue and profit if it chooses.

There are some obvious disadvantages to the profit center approach as well. Financial rigor comes at the expense of administrative overhead—rigorous financial treatment may not be worth the price. Detailed knowledge of all expense ingredients, however small, and soundly based prices may not yield improved performance overall. In addition, the firm is somewhat dependent on successful managment of the interface between IT and user organizations. There is no guarantee that prices will be properly established or that users will anticipate fully the costs of their IT services during their budgeting process. The IT organization may not earn the anticipated profit, and high-level adjustments will then be required. If IT is highly profitable, users perhaps paid more than required and alternative investments were sacrificed. The degree of financial coupling between IT and its clients may be insufficient to avoid these difficulties. Profit centers that are carefully managed avoid these difficulties and achieve success.

The Cost Center Method

Another method for handling IT cost recovery is the cost center approach. The cost center methodology is based on a process in which each user organization budgets for its anticipated IT services during the planning and budgeting process. The sum of all the user-budgeted amounts for IT is the cost center support from the users. It is expected that this amount will approximately equal the anticipated IT expenses for the planned period. Because there may be mismatches between planned IT expenses and user-budgeted support, this approach forces intense interaction between organizations during the planning stage. Planning and budgeting is highly iterative. It has other characteristics as well, which are displayed in Table 15.5.

In order for the annual planning process to go to completion in the user organizations, the users must know their anticipated requirements for the planning period. They must also know IT prices for the period so they can submit their budgets. Their budgeted support for IT is the anticipated usage times the price for the service. IT also prepares a budget that anticipates all the user demands and plans to satisfy them. The total of all the user budgets for IT must approximate the budget prepared by IT.

TABLE 15.5   Cost Center Characteristics

Promotes intensely interactive planning and budgeting activities.

Establishes prices in advance of known support.

Forces managers to handle variances.

Exposes the plan process to manipulation.

May lead to conflict (this may be beneficial).

Forces decision making.

Reinforces the SLA and capacity planning processes.

If the budgeted support is not close to the expenses planned by IT, the process recycles in the IT organization. This recycling effect means that the IT budget is the last to be completed. Also, because of changes in the support level, prices may change. The effect of this is to alter the budgets in the user organizations by a small amount. Usually this is handled by permitting some variance at plan time between the sum of all budgeted support for IT and the expense budget prepared by IT.

In the ideal situation, user budgets for anticipated IT services and the IT budget for anticipated expenses are approximately equal. Ideally, actual user demand for services equals planned user demand and prices from IT equal planned prices. In actual practice many variances must be handled during the execution of the plan. Demand changes generate price changes during the year with financial consequences to the users. For some services, such as production operations, if the demand increases slightly IT tends to make a profit; if demand decreases, IT starts incurring a loss. By year-end the variances must be resolved.

There are several ways to resolve these variances. One approach to resolving the overage or underage is to carry the variance at a higher level in the firm and generate a profit or loss at that level. Second, the profit or loss can be distributed to the users as a retroactive rate adjustment. This results in an overage or underage in the user budgets. Third, the rates or prices can be changed slightly during the year to keep the running variance close to zero throughout the year.

As one might expect, the cost center method is subject to manipulation by the users. For example, if a user organization underbudgets for IT computer services and later increases its demand, the extra demand, if it can be satisfied, may result in lower rates for all users. In this case the gamble pays off for that particular user. If many users underbudget, IT probably will be unable to satisfy the true demand and users will receive poor service. Those whose budgeting was accurate also suffer.

If the corporate policy is to cover variances at a higher level at year-end, users are likely to overbudget for IT. IT will acquire greater capacity than required, users will obtain better service than planned or justified, and the higher level must cover the excess expenses. Users may spend the excess in their IT budget on other items. These items are possibly not totally justified. There are other variations of this process as well. Given the above, the cost center method works well when the firm has strong financial discipline and the organization's controller takes action to protect IT and the firm from potential abuses.

The cost center approach is contentious and leads to conflict, some of which may be healthy. If the capacity planning process is not very effective, the cost center methodology tends to flush out latent demand. If capacity planning is well orchestrated, cost center planning proceeds quite well and the results can be very satisfactory. In any case, cost center planning forces decision making. This may be advantageous if other management process within the firm are not completely effective.

In addition to the complete profit center method and the total cost center approach, other alternatives combine aspects of each or utilize other methodologies for certain IT services. For example, some services may be priced separately, partial recovery of certain costs may be allowed, or costs may be based on long-term contracts with the users.

SOME ADDITIONAL CONSIDERATIONS IN COST RECOVERY

IT services are a varied lot, and no one single scheme for recovering costs is satisfactory to everyone. Consider, for example, the labor-intensive process of application development as compared with the operation of on-line mainframe application versus the data communication network operation. Application development has no economies of scale, generally has little excess or latent capacity, and suffers from a variety of people concerns and issues such as communication and motivation. The mainframe application may have economies of scale, may have large amounts of latent capacity, and is less susceptible to people issues and concerns. Telecommunications networks are different from each of these. Because of these differences, cost-recovery methods can be optimized to take advantage of the service characteristics.

Funding Application Development and Maintenance

Portfolio management and the management of application development lend themselves very well to customized or unique cost-accounting and cost-recovery approaches. For example, the costs of application program maintenance and minor enhancements can be recovered through the profit center or the cost center methods using rates established for application programmers. If a programmer works for 10 hours performing minor enhancements to an application, the owner of the program is charged for 10 hours at the established rate.

Since minor enhancements are a way of life for important applications, it is better to recognize this condition in advance and prepare a long-term contract describing the ongoing support for the application.[9] The contract for "period" support may state, for example, that the programming department will provide 20 hours per week of programming effort devoted to program enhancements requested by the user manager. The contract period will probably be for a year or more at a rate sufficient to recover the programmer's expenses. This type of contract has an inherent degree of flexibility. It avoids the problem of accounting for the programmer's hours except in a general way. It gives the user manager the opportunity to make minor enhancements without contending with other users for programming resources. The IT organization and the user manager benefit from this arrangement.

Application development projects may also benefit from alternative cost-recovery methods. Consider, for example, a variation of the pay-as-you-go method for recovering costs on a major programming project. At the beginning of each phase, a cost will be established to complete the next phase. At the end of the phase, the customer is billed for the contracted expense and IT recovers its costs for the phase. This approach closely relates the level of effort to the objectives established during the phase review. In effect, IT commits to producing the function on schedule *and* within budget. Since IT usually will not agree to changes in function during the phase, user and IT managers are motivated to produce a high-quality plan for the next phase. This approach has the advantage of reinforcing the discipline of the phase review process with financial incentives. Charging for application development by phases is generally preferred to recovering costs on a pay-as-you-go basis.

Charging for application development by phases exposes poor planning, reveals the financial impacts of changes in direction, and discloses the costs of poor implementation. In addition, this approach makes the decision to terminate the project, if that becomes a possibility, less difficult. The temptation to continue the project, making up for past excesses by trimming future planned expenses, is less likely.

Another method for handling application development is to recover the development cost from the application owner over some predetermined life of the application. This approach recognizes that benefits from the application are realized after installation. It relates development expense to benefits realized over time. The advantage of this method is that early risk is taken at some higher level in the firm, thus making application development more attractive for users. The advantage for the firm is that applications of high potential value for the firm can be developed when they might not be financially feasible otherwise. To the user, this method appears to capitalize the development effort, though it may be expensed at the higher level.

Some projects should be funded at the corporate level. Applications that have great strategic value for the firm should be managed financially by the firm and not by any single user manager. Though the application may be used primarily by one class of user, strategic decisions regarding the application should be made at higher levels in the organization. For example, an electronic design-automation program which gives advantage to the firm engaged in designing and building electronic products cannot be managed effectively by managers in the development laboratory. Product managers are likely to make short-range decisions; their view is mostly of the product and not of the firm and, as a group, their interests are diverse. The solution is to put decision making at a level where long-range considerations are preserved, where the firm's interests are paramount, and where the divergent motivations can be reconciled.

Widely used computer systems must also be managed financially at some high level. Office automation systems are of this type. Office managers should drive the requirements process, but the funding for office systems is best performed in a central administrative function. This approach recognizes the firm's office automation costs, but it avoids the difficulty of allocating costs to individual workers or departments. Some telecommunications systems and large database applications fall into this service category as well.

Cost Recovery in Production Operations

There are many algorithms for recovering costs in production operations. A common theme behind most of these algorithms is to charge for the resources used in the production of useful output. For instance, CPU cycles used, the amount of primary memory occupied and its duration, channel program utilization, pages of printed output, and other measures of production resources are common usage parameters. Generally these various parameters are linked together in some fashion to form a

---

charging algorithm. Someone in IT establishes a price for each resource so that the charging algorithm reflects the cost of work performed for the user. In some cases this type of charging algorithm can become quite complex; it may also be very accurate.

Although these measures can be made quite precise, they are not very satisfactory to IT customers. They are difficult to develop because they require a detailed knowledge of the system and the economics of its parts. Because of these complexities, they are difficult to explain and are hard for most users to understand. Users have trouble relating some of these measures to useful work produced for them by the system. Simple methods such as charging for the application's elapsed time are probably more effective.

For applications or processes that occur over extended periods, elapsed time is a common denominator of the charging algorithm. For example, an on-line application operating continuously may be priced according to the fraction of the CPU resources it consumes. If the on-line order-entry system requires one-third of a major CPU, then the service is priced to recover one-third of the cost of operating the machine. Large on-line data stores are frequently priced at so much per track per day. The idea is to recover the cost of on-line storage from the users of the storage in proportion to their usage. If these large on-line data stores are associated with continuously operating applications, then the elapsed-time charge should include both CPU and DASD costs.

Many other variants are useful in production operations. One such variant is the use of price differentials for classes of service. For instance, users of prime-shift capacity (8:00 AM-6:00 PM) may be charged more than off-prime shift users. The price differential can be justified since prime-shift time has more value to most users than off-shift time. The price differential encourages usage when there usually is surplus capacity. The movement of workload from prime to off-prime shift is equivalent to a capacity increase; it represents a real cost saving to the firm. By the same rationale, weekend work may be processed at reduced rates and priority or emergency work may be charged higher rates. These are examples of price incentives designed to encourage more effective usage of IT resources.

Services dedicated to one class of user should be charged directly to that user. For example, if one CPU and associated equipment and support personnel are devoted to one user department, that department should pay the full cost of that service. This charging methodology is equitable, easy to justify, and easy to explain. Dedicated telecommunications links fall into this expense category as well.

In production operations, it is important to plan ahead and to recognize the effects of technological obsolescence. This can be accomplished effectively if costs are based on a multiyear plan. The multiyear plan prevents wide fluctuations in prices as equipment is replaced or additional capacity comes on stream. Advanced planning recovers costs early and matches revenue and expenses in the long term. Successful IT managers use this approach to everyone's advantage.

Finally, some service items should be sold outright. End-user computing and office automation benefit considerably from this approach. Rather than billing the customer for the monthly cost of the terminal or personal workstation, IT should sell the hardware to the client and bill the connect time or network time based on usage. This simplifies the accounting process, is easy to justify, and is easily understood. Most user managers are quite comfortable with this approach. Upgrades or additions to the user's personal hardware or software are best handled on a purchase basis too.

Network Accounting and Cost Recovery

Networks significantly complicate the IT cost-accounting and cost-recovery processes. Production computers connected to networks see the network as another source or sink of information for the application operating in memory. IT accounting routines can deal with the application accounting in the manner described above. However, when applications in several computers communicate with each other and place computing load on the CPUs, the accounting process becomes very complex. To account for all the costs accurately and to do so in a way that users will understand is not possible.

The accounting problem becomes more complex when public carriers and value-added networks enter the picture. International operations add yet another level of complexity. Corporate network managers must have a solid understanding of the impact of expenses on service levels and must know the expenses to the component level. They should keep detailed trend information which can be related to trends in performance and utilization. Their knowledge of network costs should permit them to make trade-offs between various types of new technology. They must be able to relate new technology costs to capacity increases and performance and service improvements.

IT managers must be careful about charging users for telecommunications services. Some services, such as long-distance dial-up voice or data communication, should be charged to the user's account directly. Other network services, such as those supporting EDI applications, are probably best recovered through overhead. Other applications must be analyzed on a case-by-case basis. If the network expenses can be easily identified to an application, then the application should be billed for them. If the expenses are small or if they are comingled and not easily separated, it may be best to recover them through overhead.

Some network management systems contain accounting packages to help solve some of the problems indicated above. The need for an accounting solution may influence the selection of a network management system.

The process of accounting for IT expenses and recovering these expenses from users of IT services lends itself to creativity and benefits from a businesslike approach. The variations discussed above can be used with either the profit center or cost center structure. The methodology selected and the variations employed depend on the degree of maturity of information technology in the firm, on the importance of IT to the firm, on the corporate culture, and on the objectives that the firm's executives hold for IT. As these factors change over time the cost-accounting system and the chargeback system will also change. These processes will go through iterations as the organization matures.[10] In addition, the relationship between the IT organization and user organizations will develop as the accounting process matures.

RELATIONSHIP TO CLIENT BEHAVIOR

One important goal of user chargeback processes is to encourage and promote the cost-effective use of IT services. Many financial motivations result from cost-recovery actions; many motivations can be developed as well.

Consider, for example, a firm that underutilizes information technology and wants to promote increased technology adoption. What strategy should this firm adopt? For how long should the strategy be employed? One strategy for this firm would be to increase the funding for IT and increase IT services but accumulate the increased expenses at the corporate level. For example, if a firm wanted to encourage intersite communication it could install a network linking several of its sites and cover these expenses at headquarters. This, in effect, makes the telecommunications network a free utility to the users. User managers are highly encouraged to develop intersite communication applications under these circumstances.

This kind of funding approach is appropriate if the firm is lagging the industry in technology adoption or is at the initiation stage of growth with a new technology. The methodology hastens the firm's migration into the contagion stage, where financial incentives are perhaps no longer needed. When the contagion stage is reached, special incentives can be reduced and another approach installed.

There are many incentives inherent in the decisions regarding the approach to funding IT applications, some of which were evident in the earlier discussion. As the firm matures in its use of information technology and as the management team develops sophistication with the IT management system, variations can be employed that maximize the chances for attaining corporate goals. For example, the firm may abandon the cost center methodology in favor of the profit center approach when the opportunity to sell IT services outside is recognized.

If the correct degree of maturity and sophistication is present and if the firm owns attractive services, the IT organization may become a revenue generator for the firm. Many firms make their production facility available as a service bureau. Some firms sell the use of their applications—airline reservation systems are an example of this. And some firms engage in contract programming. As these examples illustrate, corporate goals for the IT function may require the adoption of alternative accounting methods.

IT accounting systems that utilize carefully constructed chargeback methods are fundamental to effective IT operation. They provide the basis for cost-effective usage of IT services. They greatly enhance communication linkages between IT and client organizations. The firm's performance is improved through the motivations present in the accounting and chargeback mechanisms. Successful IT managers fully exploit these concepts.

SOME COMPROMISES TO CONSIDER

The cost-recovery approaches discussed above are subject to many variations, and the variations can be expanded for many purposes. For example, it may not be required that the cost center operation fully recover its costs, or that overages be returned to the using organizations. The purpose of the managerial accounting system is to serve managers, not to be highly precise in the accounting processes. If the system controls, communicates, motivates, and helps measure and plan as desired by management, then additional accuracy and precision in the accounting processes may not be justified. Precision and accuracy can be sacrificed for reduced administrative overhead and improved effectiveness. Likewise, the accounting methodology can change as the firm matures in its use of information technology. Precise comparisons over time can be sacrificed for improved motivation or control as business conditions change. The purpose of the system is to serve management, not to be a pinnacle of accounting purity.

EXPECTATIONS

IT clients have expectations of the cost-recovery system. They expect to have a system that is easy to use and easy to understand. The cost-recovery system must be designed for the clients and must help them in their business relationship with the IT organization. Users expect IT costs to be distributed fairly and effectively, and they demand consistency. Frequent price changes or unusually large changes upset their plans and cause them to believe that IT is not managing affairs properly. Cost-recovery processes are valuable to the organization, but their use and administration requires sound management skills.

IT managers and the firm's executives have larger expectations for IT's accounting system. They expect to understand the manner in which IT is spending resources in furtherance of the firm's goals. Well-designed accounting techniques help executives and managers understand the IT organization as a business.[11] They also expect to be able to measure IT's value to the firm. IT accounting and chargeback systems help in this regard.

However, as the vignette noted, many IT expenditures can be measured only in terms of improved organizational effectiveness. Traditional accounting systems are inadequate for this purpose. Innovation in IT accounting systems can overcome some of these difficulties, but many challenges remain for the IT manager or the CIO and for the firm's senior executives.

CIOs can challenge the firm's accounting system. They can raise questions about the firm's ability to measure organizational effectiveness. And they can take action to correct the deficiencies in the traditional accounting system by installing innovative managerial accounting systems for the IT organization. In the process, they will have forged stronger ties between IT and the rest of the firm. They will also have strengthened their business relationships to the operating executives and to the firm's corporate staff.[12]

Thoughtful CIOs should consider improved ways to value investments in machines, software, and people. Improved valuations of these resources will lead to better ways to measure performance and organizational effectiveness. IT's role and contribution has been a leading management issue for a long time. Innovative accounting and measurement systems have the potential to partly defuse this issue.

SUMMARY

Information technology resources are a large and growing portion of the budget for many firms. The firm expects IT managers to use these resources effectively and to account for them properly. In many cases, the IT resources are dispersed throughout the firm, making the accounting process difficult. In addition, the accounting task is made more difficult by the complex and varied character of the resources themselves. For all these reasons, accounting for IT resources is important to IT managers and to the firm.

Accounting for IT resources assists in planning, controlling, communication, and performance assessment. It is essential to the effective operation of the IT organization and of the firm. But if the IT organization elects to recover costs from users, then additional elements of motivation and effectiveness come into play.

Several methods for recovering costs from users were presented in this chapter. The profit center approach and the cost center approach were discussed in detail because they are fundamental methods for recovering cost. Variations for handling production operations and application development were discussed and related to the objectives of these groups. The relationship of cost-recovery methods to expected motivations is important for increasing the effectiveness of these processes.

Chargeback methods are very important to IT managers; their implementation must be skillfully handled. A blend of psychology and accounting is required, tempered with the practical considerations of implementation. Accounting systems for IT organizations that intend to sell services outside the firm must be more rigorous than for those that do not. Accounting systems for IT are desirable, but they are not an end in themselves. The purpose of managerial accounting systems is to permit business managers to operate more effectively. The rules for their construction and the implementation techniques must be designed to meet this goal.

Accounting for IT resources raises some very difficult questions. For example, application developers are increasingly designing and installing network-based systems because these systems provide their owners with many important advantages. However, returns on investments in network-based systems can frequently be measured only by improved organizational effectiveness—measurements that traditional accounting systems cannot provide. Clearly this area calls for creativity from CIOs and other senior executives in the firm. Innovative managerial accounting systems are the first step in the process toward measuring organizational effectiveness.

Review Questions

1. Why is IT resource accountability important to the IT organization and to the firm?

2. What is meant by financial accounting?

3. How is managerial accounting different from financial accounting?

4. What trends in information handling are making IT accounting more difficult and more important?

5. What is the relationship between the IT plan and the IT budget?

6. What purposes can be served by IT accounting systems?

7. What are the benefits of an IT cost-recovery system?

8. How does a well-designed charging mechanism enhance the IT organization's effectiveness?

9. What is the primary disadvantage to IT cost-recovery processes?

10. What goals should an IT chargeback system seek to obtain?

11. What are the advantages and disadvantages of the profit center method for recovering IT costs?

12. What are the characteristics of the cost center method of IT cost recovery?

13. The cost center method can be contentious. Under what circumstances is this advantageous?

14. Why might it be best to use several methods to recover various IT costs?

15. What is meant by period support for application programs?

16. What alternatives can IT managers use to charge for application development? What are the advantages of each?

17. The costs of production operations can also be recovered in several different ways. What variations are useful in production operation cost recovery?

18. Why is cost recovery accuracy not necessarily important to IT managers? When is it very important?

19. What do clients expect from IT cost-recovery methodologies?

20. What role can the CIO play in the subject of IT accounting and chargeback?

Discussion Questions

1. Discuss the differences and similarities between financial accounting and managerial accounting for IT organizations.

2. Discuss the connections between planning, budgeting, measuring, controlling, and accounting for IT activities.

3. How would the activities mentioned in Question 2 work together for the successful implementation of a new market analysis program?

4. How does a well-designed cost-recovery system work to improve the effectiveness of the IT organization?

5. The IT cost-recovery mechanism need not be fair and accurate but must appear to be equitable to the clients. Discuss why this is so.

6. The text discusses some instances in which charging methods alter user behavior toward the consumption of IT resources. Can you give some additional examples of this?

7. Compare and contrast the profit center and the cost center methods for recovering IT costs.

8. When do you think the cost center method is preferable to the profit center method for IT cost recovery?

9. Why might the costs of application development and maintenance be recovered in a different manner than that used to recover the costs of production operations?

10. What goals in the management processes of application development and maintenance can be served by the alternative cost-recovery approaches discussed in the text?

11. Discuss the advantages and disadvantages of price differentials in the cost recovery of production operation resources.

12. Describe how a firm might price computer services based on a multiyear plan that includes the replacement of a major CPU.

13. What type of cost-recovery system would you recommend for a firm in the integration stage of growth?

14. What type of cost-recovery system would you recommend for a firm that believes that IT's role and contribution is a major issue?

15. Describe the possible changes in IT cost-recovery methods that might occur as the firm grows from a modest user of technology to the time when it becomes a data-servicing operation.

16. Discuss the expectations users may have of the IT accounting system. What expectations do the firm's executives have of the system?

17. To what extent do you think the necessary conditions for charge-outs as stated by Wetherbe are present in most firms today?

18. Accounting for networks is difficult and charging users for them is even more difficult. How can the databases of configuration management help account for some of the fixed costs of networks? Discuss the advantages and disadvantages of accounting for data networks as corporate overhead.

Assignments

1. Draw a flowchart of the budgeting process in a firm using the cost center method to recover IT costs. What prevents the iteration from continuing endlessly?

2. Visit a local firm and determine what methods are in use by the firm to recover IT costs. Why has the firm adopted the approach they are using? Why is it effective for them?

3. Analyze the accounting package for one network management system and prepare a report for the class. In your report indicate what goals for an accounting system this package accomplishes.

ENDNOTES

[1] "A Measured Response," adapted from *CIO Magazine*, January-February 1989, 52. © *CIO Magazine*.

[2] "The Premier 100," *Computerworld*, October 8, 1990, Section 2, 19. The 100 most effective users of information systems as measured by *Computerworld* spend 2.7 percent of revenue on IS; eighteen of these firms spend more than 5 percent.

[3] Richard Greene, "Inequitable Equity," *Forbes*, July 11, 1988, 83. Alfred Rappaport, an accounting professor at Northwestern University, commented, "As we become a more information-intensive society, shareholders' equity is getting further away from the way the market will value a company."

[4] Leslie Goff, "Dispersion of Computer Budgets Does Not Lower MIS Spending," *MIS Week*, September 11, 1989, 28.

[5] Paul A. Strassman, "The Real Cost of OA," *DATAMATION*, February, 1, 1985, 82.

[6] Jeanne Buse, "Chargeback Systems Come of Age," *DATAMATION*, November 1, 1988, 47. This article states that the chargeback process promotes a cost and value consciousness among users and IT people.

[7] James C. Wetherbe, "Down with Chargeout," *CIO Magazine*, December 1988, 62.

[8] David A. Flower, "Chargeback Methodology for Systems," *Journal of Information Management* (Spring 1988): 17. This article discusses the chargeback system at Prudential Insurance.

[9] Flower, "Chargeback Methodology for Systems."

[10] Buse, "Chargeback Systems Come of Age."

[11] Dennis Wenk, "Managing the Cost of IS," *Infosystems*, November 1986, 41.

[12] J. Robert Riggs, "Cost Control: Where Has All the Money Gone?" *Computerworld: In Depth*, November 17, 1986, 77.

REFERENCES AND READINGS

Allen, Brandt. "Make Information Services Pay Its Way." *Harvard Business Review* (January-February 1987): 57.

Connell, John Jr. "A Measured Response." *CIO Magazine*, January-February 1988, 52.

Deardon, John. "Measuring Profit Center Managers." *Harvard Business Review* (September-October 1987): 84.

Perry, William E. "User Chargeback Procedures for Distributed Systems," in James Hannon, ed., *A Practical Guide to Distributed Processing Management*. New York: Van Nostrand Reinhold Company, 1982, 73.

Sanders, William E. "User Chargeback," in James Hannon, ed., *A Practical Guide to Data Processing Management*. New York: Van Nostrand Reinhold Company, 1982, 83-98.

16 Information Technology Controls and Asset Protection

A Business Vignette
Introduction
The Meaning of Control
Why Controls Are Important
Some Principles of Business Control
    Asset Identification and Classification
    Separation of Duties
    Efficiency and Effectiveness of Controls
Control Responsibilities
Application Controls
    Application Processing Controls
    Application Program Audits
    Controls in Production Operations
Network Controls and Security
Additional Control and Protection Considerations
The Keys to Effective Control
Summary
Review Questions
Discussion Questions
Assignments
References and Readings

How Computer Science Was Caught Off Guard
by One Young Hacker

The surprise attack began between 9 and 10 Wednesday night. Among the first targets were Berkeley, CA, and Cambridge, MA, two of the nation's premier science and research centers. At 10:34 PM, the invader struck Princeton University. Before midnight, it had targeted the National Aeronautics and Space Administration Ames Research Center in California's Silicon Valley, as well as the University of Pittsburgh and the Los Alamos National Laboratory in New Mexico. At 12:31 Thursday morning, it hit Johns Hopkins University in Baltimore, and at 1:15 AM, the University of Michigan in Ann Arbor.

At 2:28 AM, a besieged Berkeley scientist—like a front-line soldier engulfed by the enemy—sent a bulletin around the nation: "We are currently under attack." Thus began one of the most harrowing days of the computer age.

The invader was a computer virus. Like some relentless, demonic automaton, it coursed through networks linking key university and government computers from coast to coast. Once inside, it multiplied, devouring the space that computers use to store information and slowing them to a halt. So ingenious and complex was the virus that some computer scientists didn't immediately realize what they were up against. It initially fooled many trying to neutralize it. They would devise a solution, only to find the virus spreading again.

In the end, the virus apparently didn't cause permanent damage to the 6,000 computers it attacked. Instead of wiping out data this invader was fairly benign; it merely used up empty storage space.

The virus nevertheless has stunned and frightened the computer world. If computers can be sabotaged so easily, so swiftly, experts wonder, how vulnerable is the system to high-tech terrorists? The virus is expected to prompt a full-scale review of computer security in government, corporations and universities. A post-mortem conference already is planned in Washington this week.

At about 10 PM Wednesday, Pascal Chesnais, a researcher working late at MIT's Media Laboratory in Cambridge, noticed something odd. Computer programs he was running had slowed to a crawl. Two or three colleagues noticed the same thing. "We thought it was just a runaway program," he recalls. "So we killed all the processes, started over, and the problem seemed to go away." Unconcerned, they soon went out for ice cream.

Across the continent, at 10:15 EDT, the experimental computing facility at the University of California at Berkeley was hit. Security software that monitors incoming electronic mail traffic sent alerts "that it was receiving unusual commands," recalls Peter Yee, a scientist at the center. Because of this early warning, Berkeley was able to contain the virus faster than others did. It shut off communications to most computers and established a trap to capture and study the unknown code that was causing the problem.

It hit about the same time at NASA's Ames Research Center. At midnight Eastern time, Ames cut off all communications with outside researchers, thus stranding 52,000 computer users.

Not all computers were targeted; just those that were on the Internet network and used a certain version of the Unix Master control software were hit. The virus took advantage of at least two loopholes in the software to sneak in. The first was an installation debugging device which was not needed after installation and about which users had been warned. A similar loophole in another communications program gave a second method of entry.

What made the virus all the more terrifying is that nobody knew for sure at the time that it was benign. "There's no reason why it couldn't have wiped out people's files, put subtle time bomb things in the system or sent junk mail to anyone," says Robert Logan, a computer systems manager at California Institute of Technology in Pasadena.

At Berkeley, researchers and students feverishly sought a vaccine. "There were about a dozen people working in a small room on eight computers and terminals," says Scott Silvey, a 23-year-old Berkeley senior. "It was crowded. The phones were ringing. People called from the Navy, the Air Force, and from Florida." Trying to reach some computer sites, the Berkeley crew found that many had no emergency phone contacts or contingency plans to deal with such an outbreak. "The sites without an emergency plan didn't do well," says Russell Brand, an artificial intelligence doctoral candidate at Berkeley and a researcher at the nearby Lawrence Livermore Laboratory.

Some computers connected through Internet weren't infected—although not for lack of trying by the virus. At the University of Maryland a computer equipped with a security system logged about 2,000 failed attempts by the persistent virus. The virus attacked Argonne National Laboratory outside Chicago starting at 11:54 PM EST Wednesday and throughout the night. But only one of the lab's many computers was infected. Luckily, researchers

Part Five: Controlling the Information Resource

a few months ago had modified the widely used code for gaining access to the system.

AT&T's Bell Laboratories also escaped. A year ago, Bell patched its electronic mail software. When Bell tried to warn other institutions of the potential for a breach in security, Bell found that few shared "our rather paranoid view of communications software," says Douglas McIlroy, a member of the technical staff.

Now that the crisis is over, computer scientists are expressing both awe and anger at the virus. Although viruses have plagued computers for years, the vast majority until lately have gone after personal computers. Last week's virus reached a new level. "We've never seen a virus this large and this successful," says Robert Cosgrove, the director of computing systems at Carnegie-Mellon University in Pittsburgh, one of the schools attacked by the virus.[2]

INTRODUCTION

Controlling the IT business and protecting its assets are fundamental responsibilities of IT managers. In nearly all firms, controls are becoming more important as information technology penetrates the organization, and in many instances business controls are required by law.

Operational systems in today's sophisticated, fast-paced business environment must be solidly grounded on a base of operational and accounting controls. The applications, networks, and hardware must be carefully protected against loss or damage. Successful IT managers install control and protection mechanisms and operate them effectively. IT managers must establish and maintain satisfactory business controls and must be prepared to demonstrate effective controls to the firm's senior executives.

Because information technology is utilized throughout the firm, the IT managers' responsibilities extend far beyond the boundaries of their organizations. As part of their staff responsibility, IT managers are expected to lead the firm in establishing IT controls. Their leadership is vital to the organization's security and control posture.

In this chapter, business controls and asset protection issues associated with application systems and their supporting hardware and networks will be discussed in detail. The participants in this activity are identified, their responsibilities are defined, and the details of the activities are explored in depth. Control and protection of application programs begin when the applications are in development and continue as long as they are in production. This chapter develops control disciplines for the applications and presents mechanisms for auditing and reporting control status. Auditing activities in the firm rely in part on the foundation of control and protection discussed in this chapter.

THE MEANING OF CONTROL

Control means that managers know the details of the significant activities taking place within the organization; they know what, when, where, why, how, and who for all important functions. In order to be operating under control, IT managers must first know what their mission is—what they are supposed to be doing. They must know what is expected of them—what is acceptable and what is not.

Control also means that IT managers have a routine means for knowing what their actual performance is versus their planned performance. The deviations between planned and actual performance are obvious in a well-controlled organization. Also, managers must have the means to respond promptly to plan deviations or to out-of-control conditions. They must have a way to acquire this information regularly and a process for responding to it quickly. They must also be able to detect improvements in performance resulting from variance corrections. They must be able to do this for all the activities within their areas of responsibility. Managers who can fulfill these conditions are operating under control.

WHY CONTROLS ARE IMPORTANT

Control is one of the primary management responsibilities. Successful IT managers establish and maintain effective business controls as part of their normal activities.

Business controls are becoming more important in automated organizations because lack of control or out-of-control conditions are becoming more subtle, less obvious, and potentially more damaging to the operation. For instance, insufficient controls in a manual accounts payable function may lead to the issuing of some unauthorized payments, but an uncontrolled accounts payable program can write unauthorized checks at the rate of thousands per hour. (In one well-known situation this actually happened.) An error in an inventory program can add discrepancies to the database much faster than manual reconciliations can detect and correct them.

Control is especially important to IT managers because many of the organizations supported by the IT function rely on computer-generated reports and other automated tools for their controls. For example, the audit of an inventory-control process in a manufacturing plant leads quickly to computer-produced reports presenting transaction activity and describing inventory status. Control weaknesses in the IT function can directly affect inventory control and control throughout the firm. Table 16.1 summarizes the reasons why business controls are important to the IT function.

TABLE 16.1   Why Business Controls Are Important

Control is a primary management responsibility.

Uncontrolled events can have subtle repercussions.

Out-of-control events can be damaging.

The firm relies on IT for control processes.

Control is required by law.

Environmental and executive pressures require control.

New technology introduction requires control.

Control requirements and accurate record keeping are mandated by law for publicly held corporations. Specifically, organizations must provide proper authorization of transactions and perform record keeping in conformity with accounting principles established for the firm. The firm must provide and maintain asset protection and must physically verify and reconcile assets with inventory records. Managers must document the extent to which corporate accounting principles have been followed.

Managers must evaluate the sufficiency of controls and must appraise actions taken to correct control weaknesses. The firm's officers must certify that these actions have been taken. Management performance and judgment in controlling assets are also subject to assessment. The following statement by one firm's officers is typical of those found in annual reports of publicly held companies:

> Management is responsible for establishing and maintaining a system of internal control designed to provide reasonable assurance as to the integrity and reliability of financial reporting. The concept of reasonable assurance is based on the recognition that there are inherent limitations in all systems of internal control and that the cost of such systems should not exceed the benefits to be derived therefrom.[3]

Controls are also important because of environmental and legislative pressures and executive concerns about control activities. Because of the growing complexity of the business environment, executives expect both manual and automated control mechanisms to be maintained at peak efficiency. Large on-line applications consisting of sophisticated hardware and software systems cross functional boundaries and interrelate business activities throughout the firm. Executives demand automated and controlled processes for these complex activities. Most firms are highly dependent on information systems; the systems must provide effective and efficient control mechanisms.

Adoption of new technology frequently creates the potential for greatly increased employee productivity. New technology usually has great potential for improving the effectiveness and efficiency of the operation overall. The use of computerized automation also brings increased capability to control errors and omissions and to prevent fraud. However, these benefits are accompanied by new and increased control risks with which management must contend. Effective controls must precede or at least accompany the introduction of the technology.

The introduction of new information technology to the firm usually coincides with the need for new control and protection measures. For instance, the deployment of personal computers and the use of new networking technology greatly increase the need for data security and physical inventory control mechanisms. However, it is common for the introduction of new technologies to lead the adoption of the necessary control mechanisms by a large margin in many firms. For these firms, the result is that insufficient controls or out-of-control situations usually accompany new technology. Alert IT managers anticipate these events and plan control systems prior to technology introduction.

SOME PRINCIPLES OF BUSINESS CONTROL

The primary job of managers in an organization is to take charge of the firm's assets entrusted to them; to control and protect the assets; to use the assets in the furtherance of their part of the business; and, in some cases, to grow, develop, or add value to the assets. Information assets, both tangible and intangible or intellectual, must be controlled and protected by all who have custody of them. In order to control and protect the firm's property, managers must know what assets they control and the value of the assets.

Asset Identification and Classification

Information assets consist of tangible, physical property like CPUs, communication controllers, and personal workstations. Many information assets are intangible, intellectual assets such as operating systems, application programs, and databases; they are more valuable than the physical assets. In almost all instances, program and data assets are worth much more than the devices on which they are stored. In most firms, the value of information assets generally exceeds the annual IT budget by a wide margin. In some firms, the value of the enterprise is considerably understated on the balance sheet because of the tremendous value of intangible information assets.[4] IT managers have a great deal of responsiblity for protecting all of these assets from loss, damage, or improper use. Information asset protection is an important task for them.

A manager's first step in controlling and protecting assets is to conduct an asset inventory—to develop an organized list of the assets for which he/she is responsible. Confining the list to IT items only, some examples might be: computer hardware, system software, application programs, databases, documentation, and passwords. The list is rather long because documentation can include many items such as strategies, plans, designs, algorithms, and other items. Taking inventory establishes what assets the manager must control and protect.

The second step is to establish a value for each of the inventoried items. This step will generally reveal three types of assets: assets with intrinsic value such as money, stock certificates, or checks; assets with possible proprietary value such as new product designs; and assets which are valuable because they control other important assets. The payroll program is an example of this type of asset. The appraisal step organizes the assets by value and provides a rational basis for establishing controls and protection for them. This step is called asset classification. Managers can develop and implement sound controls once this step is completed.

Most organizations have an asset classification scheme for proprietary information. One such classification structure has four categories of information: top secret, secret, confidential, unclassified. Unclassified information is public and available to anyone in the firm. The remaining categories of information are available on a need-to-know basis. This means that if an individual is cleared for secret information he/she can access such information if required for the job. Having a secret clearance does not mean that the individual has access to all information classified as secret.

In addition to the classification categories indicated above, most firms have an additional category for information of a personal nature, such as salary, performance, and medical data. Some firms classify this information under personnel, others identify it as personal and confidential. Access to this information is also restricted to those with a need to know.

The classification category should be indicated on the document or should be contained within the dataset. It should be obvious to anyone viewing the information what the classification is.

Separation of Duties

One of the most effective control measures in business operations is the principle of separation of duties. Separation of duties means that several individuals are involved in transaction processing and that no single individual processes the transactions from beginning to end. As an example of how this might work, let's consider payroll processing. In this process one person prepares the time card information; another validates the totals and transmits the information for processing; another controls the blank checks and supervises the processing; another validates the processing through check register data; and yet another distributes the checks. In order for fraud to occur, several people must act together. Separation of duties reduces the possibility of fraudulent acts.

Separation of duties is relatively easy to administer and control; managers can validate the control mechanisms at each interface. In addition, if managers periodically change the individuals responsible for the tasks, this control mechanism can be very effective.

It is also extremely important to validate the output with the input. In the case of payroll, this can be done periodically by hand delivering payroll checks. The person delivering the check must be from outside the immediate organization; this person must make positive identification of the recipient; and the check distributor must verify hours worked with the employee. There are many variations of this theme for accounts payable, customer shipments, incoming inventory, and other activities.

Efficiency and Effectiveness of Controls

Controls are most satisfactory when they are simple to operate and easy to understand. They are most effective when they are routinized and operate in a timely manner. To be totally effective, controls must cover all possible exposures. When action is required, control mechanisms must respond and produce action in a timely manner. For instance, when input-data errors are detected in the operation of an application, management must receive prompt notification. They must act quickly to understand the error's cause and to initiate corrective action. The problem management system must be invoked promptly to take final corrective and preventive action.

In all cases, the cost of control and protection processes must be related to the expected frequency of unfavorable events and to the anticipated loss resulting from the occurrence of these events. Overcontrol is possible, as is excessive expenditures on controls. Controls must be cost justified. Analysis of the application and management judgement are required to establish the proper balance between the conflicting forces.

CONTROL RESPONSIBILITIES

Business controls operate most effectively when clear responsibilities are assigned to specific individuals. For application systems several individuals or groups are important in the process:

1. The application program owner (almost always a manager)
2. The users of the application (some applications have many)
3. The programming manager for the application
4. The individual providing the computing environment
5. The IT manager (has both line and staff responsibility)

Each has definite responsibilities which must be discharged correctly for application controls to be effective.

The application owner is the manager of the function or department that uses the application to conduct its business activity. For instance, the owner of the perpetual inventory program in the manufacturing plant is the materials manager, and the owner of the accounts payable application is the organization's accounting manager.

Application owners are responsible for providing business direction for their applications. The owner/manager is responsible for establishing the functional capability of the application program and for providing the justification or benefits analysis for any expenditures related to the application. The owner authorizes the program's use and classifies the data associated with the application, stipulates proper access controls to the program and to the associated data, and authorizes use of the application.

The payroll manager, for example, is the owner of the payroll program and controls access to it and to the related payroll data. The owner authorizes the processing of the payroll program. If the program requires modification due to changes in the withholding rate or if the firm decides to obtain payroll processing from a vendor, the payroll manager provides the business case and establishes the business direction for the application. The payroll manager is completely responsible for the payroll program. Application owner/managers throughout the firm are responsible for the applications required to operate their part of the business.

Application users are the individuals or groups of individuals authorized by the owner to use the application and related data. They are required to protect the data in keeping with the classification established by the owner. Users are responsible for advising the owner of operational difficulties and functional deficiencies. Individuals in the payroll department who update payroll records and initiate the payroll process are examples of application users.

Some applications have many users. The network of personal workstations comprising the firm's office system is an example of one such application. In this instance, one manager in the administrative function is named the owner of the application and all the secretaries and administrative personnel are the users. Another example is the inventory control system in a large manufacturing plant. The materials manager owns the inventory application system, but workers throughout the plant, from production planners to shipping clerks, are the users. In these examples, the owners and the users all have responsibilities for the application and its data.

The responsibility for organizing and managing the work of development, maintenance, or enhancement on the application resides with the programming manager. He/she is accountable to the owner for meeting programming objectives. These objectives include functional capability, schedule attainment, cost control, and quality performance. The programming manager also is the custodian of the application and associated data during the development, maintenance, or enhancement activity. The programming manager who modifies the payroll program under the direction of the payroll manager is an example of this individual.

The supplier of computing services is responsible for providing the computing environment within which the application is processed. The supplier of services is responsible for negotiating service-level agreements with the owner and for achieving the agreed-upon levels of service. Maintaining a secure environment for the application and associated data is another major responsibility. For example, if the firm's payroll is processed in the central computer center, the operations manager is the supplier of service and is responsible for the computing environment.

IT managers are responsible for ensuring that the individuals defined above receive proper guidance regarding their responsibilities. IT managers must establish procedures for the development and use of applications. They must routinely obtain information allowing them to evaluate the control posture of the applications on a continuing basis. In some cases they may need to conduct audits to validate business controls.

These assignments and responsibilities are not optional; they are mandatory for controlled operation of application programs. Failure to establish one or more of these positions or to ensure effectiveness in these positions will lead to control weakness and possible failure. IT managers must ensure that these positions are established and managed properly.

Application controls are most effective when they generate documentation that validates the appropriate functioning of the application on a continuing basis. Application owners should use this information in conjunction with other control information to certify their processing. The controls themselves should be well documented in order for audits of the applications to proceed promptly and accurately. Automated and manual control mechanisms should be classified as confidential information and should be handled accordingly.

Application systems benefit from most of the principles normally associated with business controls, broadly applied. For example, the principle of separation of duties applies to applications as well as to other, more common activities. But in addition to general controls, some principles stem from the intangible nature of application programs as operating assets. However, application programs are best controlled with a combination of programmed and manual controls. In the case of application program assets, the characteristics of the assets are usefully employed in their control.

Application Processing Controls

Control and protection of applications consists of two parts: ensuring that application programs perform according to management-established specifications, and maintaining the integrity of programs and data. The first part deals directly with the operation of the programs. It is concerned with the correct functioning of the application and with handling input and output data properly. The second part deals with security and protection of the program and data assets. It is concerned with controlling access to the programs and data files and with the integrity of information in the files.

Correct functioning of the application and proper handling of input/output data requires the application to have auditability features and control points. They are part of the original system specifications and are best designed into the system at the outset. Chapter 8 pointed out that application control and auditability is a design issue. Operating within the phase review process, managers must establish requirements and specifications for control elements that later become part of the application design. They must audit the development process during the phase reviews to make sure their requirements are contained within the design specifications.

System control points are portrayed in Figure 16.1. Controls can be applied during transaction origination and when the data are input into the system. Once in the system, control can be established when the data are transmitted into or out of storage, when they move over the teleprocessing system, or when they are processed within the CPU. Lastly, control should be maintained over the data when they exit the system.

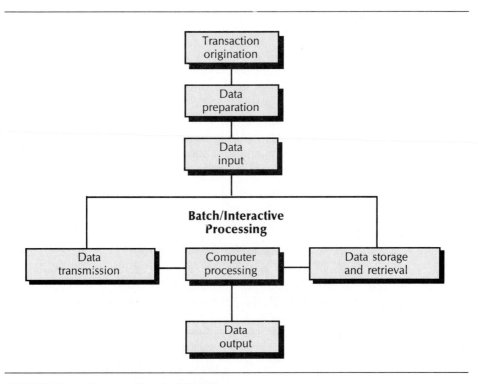

FIGURE 16.1   System Control Points

Figure 16.1 identifies points in the data processing system where control exposures exist and where controls must be exercised. Exposure points occur at transaction origination and where data is electronically input to the system. Data can be stored for later retrieval and processing or transmitted to the computer system for immediate processing. In either case, computer processing takes place and output is created for use by the application department. The results of computation may be stored for retrieval later or for further computation.

Some important control actions which must be considered at the point of transaction origination are displayed in Table 16.2. Transaction origination usually involves one or more input documents that must be controlled and monitored. Source documents are designed for the specific task. They usually have separate, printed squares for each character or number corresponding to the data requirements of the application and they are usually prenumbered for transaction control. Where possible, preprinted information should be used to reduce clerical errors.

TABLE 16.2   Considerations at Transaction Origination

Document design

Manual review of source documents

Authorization

Separation of duties

Transaction numbering

User identification

Transmittal logs between organizations

Error detection and correction

Document retention and storage

Only specifically authorized users should handle the source documents; when possible, the handling process should be divided to separate individual duties. Since the transaction documents contain identification numbers, logs can be maintained to record document transmittal between units. Because the input documents are developed in the user department and are transmitted to the computer operation, the transmission logs provide some validation that the proper documents have been processed.

Special attention must be given to error detection and correction during the input process. Resubmission of input data must be recorded and verified. Error handling must be performed according to carefully constructed procedures to prevent additional errors or fraudulent transactions. After the input process has been completed, the source documents must be stored and retained for some established period.

Batch-data input and interactive-data input controls are very important and must be clearly established and carefully monitored. Table 16.3 indicates some of the common means for establishing controls over input processes.

Batch processing includes scheduling the batch runs and ensuring that the input for the run matches the source-document data. Usually the batch application edits the input data and validates the information against the source input through control totals. For instance, the input of financial data will contain a count of the number of transactions and totals of important data such as an accounts summary. This guards against the addition of extra input or the loss of previously input data. When the input has been validated, batch processing begins and source documents are marked as entered.

TABLE 16.3   Input Data Controls

Batch Data Input	Interactive Data Input
Input processing schedules	Terminal access security
Source document cancellation	Terminal usage logs
Editing and validation	Data editing and validation
Control totals	Display and prompting formats
Batch control processes	Interactive control totals
Error handling procedures	Error detection and correction

Control over batch processing resides with the production operations department, which is responsible for advising the user department when improper batch execution occurs. Error handling is especially crucial because error correction through rerun procedures is itself an error-prone process. Errors must be carefully examined. The presence of input or processing errors is sometimes an indication of fraudulent processing. Error logs must be maintained by the operations department; the logs must be regularly reviewed by IT and user managers.

Interactive processing has many of the same control considerations as batch processing, but it is more complicated. The complications result from the use of remote devices for interactive operations. Proper physical controls over the terminal devices themselves is a mandatory first step. The devices must be secured from all but authorized users. Authorized users are identified to the system through passwords or personal identification numbers. The specifics of password construction and use will be discussed later.

Input data must be edited for reasonableness and tested for validity. Constructing the algorithms for effective editing and testing requires sound knowledge of the application and considerable ingenuity. The rules may be complex, and there may be many of them. The design of interactive displays and prompting formats can be very effective in this regard. For instance, if the field to be entered is numeric and can never exceed four digits, the screen design should not allow for more than four digits and the input editor should test for the presence of other than numeric characters.

If the dollar figure to be entered cannot exceed three figures to the left of the decimal point, screen design and prompting can control to this limit.

For some important applications, these front-end processes are performed in part by expert systems. Expert systems are especially valuable for interactive data entry where individuals can receive expert assistance and can take immediate corrective action. These considerations are part of the system design process. They must be an important element in the user's functional requirements statement.

Tables 16.4, 16.5, and 16.6 contain the control and audit tasks that must be incorporated in processing, data storage and retrieval, and data output.

Program execution validity is furthered by operating system software and by the application itself. The system software performs specialized functions that help control the use of data stores such as tapes and disk files (see Table 16.4). The system checks to ensure the correct tape file is being accessed by checking tape-label information. This helps prevent processing against the wrong file. It also verifies control totals to ensure that the correct amount of information has been processed. For all types of secondary storage, the system validates that the correct version of the data set has been used. This is especially valuable for applications that create today's output from yesterday's output and from today's transactions and where previous data sets must be preserved.

The program execution process must be accompanied by programmed routines validating that the processing was completed and that the results are correct. The processes or algorithms for doing this are highly application dependent and must be included in the program specifications when the application is developed. In some cases these algorithms are confidential. Processes for detecting errors and for correcting them should involve the application owners, who should review all error correction activity carefully for indications of improper or fraudulent conduct.

TABLE 16.4 Computer Processing Controls

Validate the input data set.

Validate the data set version.

Verify processing correctness.

Verify processing completeness.

Perform error detection and correction.

Data storage and retrieval requires full protection for classified data and application programs (see Table 16.5). Classified data must be protected so that only those with a need to know have access to it. Application program source code and object code must always be treated as classified information. Program changes are similar to data changes and are handled in much the same manner. Special procedures for reviewing classified data and for altering it will be discussed later.

TABLE 16.5   Data Storage and Retrieval

Ensure full protection for classified data.

Provide source program update procedures.

Finally, control mechanisms must be placed on the data output and distribution process (see Table 16.6). In order to be sure that the processing occurred as planned, totals should balance, input and output volumes should reconcile, and manual processes on output must be under document control. The output documents should contain control totals and transaction counts, which can be reconciled with the documented input records. Mechanisms within the application validating that the processing was accurate should produce reports for the application owner. Routine output documents and these processing reports should not be delivered to the same individual in the organization. Some of these records must be retained for future audits or for other purposes.

Output records must describe how all error-handling activity was conducted. All error-recovery actions must be documented and recorded for future review by the problem management team. Control of the data output activity is vital because it offers the opportunity to reconcile and validate the entire process.

Operations personnel must exert control over output distribution. Output must go to authorized individuals only—it must not be delivered to unauthorized individuals. Proper distribution of computer output involves issues of privacy, confidentiality, and security.

Some information is personal and must be divulged only to authorized individuals. For example, employee salary or performance records are private information and should only be seen by the employee's manager and certain other individuals. Some information is confidential, such as marketing information, new product designs, and proposed pricing actions. Disclosure of this information to the wrong people may harm the firm's competitive position. Other information must be secured because of its intrinsic value. Checks and stock certificates are examples of this kind of information. Because they describe the location and value of tangible property, inventory records should also be protected.

TABLE 16.6 Data Output Handling

Reconcile output to input.

Maintain transaction records.

Balance transaction volumes.

Control error handling.

Retain records.

Control output distribution.

Application Program Audits

Auditability features in business application programs are essential to the control posture of the firm. What makes an application system auditable?

An application system is auditable if the application owner can establish easily and with high confidence that the system performs the specified function on a continuing basis. In addition, the owner must be able to verify that exceptional conditions, discrepancy handling, and error conditions are processed according to prescribed specifications and procedures. Auditability, therefore, includes manual and automated procedures in the IT organization and in the user organization.

A system is auditable if it contains functions and features that make it easy for the owner to determine correct processing on a continuing basis. Features that provide editing capabilities on input as well as journaling or logging functions throughout the processing stages are important ingredients of an auditable system. The system must provide documented verification that processing was performed according to specifications and that the results can be compared to known or expected standards of operation. For example, the payroll program will produce a check register detailing each of the checks produced by identification number. The owner of the payroll program has physical possession of the blank checks and can account for each check processed, including any used to align the printer. Additionally, control totals are available to ensure that all elements of the payroll calculation performed as expected.

Application owners must have reliable mechanisms to determine that the system was employed by authorized users for legitimate business purposes. They must have assurance that the data owned by them and used in conjunction with the application system is protected as they prescribe. These mechanisms, assurances, features, and functions must be available to the owner in the form of documentation or reports as a regular part of system operation. A system that behaves is this manner is auditable and can be considered to be under control.

Auditability begins with sound system development techniques and practices. Well-documented programs written in high-level languages are more easily auditable than others. Sound application development techniques such as structured design, modular programming, and complete documentation are particularly important for programs that undergo maintenance or enhancement activity. Program modification activity, a part of nearly every program's life, greatly complicates the task of maintaining system integrity.

Programming disciplines and programming standards make the task of maintaining program integrity much easier. The program testing process is vital; preserving the test results and the test data makes subsequent maintenance and enhancement easier. The time to insert auditability features into an application or to add reports to improve audit trails is when the program is being developed. Maintenance and enhancement activity must preserve and, if possible, augment the original auditability features.

Controls in Production Operations

The application production process must also operate in a controlled environment. Internal application controls are only effective if the production process itself exhibits discipline. The ideas related earlier must be complemented by a well-controlled and highly disciplined production environment. In general, this means that the disciplined approach to production operations must be implemented effectively and efficiently.

The disciplined management processes for production operations are most important to application integrity. In particular, problem, change, and recovery management are essential. A well-disciplined production operation will display sound control over performance objectives. It will do a good job of maintaining sufficient system capacity for operation of the applications. Batch and on-line operations execute applications in a structured and planned manner. Detailed scheduling and rigorous on-line management provide a controlled environment in which the applications can be processed. The disciplines form an important base for a complete state of confidence regarding application control and auditability. They are required for other reasons as well; they are vital for business controls.

NETWORK CONTROLS AND SECURITY

Network managers face many challenges concerning business controls and network security. The network operations they supervise are growing rapidly with the installation of departmental LANs and interdepartmental WANs, connections to common carriers and value-added networks, and use of long-haul national and international networks. These interconnections provide faster and more direct coupling between organizations. Telecommunications facilities dispatch valuable information and assets within the firm and between it and its customers and suppliers. In the financial industry, for example, networks transfer funds aggregating a trillion dollars or more per day worldwide.[5] Because of threats to networks, the importance of controls and security to network managers in today's environment is rapidly increasing.

Networks are subject to two types of security threats, passive and active. Passive threats consist of monitoring data transmission on the net in order to read messages or obtain information on network traffic. The intruder hopes to profit from the information or to identify information sources. Active threats consist of attempts to alter, destroy, or divert message data, or to pose as a network node. In the extreme, the intruder may become an active participant on the network, exchanging information with legitimate sites and obtaining free network services.

In order to deal with these threats, network managers must control access to the system and to the data, and they must protect data in transit. The first step in controlling access is to secure the physical system. This means that access to user devices such as terminals, facsimile machines, and phones must be controlled by lock and key. The facility should be secured so that only authorized individuals are admitted. Rooms containing cluster controllers or network servers must also be tightly secured. Cables connecting user devices to the controllers should pass through restricted access areas. Transmission media from controllers and network gateways to the communication controller must also be protected. Fiber-optics works well in this link since it is difficult to tap and tapping is usually easy to detect.

The second step in securing access to the system is to establish user identification and verification processes. For most systems this means that users sign on to the system with their name and then supply a password. The user identification/password scheme can be effective if properly implemented and used. For example, users can be required to establish passwords that have six characters, two of which must be numbers, with alphabetic characters separating the numbers. In this case, a2gj3b would be a valid password; cwfa12 would not. Even with these restrictions, users tend to develop passwords that are easy to remember and therefore easier to duplicate. Well-protected operations have system-generated passwords that are changed frequently.

Terminal systems should have automatic disconnect. If password validation fails after five attempts, the terminal is disconnected from the system and the event is logged. This prevents the use of automatic password-creation programs by hackers. All attempts to gain unauthorized system access or to use the system in an unauthorized manner should be recorded and reviewed by management.

Network and user managers should not hesitate to withdraw terminal access to employees who abuse system controls or who use the system for unauthorized purposes. In some firms, employee use of information systems for personal or unauthorized activities is cause for dismissal.

The third step toward system access security is data security. Properly secured data means that each data set in the system has an owner and that the owner's identification and the data set classification are part of the data. Each owner must specify who can access the data set and what kind of access the user has. The types of access are read, write, alter, delete, and execute. For data security to work properly, each valid user is authorized by the owner to access the data set and the user's authorized actions are identified. When a user tries to open a data set the system validates the user and permits only the type of access authorized by the owner. Unauthorized attempts to access data are recorded and investigated by management. Since application programs are also data sets, the execute authorization permits the user to operate the application.

For several reasons access security can never be perfect. For example, common carrier links can be tapped, microwave transmissions can be disrupted, and satellite transmissions can be intercepted. Therefore it is necessary to protect date en route. The most important tool for protecting data in transit and for maintaining network security is message encryption. Encryption changes message characters by means of an algorithm and a key into a different stream of characters prior to transmission. The message must be decoded after reception using the algorithm and key. The encryption process may operate at the character level but usually is applied to the message bit stream.

In 1977 the National Bureau of Standards established a data encryption standard (DES) which has been widely adopted and used in commercial applications. Proper use of data encryption can make telecommunications very secure. There are means for authenticating transmissions and for validating signatures. For especially sensitive traffic, a third party can be used to validate the traffic and to guard against lost or stolen keys. In addition, research is underway to improve the encryption algorithms by increasing their speed of operation and by making them more secure. Network managers must secure the firm's vital network traffic against passive and active threats by using encryption and authentication techniques.

Control of today's business operations and the security of the firm's intangible assets depend heavily on control and security of telecommunication systems. They are the link between users and applications; they link systems and applications with customers and suppliers; and they integrate the firm's far-flung operations. They must operate under control, and they must be secure. The network manager must make sure that this is the case. Table 16.7 lists the network security considerations we have discussed.

TABLE 16.7   Network Security Considerations

Physically secured terminal devices.

Physically secured controllers and cables.

User identification and verification.

Processes to deal with unauthorized usage.

Data set protection mechanisms.

Data encryption and authentication processes.

Protection of information assets tied to networks requires special attention. The business vignette that opened this chapter described what can go wrong when this attention is lacking.

There are several clear lessons to be learned from this vignette and from other incidents like this one. First, there are a small number of people who will try to intrude on your network and invade your systems and applications for pleasure or profit. They are intelligent and persistent; your network and your systems and applications must be protected from them. Second, networks and the systems connected to them can be secured from most intrusions; the degree of protection depends on the amount of resources expended and on the type of protection obtained. Third, despite the obvious threats, many organizations fail to take the simplest of precautions.[6] IT managers have considerable responsibility for the firm's assets; they must actively protect the vital assets entrusted to them.

Network control and security are disciplines like others discussed earlier. They relate directly to controls and audits in applications because networks are the sources and sinks of much of the application data. Network control and security relate directly to the other network management disciplines such as problem, change, and recovery management because effective network control reduces network problems and because network security reduces damage to information assets. The disciplines of network management are an integral part of the IT management system.

ADDITIONAL CONTROL AND PROTECTION CONSIDERATIONS

IT managers have many security and control responsibilities. Some of the most important ones concerning applications and networks have been discussed above, but others must be addressed. In particular, physical protection of major processors, control and protection of critical applications, and security of unusually important data sets are issues that IT managers must handle.

The data center containing the firm's central CPU complex and stores of information requires extraordinary physical protection. Some practices that help data center managers secure their operations are the following:

1. Only people who work in the data center should be allowed routine access.

2. Data center workers must wear special badges that identify them on sight.

3. The identity of all visitors to the center must be validated and they must sign in and sign out.

4. Duties within the center should be separated so that operators who initiate or control programs cannot access the data stores.

Under certain circumstances additional actions may be necessary. For example, in some critical centers visitors are prohibited entirely and the center itself is protected by specially constructed floors, walls, and ceilings. In all cases, however, the degree of protection must be consistent with the value of the assets.

Downsizing the mainframe means that all of these considerations may apply to the distributed operations. Users who operate their applications on their own processors must take many of the security and control precautions required of mainframe operators. Managers of distributed operations assume considerable control and security responsibilities.

Systems programs such as operating systems, file handling utilities, password generation programs, and data management systems must also be specially protected. Managers must take careful precautions with system programmers and network specialist who have access to restricted utility programs and control elements. For example, the file of authorized users and their passwords is available to system programmers. Utilities that can copy or rename data sets or that can alter load modules or tape labels are used by system support personnel. Most systems also have a superuser capability—one or more privileged passwords can be used by system programmers to obtain access to any data set on the system. System programmers need these capabilities in order to do their jobs.

Managers can take several routine precautions with the few individuals in the data center who have these special privileges. People in these responsible positions should rotate or change duties frequently. Their actions should be routinely recorded and managers should review the records frequently. When system programmers use privileged passwords or access restricted utility programs, they should do so only after obtaining clearance from the center manager. The manager should validate their actions upon completion.

Privileged network technicians have many of the same capabilities that system programmers have, but they may also have access to codes, keys, utilities, and passwords for remote operations. They must be controlled in the same manner as system programmers.

In addition to the controls and procedures required for the operation of the applications, some programs in the portfolio require special handling. Application programs that permit or authorize the transfer of cash or valuable inventory items need careful and unusual controls. For example, payroll, accounts payable, and inventory control programs need special protection. Managers must identify these programs and maintain an inventory of them. Application owners must prescribe necessary protection and security conditions for these programs to the custodian or supplier of service. The owners must also ensure these special considerations are implemented satisfactorily. The owner must prescribe controls on program storage, operation, and maintenance.

First, program source code, load modules, and test data must be identified as classified or sensitive information and protected accordingly. In most cases the protection should be the highest available on the system. For instance, the owner of the accounts payable program may classify the source code as confidential and restrict access to one maintenance programmer only. Load modules used for program execution may be restricted to another individual. This ensures that maintenance programmers do not have the ability to operate the program with live data. The test data sets should be entrusted to yet another individual who will operate the modified program with them and deliver the results to the owner. When program testing is complete, the executable load modules are updated and protected. Change control for these special applications must be carefully managed.

Second, operation of these critical programs usually proceeds differently from routine applications. For example, control over input and output documents is tighter for accounts payable than for most other applications. Blank checks are hand carried to the computer center after the sequence numbers have been recorded and verified. When processing is complete, the output is returned to the accounting manager who verifies the check count, returns unused checks to the safe, takes the checks and stubs to the distribution center, and gives the check register and other control information to the accounting department. These operations are verified in the accounting department and are recorded for later reference.

Some data sets for these applications are also highly sensitive, for example, the vendor name and address file for accounts payable. Anyone wishing to create a fictious vendor to whom payments will be sent must access it. Accounting managers must verify all transactions against this file.

Application owners must be especially vigilant during periods of maintenance or enhancement. Only authorized changes can be made and all modifications must be reviewed by the owner. The owner must control maintenance through close supervision and through documentation and testing procedures. It is important that the normal audit and control features in these programs function flawlessly and without modification. Additions and changes to these applications must incorporate additional audit and control features. Applications of this type must be guarded more carefully than the assets they control.

THE KEYS TO EFFECTIVE CONTROL

In order to operate under control, managers must have a complete understanding of their control responsibilities. They must know what their assets are, they must know the value of the assets, and they must classify and protect them accordingly. Managers must be actively involved in the control process. Their involvement must be timely and responsive to changing conditions; they must follow up to ensure their actions have been effective.

The operation of information assets such as systems and applications must routinely produce measurements and reports that reveal the state of control. Managers must review these reports frequently. They must be able to determine that specific operating procedures have been followed and that the operation is in compliance with defined control practices.

Managers should ensure that separation of duties is an established practice and that employees in critical positions are rotated to different jobs frequently. IT managers must conduct audits of their operations periodically, and they must use their findings to improve their business controls.

Business controls in application systems are of interest to user managers, IT managers, and also to other senior executives in the firm. The capstone discipline—management reporting—must function effectively. IT managers, their peers, and superiors require knowledge of correct application performance periodically. Astute IT managers will take extra steps to keep all interested individuals informed of the sound control posture surrounding the application assets. For example, the IT manager may summarize problem management actions for senior executives, or they may present trend information from internal audits and reviews of their operations.

These periodic reports should highlight the major routine actions that ensure correct and valid performance of the applications. A summary of the deficiencies in operation, if any, and the corrective actions taken must also be reported. The results of routine tests and audits of the control mechanisms performed since the last reporting period should be included. Additional steps taken to augment existing controls or to secure further the development or operations areas will be important to these managers. Comprehensive reporting is important because, in its absence, senior executives are likely to seek information through outside audits, independent reviews, or other less welcome means.

SUMMARY

Business controls and asset protection are fundamental to business operations. They are part of every manager's primary responsibilities. Controls are more important now because of changing business conditions, new technology introduction, and increased attention demanded by law. Controls in IT are especially important because the firm depends on computerized systems for control throughout its operations.

In order to control and protect assets, managers must know what their assets are and must have an estimate of their value. Managers must classify their assets and must specify protection in keeping with the assets' classifications. For application program assets, individual responsibilities are assigned to owners, users, programming managers, and suppliers of the computing environment. IT managers must ensure that these individual responsibilities are discharged effectively.

When applications are processed, control must be exerted over transaction origination, data preparation, data input, computer processing, and data output. The operation of application programs generates reports that provide evidence to the owner that the program operated as specified and that data is handled in accordance with its classification. The production environment is controlled through the disciplines of production operations and by separating duties within the center.

Networks are especially critical in today's business operations. They are subject to various threats. Managers must protect the physical network elements; they must control access to network devices and to data stores; and they must protect data in transit. User identification and passwords, data set classification and access protection, and data encryption all help secure the network from intrusion or protect data from unauthorized viewing or use.

Some information assets require special care because they directly manage other assets or because they control access to files of critical information. System control programs, utility and data management programs, and cash dispensing programs fall into these categories. These critical programs must be tightly controlled during storage, operation, and maintenance. The manual operations surrounding these applications are critical too, and they must be carefully controlled.

IT managers must evaluate the effectiveness of their control and asset protection processes. Their evaluations should be reported to the firm's executives who need to know the adequacy of controls in the firm. Corporate information of all kinds is valuable: IT managers must ensure that it is not vulnerable to loss or damage.

Review Questions

1. What lessons did you learn from the business vignette?

2. What is the first thing that managers must know in order to establish effective controls? What are some other things they must know?

3. Why are business controls important to IT and user managers?

4. Why does new technology usually require new and different business controls? What new controls might be required for a firm introducing office automation?

5. What staff responsibilities do IT managers have regarding business controls?

6. What are some physical information assets that must be controlled? What are the most important intangible assets that must be controlled and protected?

7. Intangible assets usually have great value to the firm. Can you give an example of a firm in which intangible assets are more valuable than all other assets?

8. What is meant by separation of duties? How might this work in managing inventory?

9. What are the characteristics of efficient and effective controls?

10. Identify the participants required for complete control over the development and use of application programs.

11. What responsibilities does the application owner have? How are these related to user responsibilities?

12. What are the duties of the IT manager to all the other participants in the business controls process?

13. Identify the control points in computerized data processing.

14. What items comprise control features of transaction origination?

15. What are the similarities and differences of controls for batch and interactive data input?

16. Who specifies control features in applications? What management processes ensure the correct and complete implementation of these features?

17. Identify the main elements in computer processing designed to provide protection and control.

18. The final control point in data processing occurs in data output. What control activities take place at this time in the processing cycle?

19. What is meant by system auditability? Why do application owners require auditability features as part of system controls?

20. Why is a disciplined production process essential to a well-controlled application portfolio? What constitutes a disciplined production process?

21. To what kind of threats are networks exposed? What kinds of action can managers take to contend with these threats?

22. What actions can managers take to protect data center assets? Downsizing changes the physical arrangement of data center assets and rearranges the control and security responsibilities. What does this mean to user managers? What are some considerations that might govern their actions in these circumstances?

23. What protections must be taken with system program assets? System programmers have almost unlimited access to information system assets. How do managers control this situation?

24. What special precautions must be taken with critical programs?

25. What are the keys to effective system controls and asset protection?

Discussion Questions

1. Why are application business controls not listed as a critical success factor in Chapter 1? In what instances do you think they should be?

2. Why are control issues in applications more important now than 10 years ago?

3. The business vignette illustrated some of the many things that can go wrong in networked systems. Considering all that you learned in this chapter, discuss the actions you would take to prevent this type of incident.

4. Discuss the control issues that accompany the introduction of facsimile machines to the firm. Who has responsibility for resolving these issues?

5. Intangible property is very valuable in most firms. Suppose you were the Merrill Lynch manager who controlled the name and address file for the CMA program discussed in Chapter 2. Discuss how you might protect and control this asset from loss or damage.

6. Along with separation of duties, the text mentioned the need to rotate employees regularly. Discuss why these two actions work together to improve security and controls.

7. In some firms the IT manager has ownership and control responsibilities over the application programs and the databases. Discuss the advantages and disadvantages of this arrangement.

8. Discuss the changes in control responsibilities accompanying end-user computing and downsizing. What changes occur in the provider-of-services role? "With end-user computing the owner, users, development manager, and provider of services are all in one department." What are the implications of this statement?

9. What are the IT manager's responsibilities to departments that have adopted end-user computing?

10. Discuss the special precautions you think might be advisable regarding controls during the application enhancement process.

11. If you were the payroll manager, what control actions might you take when the payroll program is being altered to handle changes in the tax law?

12. System control and auditability must exist across the automation boundary. Using the accounts payable program as an example, discuss the shifts in responsibility at this interface.

13. Discuss why the disciplines of problem, change, and recovery management are also business controls issues.

14. What, if any, ethical issues might arise in testing business controls in applications?

15. Discuss the sometimes conflicting issues surrounding ease of use and business controls. How can an analysis of risk help resolve these issues.?

16. Develop a list of critical applications that you believe exist in many firms. Why are these applications critical? How should they be protected?

17. What special control considerations are involved when considering purchased applications?

18. Discuss the important business controls issues that arise when a service bureau is employed to process payroll.

19. What do you think the role of the firm's controller should be in the subject of controls and audits in application systems? If you were the controller in a firm contemplating downsizing the firm's data processing operations, what would you want to know before you approved the plan?

Assignments

1. Assume that you are the manager of the workstation store in a firm starting to implement end-user computing. Develop a list of business controls actions for the store's operation. As store manager, what reports and audit trails do you think are necessary, when should they be produced, and how would you handle separation of duties among your three employees?

2. One of the applications serving your department is being replaced by a newly developed program. The prototyping methodology is used for the development process. Devise a process by which you, the manager, can remain assured that business controls issues receive sufficient attention during development.

ENDNOTES

[1] Exerpted with permission from "Spreading a Virus," *The Wall Street Journal*, November 7, 1988, 1.

[2] In this case the perpetrator was caught, prosecuted, and convicted of felony acts. He was sentenced to a $10,000 fine, three years' probation, and community service.

[3] *Annual Report, 1989*, Baxter International Inc., 33.

[4] Richard Greene, "Inequitable Equity," *Forbes*, July 11, 1988, 83.

[5] Deepak Sarup and Andrew Davies, "EFTS International Networking: Addressing the Global Security Problem," *The EDP Auditor's Journal*, vol. 4, 1988, 51.

[6] Ali Farhoomand and Michael Murphy, "Managing Computer Security," *Datamation*, January 1, 1989, 67. This article reported a survey of *Fortune 500* firms which indicated that many spend less than 1 percent of their IS budgets on computer security.

REFERENCES AND READINGS

Bruns, William J., Jr., and F. Warren McFarlan. "Information Technology Puts Power in Control Systems." *Harvard Business Review* (September-October 1987): 89.

Cerullo, Michael J. "Controls for Data Base Systems." *The CPA Journal* (January 1982): 30.

Coughlin, John W. "The Fairfax Embezzlement." *Management Accounting* (May 1983): 32.

Dasher, Paul E., and W. Ken Harmon. "Assessing Microcomputer Risks and Controls for Clients." *The CPA Journal* (May 1984): 36.

FitzGerald, Jerry, and Ardra FitzGerald. *Fundamentals of Systems Analysis*, 3rd ed. New York: John Wiley & Sons, 1987.

Gregory, Ed. "FBI Arrests NT Employee." *MIS Week*, May 30, 1988.

Mason, Richard O. "Four Ethical Issues of the Information Age." *MIS Quarterly* (January 1986): 5.

Perry, William E. "Evaluating Data Center Controls," in Robert E. Umbaugh, ed. *Handbook of MIS Management*, 2nd ed. Boston: Auerbach Publishers, 1988.

Part
Six

17 People, Organizations, and Management Systems

18 The Chief Information Officer's Role

Preparing for Advances in Information Technology

The work people do and the organizations they are in often undergo significant change because of technology introduction. Astute managers deal skillfully with the people issues as transitions occur for individuals, departments, and organizations. Organizations and effective leaders facilitate people working together to achieve the benefits of new technology. The chapters in this Part are People, Organizations, and Management Systems and The Chief Information Officer's Role.

Managing human resources effectively is the most important success factor for IT managers. Effective information technology managers display an abundance of people management skills.

17

People, Organizations, and Management Systems

A Business Vignette
Introduction
Technology Shapes Organizations
  Organizational Transitions
  Centralized Control—Decentralized Management
  The Impact of Telecommunications
  The Span of Communication
People Are the Enabling Resource
  People and Information Technology
Essential People Management Skills
  Effective People Management
  Attitudes and Beliefs of Good People Managers
  Achieving High Morale
The Collection of Management Processes
  Strategizing and Planning
  Portfolio Asset Management
  The Disciplines of Production Operations
  The Management of Networks
  Financial and Business Controls
  Business Management
The IT Management System
Summary
Review Questions
Discussion Questions
Assignments
References and Readings

A Business Vignette

In 1854 my great-great-grandfather founded a small glass manufacturing business, the Union Glass Company. Today it is a global corporation known as Corning Inc.[2] Until a few years ago, like most American companies, Corning had a traditional business structure more appropriate to the 19th century than the intense competitive environment of the 20th. The corporate hierarchy, with tasks strictly defined by organization charts and tightly drawn job descriptions, was deeply rooted in our history.

But as a company, we were also quick to realize the importance of joint ventures and other types of business alliances and formations. In 1924, for instance, we entered into our first joint venture to make cartons for glass products. Later, a handshake between my father and Dr. Willard Dow in 1941 led to the formation of the Dow-Corning Corporation to produce silicones, which our scientists invented.

Since 1983, we have been experimenting with ways to make our management structure conform better to our business structure. As a result, Corning is now what we call a "global network." We use the concept of the network in this broad way: A network is an interrelated group of businesses with a wide range of ownership structures. Although diverse, these business are closely linked.

Over the years, our network strategy evolved from our basic businesses, which we visualize as four segments of a wheel: consumer housewares, specialty materials, telecommunications, and laboratory sciences.

At the center of the wheel are the formal linkages that bind the network and establish priorities. For Corning, these are technology, common values, and shared resources. Within each sector there are a variety of business structures that range from traditional line divisions, to wholly owned subsidiaries, and to alliances with other companies. For example, in the telecommunications sector we manufacture optical fiber as a line division here in the United States and as a joint venture in Europe; produce fiber-optic components through a subsidiary in France and joint ventures here; and make optical cable in alliances with Siemens A. G.

Alliances or joint ventures are a major component of our structure. They contribute about half of our earnings, which I believe is unique among Fortune 500 companies but is likely to become more commonplace. Our specific objective, however, is not to creat joint ventures or acquire subsidiaries but to select a business structure that will provide the best chance of success in a particular market. Unfortunately, forming partnerships is still often construed as a sign of weakness outside of our company.

Not only is this attitude outdated, but it can also be harmful. As an example, after a company makes an invention it no longer has the luxury of nurturing that discovery over many years. Speed is the key, and a partner may add expertise in manufacturing or marketing that will bring the product to market much faster.

Our people have been trained to look at alliances as strengths. We are also promoting the idea that skill at creating and growing successful joint ventures is needed throughout the business network. We have found that the successful operation of a global management network requires a new mindset. A network is egalitarian. There is no parent company. A corporate staff is no more, or less, important than a line organization group. And, being a part of a joint venture is just as important as working at the hub of the network.

Ralph Kilmann, a professor at the Katz Graduate School of Business at the University of Pittsburgh, has written that "the hub will assume some of the functions of a broker, conflict negotiator, facilitator and think tank" and that its major role will be to provide the information, resources, and guidance to keep the network functioning.

Our network is still a "work in progress." We need to spend more time, for instance, on how we can use the network to enhance career opportunities for people who want to move between diverse businesses. We also must decide what information must be disseminated and what policies carried out through the network to break down walls between businesses.

I am sure there are many organizational surprises ahead. But I am also convinced that we are on the right track. We expect international competition to grow rather than abate, and by the year 2000 we will face a dramatically different business climate. The global network will give Corning—and other companies as well—the flexibility and strength to prosper in the future.

INTRODUCTION

The emphasis in this text has been on the management of information technology assets and on tools, techniques, processes, and procedures required for effective control and use of these assets. The processes of strategizing and planning set the stage for understanding and embracing software and hardware technology trends. Management and control of application assets, and the production facility in which applications are processed, followed technology trends. This book turns now to the most important information technology assets, IT people and their organizations and structures.

People and their organizations are critical to the successful functioning of the modern business firm. But the effective use of people within the firm requires a corporate culture within which they can thrive and be productive for themselves and for their organizations. Employees and managers need to know "how we do things around here." To a large extent, that is a function of the management system. The management system provides an intellectual framework not only for management actions but also for employee actions.

As the quotation from Peter Drucker in Chapter 1 and the business vignette in this chapter indicate, dramatic changes are taking place in conventional business organizations. These changes are not confined to the firm, but involve its customers and its suppliers. In many cases, changes within the firm result from collaboration with others. In all cases, information technology enables new structures of business activity and is itself a change agent.

TECHNOLOGY SHAPES ORGANIZATIONS

We are witnessing immense, fundamental, pioneering changes in the business environment. The environment is rapidly becoming much more complex; the complexity is driven by tremendous advances in communications and information processing capability. These driving forces are significantly affecting our business structures.

> There is no question that the new information technologies (IT) are having a major impact on the range of strategic options open to an organization. IT is not only creating an environment that is making it imperative for organizations to evolve, it is also helping them adjust to the enormous changes outside their own boundaries.[3]

Organizational Transitions

The decade of the 1980s was one of major organizational transition for many corporations worldwide—the decade of the 1990s promises more of the same. In the United States, mergers, acquisitions, and leveraged buyout activity reached a frenzied pace. "Before the decade had ended, a towering $1.3 trillion was spent on shuffling assets—an amount on a par with the annual economic output of West Germany."[4] More than 2800 companies were involved. The largest deal was the $24.7 billion LBO of RJR Nabisco by Kohlberg Kravis Roberts. The phenomenon was not unique to the U.S., as the purchase of Jaguar by Ford and many other such purchases illustrate.

Mergers, acquisitions, and leveraged buyouts caused turbulent times for IT executives and for employees, too. In a study by Steelcase, Inc., nearly 50 percent of office workers and top executives in Canada and the U.S. indicated that their companies had reorganized. Thirty-nine percent revealed the reorganization resulted from mergers or acquisition.[5] About 25 percent of the restructured firms reduced the number of white-collar jobs. Corporate reorganization frequently leads to reductions in employment—in many cases workforce reductions are the motivation for restructuring. Information systems professionals are not immune to this.

However, for many IT professionals, mergers, acquisitions, and corporate reorganizations represent opportunities. One example of this occurred when Baxter-Travenol Laboratories, Inc., purchased American Hospital Supply Corp. in 1985. Baxter was attracted to American in large part because of American's highly successful automated order-entry system. Michael Heschel, American's IT chief, became Baxter's IT executive and immediately began consolidating the MIS organizations. The combination included 800 employees and a budget of about $100 million.[6] In 1989, Baxter became the *Computerworld Premier 100's* number one IS user. In the meantime, the consolidation continued and Heschel advanced to a more responsible job with Security Pacific.

Through purchases, joint ventures, and planned internal expansion, many firms are attaining a significant international presence. Some of the world's largest and most successful corporations are becoming truly international businesses. Some examples illustrate the point: Honda manufactures in several countries and sells in many. IBM manufactures in many countries and sells globally. Phillips, the huge Dutch electronics firm, has 10 percent of the world's color TV market and is the world's largest manufacturer of light bulbs. Royal Dutch-Shell, the world's largest industrial corporation, is of Dutch and British origin but must be cognizant of the interests of many nations in its worldwide operations. The trend toward dispersed national, transnational, and international business enterprises is certain to accelerate as communication technology and political considerations provide the impetus.

Centralized Control—Decentralized Management

Not only are firms growing in size through mergers and acquisitions, but alliances between firms are increasingly common. This text gave many examples of alliances. They were discovered in connection with the study of strategic systems. The semiconductor and telecommunications industries demonstrated the value of cooperative efforts. And alliances were found in connection with application program development and computer center operations. Alliances and partnerships, once considered a sign of corporate weakness, are becoming necessary to speed new products to market, to improve operational efficiency, and to obtain the skills necessary to meet business goals.

The adoption and use of alliances and joint ventures to capture the advantage of time, to employ critical skills, and to obtain access to distribution channels or to new markets adds complexity to the firm's value chain. Value is added to the firm's product through the processes of inbound logistics (the firm imports raw materials or subassemblies from suppliers), internal operations (assembly, manufacturing, or process activities), outbound logistics (distribution activities), sales and marketing, and service. These processes are supported by activities such as procurement and technology development, and they depend on human resources and the firm's infrastructure.[7] Firms operating joint ventures or firms with alliances have a product value chain several parts of which may be outside the firm itself.

For example, consider the five firms in five countries that manufacture jet engines, mentioned in Chapter 14. Value is added to the final product as the five manufacturers produce engine subassemblies, develop engineering and technical documentation, and produce maintenance spare parts for delivery to final assembly locations. There are many other instances in many industries; the auto industry and the computer industry are some well-known examples. In all these examples each firm's infrastructure and information technology architecture are used cooperatively by managers to bring added value to the end product.

Information technology is essential and in some cases absolutely critical to the success of firms engaged in forming partnerships and alliances or in internationalizing their businesses. We have seen that Corning required technology, common values, and shared resources to bind interrelated businesses together. Information and guidance from the hub of the corporation are the intellectual resources that allow firms in five countries to develop and build jet engines or enable others to produce oil on several continents and to refine, distribute, and market it in many countries. Executives use information technology extensively to keep the networked business operating as a cohesive unit.

Centralization versus decentralization has been a longstanding corporate strategic issue, one that looms larger now because of the recent trends toward corporate restructuring and global strategies. As firms grow in size, there is a tendency to decentralize operational control and to put decision making closer to operational centers. For many firms, especially conglomerates, growth via acquisition accompanied by decentralized profit centers was the norm. Decentralized operations with limited centralized control were usual. However, information technology, especially telecommunications systems, offers executives the potential to have the best of both worlds. Decision making and operational control can be delegated to operational units; control information can be available to headquarters on a real-time basis. Information technology permits the organizational structure described by Houghton in the business vignette to operate effectively and efficiently.

The Impact of Telecommunications

Advances in telecommunications technology and rapid implementation of these advances to achieve important business goals are the hallmarks of information technology today. The union of information systems and networks gives business executives enormous capability to expand their operations, to form alliances and joint ventures, and to restructure their assets for greater effectiveness. Telecommunications systems shrink time and distance and help establish new relationships. They improve organizational efficiency and effectiveness, and they promote innovation.[8] The following examples illustrate how companies capture these advantages.

Volvo uses an international computer-aided design network to design and manufacture heavy trucks. IBM uses its engineering design system and programming support systems through national and international networks to design and manufacture computer systems for a global market. Network systems allow GM to sell cars in the U.S. that were built in Korea but designed in Germany. More than 12,000 companies in North America now use electronic data interchange (EDI) to transmit information from one computer system to another.[9] The electronic communication can take place between the decentralized operations of the firm or from the firm to the offices of a supplier or a customer. The concepts of just-in-time manufacturing and the paperless factory are critically dependent on telecommunications systems. Adoption of sophisticated telecommunications technology is a critical success factor for most firms today.

The Span of Communication

Telecommunications systems greatly increase the span of communication and remove the need for filtering layers of management, a development that has made the traditional span of control irrelevant.[10] Executives who communicate electronically throughout the organization can understand details of their operations essential to maintaining control. The reduced number of middle managers must expand their spheres of influence in order to facilitate the attainment of corporate goals and objectives. They must also use advanced communication techniques to increase their effectiveness over a broader range of organizational activity.

The span of communication can be large in modern industrial firms. For example, technicians in California, working with technologists in New York, help manufacturing engineers in Germany solve difficult production problems on advanced disk drives for European customers. On a smaller scale, application software specialists diagnose customer problems and provide solutions electronically. The customers are located throughout North America, but the specialists are located in Colorado. On yet a smaller scale, employees at one computer development and manufacturing facility unite product designs with manufacturing technology and procurement operations to build customer solutions when the customer needs them, just in time. Telecommunications technology enables these operations. Indeed, telecommunications enables parallel processing at the firm level whether the firm is local, national, or international in scope.

Organizational impacts of technology are not limited to user organizations; the IT organization itself is undergoing substantial change in most firms. It must adapt even further if it is to be effective. The days of a centralized IT organization are rapidly disappearing and, in its place, a new structure is emerging. The new structure recognizes that some information processing functions are best performed centrally but that others must be performed locally. IT organizations must have strong leadership to develop, operate, and maintain systems vital to the firm's centralized operations. The IT leader must ensure that both central and dispersed IT activities conform to and support the firm's strategic direction.

The new IT structure must also form partnerships and develop alliances with users throughout the firm to ensure effective use of the technology at all levels. IT must facilitate user adoption of appropriate information systems and must support their growth and development. Important user tools such as computer-aided design systems, computer-integrated manufacturing systems, office systems, and end-user systems for marketing, sales, or service are all part of the firm's information infrastructure. IT managers are responsible in part for the success of these user systems. IT organizations must be tightly coupled to the using organizations; shared values, goals, and objectives must be the norm.

Cooperative endeavors between individuals or groups are the life-blood of business enterprise. Improved technology, sophisticated management systems, and effective people management enable employees, managers, and the enterprise to adapt to the steady stream of changes required for success. But as the following passage illustrates, the difficulties for the innovator are an age-old problem.

> It should be borne in mind that there is nothing more difficult to arrange, more doubtful of success, and more dangerous to carry through than initiating changes. The innovator makes enemies of all those who prospered under the old order, and only lukewarm support is forthcoming from those who prosper under the new. Their support is lukewarm, partly from fear of their adversaries, who have the existing laws on their side, and partly because men are generally incredulous, never really trusting new things unless they have tested them by experience.
>
> In consequence, whenever those who oppose the changes can do so, they attack vigorously, and the defense made by the others is ineffective.
>
> So both the innovator and his friends are endangered together.

A. Machiavelli, *The Prince*, 1513

PEOPLE ARE THE ENABLING RESOURCE

Managers universally recognize that human resources in a business enterprise comprise the most important assets on which business success depends. Most CEOs pay tribute to this fact by noting the importance of people resources in letters to stockholders or in annual reports. It is fairly common to find a paragraph in the letter to stockholders thanking the employees for their contributions to the success of the business and making the point that "people are our most important asset."[11]

When R. J. Stegmeier assumed the position of CEO at Unocal he published the following in his first letter in April 1989 to employees and stockholders:

> In my view, this company's greatest asset has always been its human resource. Without the talent and teamwork of Unocal's 18,000 employees, our oil would stay in the ground and our facilities would stand idle. We face some tough challenges in the months and years ahead, but together—by our creativity and hard work—I'm certain that we can meet those challenges and turn them into opportunities for productivity and growth.

This is a good example of CEOs' regard for human resources.

Although most of the corporation's functional areas are vitally dependent on a base of skilled employees, the overall performance of some organizations may be limited by their ability to recruit, train, and retain them. Individual performance relates directly to the performance of the unit. For example, the revenue of a marketing unit may be limited by the availability of people trained to sell the firm's products. New product development depends on the creativity and ingenuity of the engineering force engaged in development activities. The firm's ability to grow and prosper hinges on the skills, abilities, and energy of the management team in establishing and implementing strategies and plans for success.

Not only is it generally agreed that human resources may limit unit performance, but in many instances significant contributions by a small number of individuals or by an individual working mostly alone materially improve the bottom-line performance of the organization. An engineer whose brilliant idea spawns one or more new products or a salesperson who clinches a big sale are very important to the firm. Thomas Musmanno, who invented the cash management account, and the managers at Merrill Lynch who led the implementation of the CMA are truly valuable assets.

People and Information Technology

Since the beginnings of electronic data processing in the 1950s, the effective implementation of information systems in business and industry has been limited to some extent by the availability of highly skilled people. Recently, the shortage of trained and experienced managers to lead the introduction of complex and rapidly evolving technology has limited progress. The job opportunities for programmers and systems analysts has remained firm for 30 years; these skills are predicted to be in short supply during the near future. Numerous articles have detailed the consequences of inexperienced or ill-prepared individuals leading organizations to increasing automation in the face of high corporate expectations. Although 30 years of experience have taught us much about coping with these difficulties, there remain plenty of examples to remind us that we still have a long way to go.

Not only have we not closed the skills gap in the field of information handling, but current trends in telecommunications, end-user computing, downsizing, and out-sourcing are placing increasing demands on skills currently in short supply. Talented individuals who are able to capitalize on advances in hardware, software, and telecommunications are in great demand. Additionally, the new skills required to implement strategies in these growing and potentially profitable areas differ in many ways from the current skill base. The task of managing people in the IT organization has always been a difficult one, but emerging challenges promise to keep IT personnel management exciting.

Prudent IT managers and their superiors are planning to cope with personnel shortages lest they find themselves on the critical path of their parent organization's roadmap to success. Some alternatives are available. Managing the present employees, who are familiar with the organization and its mission, in an effective manner must be the highest priority. If the organization fails to manage its present staff in a superior fashion, it stands little chance of obtaining and retaining outstanding individuals.

ESSENTIAL PEOPLE MANAGEMENT SKILLS

The key to effective utilization of information processing technology is management's ability to employ skilled individuals. These individuals must be knowledgeable about the technology itself and the use of the technology within the firm. Talented people enable the firm to capture the benefits and capitalize on the opportunities offered by advancing technology. Employing skilled people, managing them with sensitivity, and providing effective leadership are necessary conditions for success. Managers who fail to capitalize on their people set the scene for mediocre or marginal performance in other endeavors. Successful managers employ solid people management skills as the cornerstone of their success.

What do we mean by solid people management skills? What illustrates good people management traits in the minds of employees, and what reasonable expectations can employees hold of their managers? What distinguishes managers possessing these skills from those who do not?

The answers to these questions depend on the assumptions one makes about organized human effort and the values individuals derive from participation in organized activities. Abraham Maslow attempted to explain individual needs as a hierarchy with basic physiological needs such food and drink at the bottom and self-actualization needs at the apex. Physiological and safety needs, the most essential, are followed in order of importance by social needs and esteem needs, which precede self-actualization needs.[12] Individuals will try to satisfy the most basic needs first; when these are satisfied, they will be motivated to satisfy the next most important needs. For example, an individual whose physiological and safety needs have been met will be motivated to satisfy social needs such as love or a sense of belonging. When these needs are met, the individual will be motivated to attain esteem and recognition. When all other needs are fulfilled individuals strive for self-development, according to Maslow.

Most IT employees have satisfied their physiological and safety needs and are motivated toward higher needs. Organized activity in the firm provides the means for managers to appeal to these needs and to accomplish objectives for the firm. Skillful managers believe that most employees prefer to be engaged in meaningful work in which they also attain self-satisfaction and have a chance for self-development. The ideal situation is one in which the individual's and the organization's goals are congruent. Managers who have a good understanding of their employees as individuals and who can arrange for this congruence have taken a giant step toward improving morale and increasing productivity.

Effective People Management

It follows that effective people management centers around the concept of respect for the dignity of the individual. Good managers recognize the enormous, frequently unrealized potential in each individual. Effective people managers work hard at understanding each individual's preferences and motivations and displaying respect for each employee's unique personal characteristics. Talented professionals expect to be treated as special individuals; good managers meet this expectation.

Good people managers make the right assumptions about individuals and act with these beliefs in mind, unless proven wrong. These assumptions form the basis for managers' behavior toward employees and, in turn, influence employees' attitudes about managers.[13] The assumptions are very important, not only because they govern attitudes, but because there is a certain self-fulfilling character about them. Therefore, the assumptions and the attitudes derived from them have an important bearing on individual and group performance.

Attitudes and Beliefs of Good People Managers

Good people managers believe that employees are honest and industrious, that they act with the best interests of the firm in mind, that they are intelligent and willing to learn, and that they desire self-fulfillment. In keeping with these beliefs, good managers establish an environment of high but reasonable expectations for their employees. Based on the belief that professionals desire to make meaningful contributions to the organization, managers work with individual employees to establish challenging goals. The manager and employee agree on objectives that are aligned with the needs of the organization and that lead to self-fulfillment for the employee. Employees who desire to achieve challenging and self-fulfilling goals that are congruent with organizational goals

---

are ideally positioned to make significant contributions to the firm. In the process, their accomplishments achieve self-satisfaction for themselves. They are productive and they are self-fulfilled.

Management equals leadership within an organization. Employees expect their managers to lead with clear goals and objectives for the organization based on a vision of the future. Good managers must clearly articulate this vision to the organization, establish organizational goals and objectives, provide pathways, and set directions that lead employees toward accomplishing challenging goals for the organization and for themselves. Good managers provide good leadership through vision and goal setting.

Managers must also set challenging performance standards for themselves. By setting good examples for others to follow, effective managers establish an environment of high performance and productivity for the organization. High-performance organizations are led by managers who have high expectations for themselves and who have attained agreement on challenging goals for their employees. Managers achieve self-satisfaction and self-fulfillment from attaining difficult goals just as employees do and their morale and productivity rises with that of their employees.

When managers and employees have reached mutually acceptable standards of accomplishment, employees should assume responsibility for meeting the standards and should be given the authority to accomplish the tasks. They should be provided with tools to do the job and trained to use the tools effectively. In addition, good managers delegate responsibility to individual employees and give them the authority to carry out the objectives. Through delegation, managers empower employees to meet objectives for the organization. Managers must also pinpoint accountability and responsibility so that employees are clear about what is expected of them and so that they can be rewarded individually for their accomplishments.

All organizations contain obstacles that retard progress. Frequently bureaucracy appears to stand in the way and meaningless rules make it difficult to get the job done. Sometimes the bureaucracy is used as an excuse for lack of progress. One of the tasks of good people managers is to remove or reduce the barriers to accomplishment. If the rules are truly meaningless, managers should remove them. If the rules are required, managers should explain why they are needed and should assist employees in complying with them. Managers represent the firm to employees; they should enforce effective regulations and eliminate ineffective ones.

Good managers expect employees to be thoughtful about their work, but they also expect action. In most situations there is a balance between exhaustive analysis and thoughtless action. Frequently, in the hope of obtaining all the facts surrounding a decision, employees and managers spend far more time on analysis than the reduction in risk is worth. It may be impossible to understand all the minute consequences of certain actions, but some action must be taken in the face of incomplete information. That's the essence of decision making, after all. Effective decision making by both managers and employees means striving for the middle ground, which balances expended effort against remaining risk.

Good managers expect employees to solve problems for the firm, but they also expect employees to prevent them, since solving problems usually requires more time and energy than preventing them in the first place. Good people managers encourage and reward problem prevention. In many cases this is a difficult task because visible external signs such as distressed operations and frenetic activity accompany problem solution actions, but problem prevention usually consists of thoughtful actions taken without fanfare while operations are proceeding smoothly. Good managers must take special care to acknowledge and reward the subtle but valuable actions of those who work to prevent problems in the first place or who have the ability to keep small problems from becoming large.

Through their actions and their attitudes, good people managers establish an environment of high productivity and good morale. In this environment, individuals can accomplish challenging objectives and experience high degrees of self-fulfillment. The environment encourages creativity, innovation, and invention. By providing an atmosphere of productivity filled with opportunities for self-satisfaction, managers create an environment where innovation can occur (for many, innovation is the highest form of self-fulfillment). But managers' actions must also provide support to those whose innovations were less than expected or whose attempts at invention failed. Managers know that not all innovative endeavors result in success, and they must remember that the surest way to stifle innovation is to belittle or criticize individuals whose innovative ideas were less than totally successful.

Achieving High Morale

Studies have shown that employee morale and employee opinions of their immediate manager's performance are strongly correlated with the level of trust and confidence employees have in their manager.[14] Employee trust and confidence is granted to managers whose behavior is reliable, based on predictable norms for the organization, and consistent across similar situations. Managers who maintain steady and consistent patterns of behavior inspire trust and confidence in their employees and are viewed more favorably by their employees. Employee morale in departments headed by these managers is also predictably higher.

High employee trust and confidence in management stems from many factors of managerial behavior; factors that are under manager's control and that can become part of the practice of management. Some of the actions that managers can take to increase employee trust and confidence are:

1. Maintain two-way communication with employees to understand their needs and desires and to share company information.
2. Provide training and complete information so that employees can do their jobs effectively and efficiently.
3. Inform employees of opportunities for promotion and career advancement.
4. Listen to employee suggestions for improving the work environment and respond to all good suggestions.
5. Sponsor teamwork and cooperation among department members.
6. Be available to employees when they desire consultation.
7. Understand the amount and quality of each employee's work.
8. Use the knowledge of each employee's work to grant fair salary increases and promotions.
9. Provide enthusiastic leadership to the department in achieving its goals and objectives.

Managers who inspire trust and confidence in their employees are good people managers. They are effective communicators who share information candidly and consistently and they welcome dialogue intended to improve business operations. Effective managers treat individuals with respect and dignity. They understand individual wants and needs and individual contributions. Good people managers understand good performance and reward it both publicly and privately. Good people management is the hallmark of effective executives and is a critical success factor for IT managers.

THE COLLECTION OF MANAGEMENT PROCESSES

Effective operation of the IT management system is also a critical success factor for IT managers. It provides a framework or background within which employees and managers operate and from which the norms for the IT organization are in part derived. The management system is comprised of tools, techniques, and processes that are exercised periodically.

For example, the strategic planning process, which takes place annually, establishes long-term corporate directions for the use of information technology; the problem management system, which operates daily, verifies managers' intentions to achieve service levels. The IT management system must be aligned with the firm's management system: it must support and augment the firm's management system, and embrace the firm's values and basic beliefs.

IT managers must have systems that guide them in achieving critical goals for themselves, their organizations, and their firms. As we learned in Chapter 1, these critical factors can be organized into strategic and competitive issues, planning and implementation concerns, operational items, and business issues. The management systems presented in this text are designed so that managers can achieve their critical goals and deal effectively with the issues facing them and their firms.

Strategizing and Planning

The management tasks of building IT strategies and developing long- and short-range plans to implement IT's strategic direction are critical first steps for IT managers. Strategizing and planning are the cornerstones of the IT management system, activities that link the IT organization to the firm's management system. Their outputs align the IT strategic direction with the firm's strategic business direction. The linkages and alignments are critical to IT and to the firm. The relationship of these activities is shown in Figure 17.1.

Information technology planning and control begin with the firm's business strategy development process. The firm's mission statement and its goals and objectives are the foundation on which the firm's business strategy is built. The firm's business strategy and the IT strategy are intimately linked through shared goals and objectives and through shared processes. Business objectives for the firm that require IT resources and actions are translated into IT strategy directions. Thus the process of IT strategy development occurs in conjunction with the business strategy development process. Since the IT strategy is the basis for IT plan development, shared strategic directions and plans are developed within the firm, forging links between IT and the remainder of the firm. It is this interweaving and sharing of goals, objectives, and processes that ensures alignment between the IT long-range plan and the business strategy.

These cohesive strategizing and planning activities deal directly with many of the issues facing senior executives, IT managers, and their peers throughout the firm. Specifically, these processes force alignment of IT and corporate goals, educate senior managers on IT's role and its potential, and demonstrate IT's contribution to the business. If properly implemented, these processes eliminate strategic planning as an issue. They provide a mechanism through which the firm can exploit information technology for competitive advantage in a systematic manner.

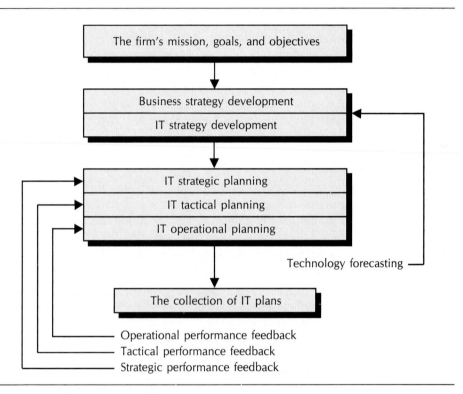

FIGURE 17.1 **IT Planning and Control**

IT strategizing and planning processes offer excellent opportunities to coach others in the firm, including executives, about how technology can be used to achieve objectives. Forecasts of technology capabilities must be provided as input to the firm's strategy and to the IT strategy. Technology capability is matched with the firm's requirements for new technology and plans incorporating it in business processes are developed. Planning for the adoption of new technology and improved use of current technology can reduce costs, improve efficiency, or provide and sustain competitive advantage. If properly handled, these actions will also ensure realism in long-term expectations regarding technology.

There are several reasons why the process described above may not be successful and the resulting IT strategic plan may not be aligned with the firm's business plan. For example, the process can only work when the firm itself has a well-defined mission statement and a business strategy. In their absence, an IT strategy may not be well correlated with the intended business direction. In some firms, IT is not considered critical to the firm's success and is not brought into the firm's planning process. In these firms, strong support for corporate goals and objectives from information technology is not required. In other cases, IT managers share in the process but lack the necessary business skills and knowledge to construct strategies and plans aligned with the the firm's plans. To mitigate these difficulties, IT managers must ensure that they understand what their role is and that they contribute to the firm in substantive ways.

Control in plan execution stems from operational, tactical, and strategic performance information. Performance information describes the variances between actual results and planned results. It also includes variances between environmental and business assumptions in the plans and actual business and environmental conditions. For example, if competition threatens some of the firm's markets the IT development team may need to improve the planned schedule of a marketing system under development. Depending on the nature of the variances, feedback may generate course corrections in operational, tactical, or strategic planning. Also, major environmental perturbations may cause the firm to adjust its strategic direction. In the extreme, the firm may adjust its mission to capture perceived opportunities or to avert potential threats.

Effective strategizing and planning are critical activities for the firm and its managers. These vital activities direct the firm's exploitation of its current and future assets and form the foundation on which its future depends. The management systems described in this text provide an excellent framework for attaining success in this crucial activity.

Portfolio Asset Management

Managing the application portfolio assets is an important activity; IT managers must deal successfully with many issues in this task. In earlier chapters, this text described methods for prioritizing the application backlog, managing the development process, and finding alternatives to local application development. Figure 17.2 describes the relationships between these activities.

The application portfolio is a large asset that is growing in size and complexity. Most firms spend a significant percentage of their IT budget maintaining and enhancing the applications and managing the associated data resources. The management systems for dealing with these challenges begin with IT strategizing and long-range planning activities. These processes establish the strategic direction for enlargement and enhancement of the application portfolio and lay the foundation for incorporating technology advances. They help managers add value to the portfolio assets.

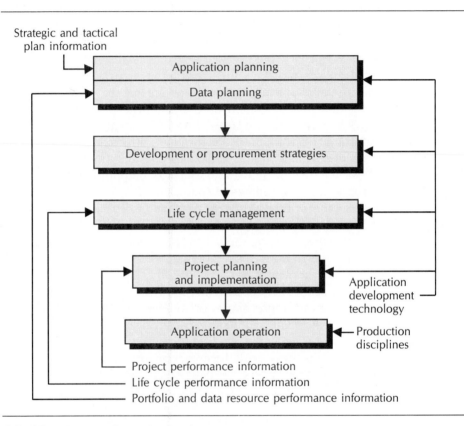

FIGURE 17.2 Application Management

Application and data planning rely on information gathered during the firm's strategy and plan development activities. Application performance and capability are compared with the firm's requirements and the sufficiency of the data resources is analyzed. Using this information, the prioritization process identifies the applications that qualify for investment and specifies what acquisition methodology is preferred for each application. This decision-making process combines the technical knowledge of computer experts with the business knowledge and vision of senior management.[15] With information derived from this analysis, IT managers build an acquisition plan and establish an installation schedule.

Life cycle management and project planning and implementation processes govern the investment of resources in the selected applications. These management-oriented processes consist of business case development, phase reviews, resource allocation and control, risk analysis, and risk reduction actions. The goal of these processes is to ensure that the objectives embodied in the strategic and tactical plans for the portfolio are achieved in a controlled manner. Skillful implementation of these management processes results in portfolio enhancements and enlargements that satisfy functional and business objectives.

Application planning and management processes account for new application technology that may be valuable to the firm's business. Current examples include such developments as hypertext or imaging systems and new data technology such as CD-ROM or important data transfer technology such as EDI. In addition, the management process must also incorporate new application development technology such as advanced languages (e.g., fourth-generation languages), new methodologies (e.g., object-oriented design), and new development tools (advanced CASE tools). Input regarding technological developments is a vital part of the application management system.

Successful application asset managers must consider alternative acquisition methods such as alliances, joint ventures, and purchased application packages and alternative strategies such as end-user computing, downsizing, and out-sourcing application development and operation. Businesses are critically in need of modern applications and rapid IT solutions to problems. IT managers must use a variety of techniques to improve productivity in bringing new solutions to the user community.

Management information is available during the implementation of these processes to assess whether the activities are proceeding satisfactorily and to provide data for course corrections that might be required. Information derived from phase reviews, for example, serves to keep application development projects on schedule or to make schedule corrections if required.

Application and data planning performed in conjunction with strategic and tactical planning for the firm lay the foundation for developing an information architecture. Information architecture, an enterprisewide configuration describing the sources and sinks of information mapped against the information requirements for the firm, is an important concept for IT and the firm. Strategic plan objectives coupled with assessments of present application and data performance measured against requirements allow an information architecture to develop from application and data planning. Developing an information architecture is a critical issue in most firms today.[16]

The management system governing the application portfolio assets attacks many other issues of importance to executives, including software development, distributed systems, CASE technology, end-user computing, and systems integration. The systematic application of portfolio management processes causes executives to focus on the issues and gives them tools and techniques to address these concerns in their firms.

Through effective use of the portfolio management systems, the firm achieves the acquisition of applications and functions it needs, on schedule, and within the cost targets justified in the business case. New applications and enhancements to present applications become available for productive use within the firm in a controlled and optimal fashion.

The Disciplines of Production Operations

The portfolio management system delivers new applications or enhanced current applications for productive use in the firm's business. These applications support most of the firm's internal functions and interact closely through telecommunications systems with customers and suppliers. They may provide competitive advantage to the company. Successful operation of the applications is a critical success factor for the firm's managers.

IT customers relate to the production operations department through the disciplines of service level agreements, batch operations, and on-line operations. These relationships are displayed in Figure 17.3.

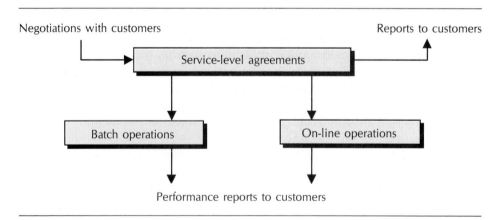

FIGURE 17.3   **Production Operations—Customer Performance System**

Customer expectations are developed in the process of establishing service-level agreements between the customers and the IT organization. Levels of service are negotiated that are achievable technically and financially and that are required to meet the needs of the business. Production service is delivered through batch processing, on-line processing, or a combination of both. The ingredients of production service required by customers and described in service-level agreements is factored into the batch or on-line processes. These processes are designed to deliver committed service levels to customers.

For example, routine inventory transactions are entered continuously during the day via on-line applications and records are accumulated for the nightly batch run. Nightly inventory processing is scheduled as a batch operation and the results are delivered to the inventory department before 7:00 A.M. daily. In each activity, the inventory department achieves agreement with IT on the parameters of service that it needs and can afford. For these activities, IT performance in meeting service levels is measured and reported to its customers on a regular basis.

These processes form the basis for achieving customer satisfaction in production operations. They establish customer requirements in a rigorous manner and provide the means for IT to address these requirements systematically through batch and on-line activity. These management systems meet many, but not all of the necessary conditions for succes in application operation.

During the operation of application programs, problems that impact service levels occur and changes must be made to the applications or to the environment in which the systems operate. Disciplined processes to manage these activities must be established. Achieving service levels also depends on the capacity of hardware and software system components and the performance of these components. The relationship of management systems to deal with these complexities is displayed in Figure 17.4.

Production operations occasionally experiences problems or defects that affect service or have the potential to do so. These incidents are handled as described in Chapter 12 through the problem management discipline. Significant activities of problem resolution are reported to management for monitoring and control purposes. The resolution of problems initially reported by customers is also confirmed.

System changes to complex data processing or telecommunications operations must be carefully controlled. Changes originate from two primary sources: planned changes and changes required to correct problems. The change management discipline obtains input from these sources, manages the changes through implementation, and reports the results to appropriate managers. Working together, the disciplines of problem management and change management provide a systematic way for IT managers to correct the inevitable operational faults that occur and to implement changes in hardware, software, or applications required to correct problems or implement planned system alterations.

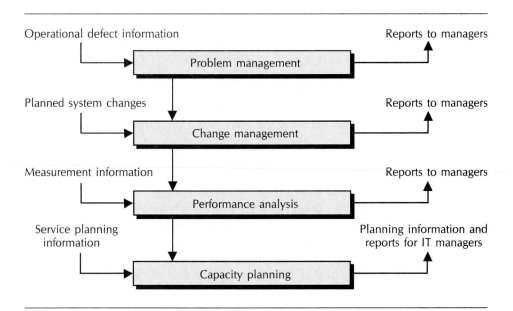

FIGURE 17.4   Production Operations—Internal Systems Management

The management system for problems and changes mostly involves IT people, but it directly affects service-level attainment and therefore is critically important to both IT and its customers. Although these processes are critical, they are not sufficient for success.

Performance management and capacity planning round out the management system for production operations. Since service levels are based on known or anticipated hardware, software, or application performance factors, ongoing measures of performance must be performed to validate system productivity. Analysis of system productivity is reported to IT managers who can take action if required to improve performance through system tuning or capacity additions. Therefore, capacity planning receives input from performance analysis and from service-level planning. Plan input, customer requirements, and performance analysis generate future capacity requirements that become inputs to tactical and operational planning.

Production operations management is a critical success factor for IT managers. The management systems governing application operations builds an IT infrastructure for dealing successfully with the issues and forms a framework for effective management actions. This operational framework combined with the tactical and strategic management systems discussed earlier forms an important base for the successful operation of the IT organization.

The Management of Networks

Advances in various technologies for processing, transmitting, and storing binary data have led to systems that handle voice, data, and image information interchangeably. Consolidation of the hardware and software for processing all types of information has been followed by consolidation of the organizations supporting these operations. In most cases, these merged organizations have found it advantageous to consolidate previously separate management systems. This means that the management systems for production operations also support network management.

For example, network faults (defects that impair or have the potential to impair network service levels) are processed by the problem management system and network changes are handled by the change management system. Network performance and network capacity planning are also consolidated with application systems performance and capacity planning. This works well for the IT organization but also makes a great deal of sense for IT customers. Customers are much less able to distinguish network components from application components in their systems than IT people and they are much less likely to care about these distinctions. Customers are interested in obtaining the service they need and they appreciate a management system that provides good service without complications or bureaucracy.

In addition to the management systems for networks discussed so far, IT needs network configuration management to provide data for other disciplines. Figure 17.5 displays the sources and uses of the data developed and stored by configuration management.

Configuration management maintains several databases that are useful to managers as they manage problems and changes and plan recovery actions. Managers also are able to obtain information on the status of the network and its components from the system. Automated network management systems use configuration data to optimize network operations and to update the databases. The management system for networks depends heavily on configuration databases.

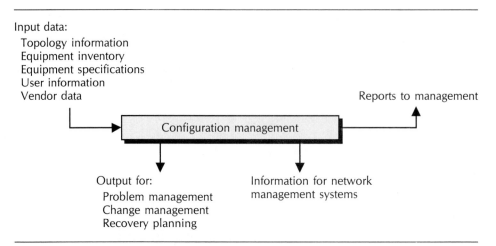

Input data:
  Topology information
  Equipment inventory
  Equipment specifications
  User information
  Vendor data

Reports to management

Configuration management

Output for:
  Problem management
  Change management
  Recovery planning

Information for network
management systems

FIGURE 17.5   **Configuration Management System**

Financial and Business Controls

Effective operation of the IT function requires that managers exercise control over all aspects of the IT business. This means that they understand the goals and objectives of the organization and that they are managing IT to meet organizational objectives in a planned manner. This is facilitated by strategizing and planning processes. Plans to meet established goals are implemented and disciplined processes for managing the acquisition and operation of applications are installed. The effective implementation and utilization of application assets is guided by operational management systems. In addition to these processes, IT managers must maintain control over the financial and business aspects of their operations. This means that systems of financial and business controls must be in operation.

The mechanism for monitoring IT finances and maintaining financial control begins with the tactical plan and its accompanying budget. Whether IT is organized as a cost center or a profit center, or operates as corporate overhead, the annual planning process gives IT a budget within which it is expected to operate. Usually each organization in the firm receives monthly financial statements that describe the actual financial position versus the budgeted or planned position. IT managers review actual expenditures as compared with planned expenditures and take action on the variances. Figure 17.6 displays this process.

If IT is organized as a cost center, the financial plan will contain planned cost center support and expenses, both of which are subject to variances. IT managers must analyze each and take corrective action if

Part Six: Preparing for Advances in Information Technology

the variances are large or if the trend is unsatisfactory. In some cases this will lead to negotiation with users and may result in changes to the planned rates for some classes of services.

If IT is organized as a profit center, both revenue and expense must be analyzed for discrepancies. Variances can be corrected by changing revenue, altering expenses, or permitting changes in profit margin. Revenue can be altered by increasing or decreasing business volumes or by changing prices. An IT profit center operates as a business within a business and managers have many opportunities to make adjustments that affect revenue, expenses, or margins.

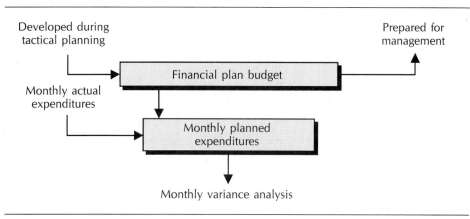

FIGURE 17.6   Budget Management

IT financial management systems that control to a budget developed in coordination with tactical and operational plans and that account for user services and recover costs from users attack several of the issues of concern to top management. IT organizations using these techniques operate in a businesslike manner and are spending resources on activities that the firm's managers agree are required. Mechanisms to control costs and expenses in a visible manner are important because, in most firms, IT expenses are significant and sound financial management is a critical activity for IT managers.

Business Management

Sound management systems are essential to IT managers because they provide a framework for accomplishing what is important. Properly functioning systems guide managers in the execution of their responsibilities and help them carry out their tasks in a logical and organized manner.

But systems operate through people and require good people relationships to be effective. Not only must IT managers execute their management systems, but they must do so in harmony with the other managers in the firm with whom they interact if they are to achieve maximum effectiveness.

There are several ways for IT managers to foster improved interaction with users to increase coupling between customers and suppliers and to provide input for improving service. Many firms form an IT steering committee consisting of executives from user organizations who work with key IT executives. The steering committee acts as a high-level sounding board to help establish IT strategies and develop actions for the IT organization. In many ways the steering committee acts as a sponsor for IT and helps build awareness of IT benefits to the firm. Through its actions and deliberations, the committee influences service levels by recommending and approving spending levels. In firms that have a cultural bias toward participative management, the steering committee can be particularly effective in communicating with top management and with functional managers. This communication is vital to IT.

Another effective process for linking IT with users is to assign individual IT managers liaison responsibility to user organizations. For example, an application programming manager may be assigned as the business service representative to manufacturing or an operations manager to product development. The manager represents IT to the individual function and represents the function to IT.

The service representative responsibilities include:

1. obtaining information on new or changing user needs
2. communicating this information to the right IT people
3. answering questions about IT procedures from users
4. facilitating problem resolution for users
5. fostering good relations between IT and users

In short, the IT service representative provides a high-speed communication link between IT and other functions in the firm to everyone's mutual advantage. In a large, diversified firm there may be many service representatives interacting with product development, manufacturing, marketing, sales, service, finance, administration, and possibly other departments.

Effective IT executives use every means at their command to maintain and improve communication between IT and the rest of the organization. They understand the high value of effective communication and work to maintain close contact with their peers throughout the firm. These formal and informal communication processes are an important ingredient of their management system.

THE IT MANAGEMENT SYSTEM

The management system described in this text is designed to cause IT managers to focus on activities that improve their effectiveness and increase the firm's competitiveness. IT managers must do the right things, things that the firm needs done for its success. IT management processes cause managers to sort through a list of possible activities and, in an organized manner, select the activities with highest payoff for the firm. Through effective operations IT improves the firm's competitive position.

The management system also causes IT activities to be conducted in an efficient manner. The system forces IT managers to focus on operational details and on control and financial details. This focus provides the means to improve efficiency and to increase productivity. Productivity increases in IT business processes will alter the work people do and change the organization in which they do it. When businesses are able to utilize information more effectively, decision making will improve and the performance of IT and of the firm will increase dramatically.

SUMMARY

Today's business environment is undergoing rapid and fundamental changes and is becoming much more complex. Information technology is one of the driving factors behind these changes because it provides the means for altering competitive positions among firms and gives firms opportunities to restructure in response to the changing environment. The restructuring of firms enables them to reduce costs, improve their value chain, or compete globally. It presents both problems and opportunities.

Information technology, particularly telecommunications, offers modern firms many opportunities to improve or enlarge their operations and to maintain control of widely separated organizations contributing to their value chain. The organization may take the form of alliances or joint ventures and may be national or international in scope. Centralized control of decentralized, geographically dispersed management is facilitated by IT; global operations are well supported through information and telecommunications systems.

Organizations are comprised of people working together toward common goals, directed by managers who provide leadership within the organization. People, the most important assets of the firm, must be managed with skill and sensitivity. This means that effective people managers respect the dignity of their employees and make assumptions about people that reflect their beliefs about them. Productive organizations operate in an environment of high expectations supported by managers and employees who encourage innovation and who recognize and reward accomplishment. Productive organizations are filled with people who are achieving self-satisfaction and self-actualization through accomplishing challenging tasks for the firm.

Well-managed organizations display high levels of trust and confidence between employees and managers and exhibit high morale and productivity. Trust and confidence in management is fostered by open and candid communication on topics of importance to employees and by consistent behavior on the part of managers. Managers who inspire trust and confidence in their employees are effective leaders and good people managers. Good people managers lead productive departments where morale is high and teamwork is effective. Good people management is a trait of effective executives and a critical success factor for IT executives.

IT executives are supported by management systems consisting of tools, techniques, and processes that guide them and their organizations in achieving corporate and organizational goals within the cultural norms for the firm. IT management systems begin by developing strategies and plans that support the firm's goals and objectives and that are congruent with the firm's strategic direction. IT planning processes are linked with the firm's processes, thus ensuring alignment between IT and the firm's goals. New technology begins to enter the firm during the planning process when it supports needs or lays foundations for new directions.

The application portfolio direction also originates from strategic and tactical plan information. Application development, enhancement, and maintenance are controlled through acquisition strategies and life cycle management techniques. Projects are controlled through phase reviews and associated techniques such as business case development, risk analysis, and resource allocation and control. Implementation of portfolio management techniques leads to applications operating in support of the firm's objectives.

Management systems for operations are based on service-level agreements and consist of techniques to handle problems, changes, and recovery actions. Performance analysis and capacity planning ensure that managers know what resources are needed to meet service objectives. Network management depends on the same disciplines but requires configuration processes and databases to interface configuration data to the management system. The management systems for operations establishes an environment in which service levels are achieved, costs are lowered, and planning takes place. The system enables operational success for IT managers.

Financial management and control is accomplished through planning, budgeting, and review activities. IT managers must analyze financial variances and take action to maintain control through pricing actions, rate adjustments, expenditure alterations, or other means. Although these activities form an effective basis for IT management, their effectiveness is increased by sound communication channels between IT and users. Steering committees and IT service representatives can play an important role in fostering improved communication and increasing understanding and cooperation. Effective IT and user managers promote these communication techniques.

Review Questions

1. What role does information technology play in the dramatic changes taking place in the business environment today?

2. In what ways are alliances or joint ventures important to topics in this text?

3. How does telecommunications technology relate to the issue of centralization versus decentralization in business operations?

4. How do Volvo, IBM, and GM use telecommunications in their businesses?

5. What is meant by span of communications? What is the importance of span of communications to middle managers today?

6. Why does this text state that people are the enabling resource in a business enterprise?

7. What IT trends are causing people skills to be more important now than formerly?

8. Describe Maslow's hierarchy of individual needs. Where on the hierarchy are most IT people today?

9. What are some of the assumptions that good people managers make about people, according to the text?

10. What are some of the reasons why managers and employees should establish high expectations for themselves?

11. How can managers and employees achieve the required balance between thoughtfulness and action when making decisions?

12. The text states that managers should expect employees to solve problems but that they should expect employees to prevent problems too. What possible difficulties arise when managers attempt to have their employees accomplish these goals?

13. Why are strategizing and planning activities so important for IT managers?

14. How are new technologies woven into the firm's strategic plan? In what other ways can new technology enter the firm?

15. Why might the management system for IT strategy and plan development fail in some firms? What are the consequences of failure?

16. Why is the management system for the application portfolio important? What critical issues does it address?

17. What possibilities exist for making additions to the applications portfolio?

18. What are some important steps in application life cycle management?

19. Describe the management system for service-level agreements and indicate its role in production operations.

20. What are the processes through which IT budgets are developed?

21. What are some activities that steering committees can perform? Why are these activities valuable for IT and for users?

Discussion Questions

1. Describe the features of the global network used by Corning.

2. Discuss the connection between Drucker's comments in Chapter 1 and the business vignette by Houghton in this chapter.

3. Describe how alliances or joint ventures can be used by firms to improve their value chain or expand their operations.

4. Discuss how the issue of centralization versus decentralization is related to the use of information technology.

5. Discuss how partnerships and alliances apply to IT organizations today. Describe the relevance of these concepts to the IT activities of downsizing and out-sourcing.

6. Describe how effective people managers use their knowledge of human needs to improve their managerial performance.

7. Why do you think that managers' assumptions about people will impact employees' performances?

8. List the characteristics that good people managers display in dealing with their employees. Which of these characteristics is most important, in your opinion? Which characteristic do you think is most difficult to attain in practice?

9. Describe the environment that good people managers attempt to attain. What are the advantages to managers and employees who work in this kind of environment?

10. What kinds of managerial behavior patterns improve employees' perceptions of managers' performance and inspire trust and confidence in them?

11. Describe the role that management systems play in managing the IT function. How do management systems relate to the corporate culture?

12. Describe the management system that results in the collection of IT plans. What issues noted in Chapter 1 are addressed by these processes?

13. Describe how IT might implement control processes to ensure that plan implementation is proceeding correctly.

14. Describe the system for portfolio asset management. What processes provide input data for the system? What is the desired outcome of effective portfolio asset management?

15. How can the application management system discussed in the text help manage the large data stores associated with the portfolio?

16. Describe the management system for production operations and tell how it relates to other systems.

17. Discuss the advantages and disadvantages of having separate operational processes for applications and for networks.

18. How is financial control maintained in an IT profit center?

19. Describe how steering committees and service representatives augment the IT management system.

20. How would the service representative concept differ between an organization that operated as a cost center for the firm and one that sold IT services to other firms for profit-making purposes.

Assignments

1. Read "The Changing Value of Communications Technology" by Hammer and Mangurian (*Sloan Management Review*, Winter 1987, p. 39) and prepare a two-page summary for the class.

2. Interview a human resources manager at your university on his or her views of people management. Compare and contrast your interview observations with the approach toward people management advocated in the text.

3. Visit a computer center near your school and review its written operational procedures. Evaluate them in light of the management systems described in the text. Did you find any processes or procedures unique to the center you visited? If so, why are these important?

ENDNOTES

[1] Copyright © 1989 by The New York Times Magazine Company. Reprinted by permission.

[2] Corning ranked 337 in the *Forbes 500* for 1989. It had sales of $2.4 billion on which it earned $259.4 million. Corning employs 26,900. "Ranking the Forbes 500s," *Forbes*, April 30, 1990, 286.

[3] Michael S. Scott-Morton, "Information Technology and Corporate Strategy," *Planning Review* (September-October, 1988): 28.

[4] "The Best and Worst Deals of the '80s," *Business Week*, January 15, 1990, 52.

[5] "The Shape of Things," *MIS Week*, July 10, 1989, 24.

[6] Charles Pelton, "MIS Masters a Merger," *Information Week*, August 28, 1989, 27.

[7] Michael E. Porter, *Competitive Advantage: Creating and Sustaining Superior Performance* (New York: Free Press, 1985): 12.

[8] Michael Hammer and Glenn E. Mangurian, "The Changing Value of Communications Technology," *Sloan Management Review* (Winter, 1987): 39.

[9] Eric E. Sumner, "Telecommunication Technology in the 1990s," *Telecommunications*, January 1989, 38. Data traffic on U.S. telecommunications networks is growing about 20 percent per year and voice traffic at about 5–6 percent per year. The result of these growth rate differentials is that data will account for about half of the business telecommunications traffic in the early 1990s.

[10] Peter Drucker, *The Frontiers of Management* (New York: Truman Talley Books, E. P. Dutton, 1986): 204.

[11] For examples see *Annual Report*, Montana Power Company, 1987, 3 and *Annual Report*, Mobil Corporation, 1987, 3.

[12] Abraham H. Maslow, *Motivation and Personality* (New York: Harper & Row, 1954): 80.

[13] Douglas McGregor, *The Professional Manager*, edited by Warren Bennis and Caroline McGregor (New York: McGraw-Hill Book Company, 1967): 16.

[14] M. L. Pesci, personal communication.

[15] Thomas H. Davenport, Michael Hammer, and Tauno J. Metsisto, "How Executives Can Shape Their Company's Information Systems," *Harvard Business Review* (March-April, 1989): 131.

[16] See information systems management issues in Chapter 1.

REFERENCES AND READINGS

Cash, James I., Jr., and Poppy L. McLeod. "Managing the Introduction of Information Systems Technology in Strategically Dependent Companies." *Journal of Management Information Systems* (Spring, 1985): 5.

Cougar, D. "Key Human Resource Issues in the 1990s: Views of IS Executives versus Human Resource Executives" *Information & Management*, Vol. 14 (1988): 161.

Drucker, Peter F. "The Coming of the New Organization." *Harvard Business Review* (January-February, 1988): 45.

Hartog, Curt, and Robert A. Rouse. "A Blueprint for the New IS Professional." *Datamation*, October 15, 1987, 64.

Sutherland, Vernell M. "Managing Change." *CIO Magazine*, April, 1988, 8.

Walton, Richard E. *Up and Running: Integrating Information Technology and the Organization*. Boston, MA: Harvard Business School Press, 1989.

Winkler, Connie. "McGraw-Hill Taps One Unified System for Many Markets." *Computerworld*, September 12, 1988, 47.

18

The Chief Information Officer's Role

A Business Vignette
Introduction
Corporate Challenges for the Senior IT Executive
The Chief Information Officer
    Organizational Position
    Career Paths
    Performance Measurement
Challenges Within the Organization
The Chief Information Officer's Role
    CIOs Manage Technology
    CIOs Introduce New Technology
    CIOs Facilitate Organizational Change
A Business Vignette
    CIOs Must Find New Ways of Doing Business
Successful CIOs Are General Managers
A Business Vignette
What CIOs Must Do for Success
Summary
Review Questions
Discussion Questions
Assignments
References and Readings

A Business Vignette

CIO: Misfit or Misnomer?[1]

According to the new IS mythology, a vital ingredient has been added to the churning mass of assets and people that is corporate America—a new breed of manager with unusual gifts; a bridge, it is hoped, between the alien cultures of the executive suite and the computer room. A torrent of books and magazine articles have painted an idealized portrait of these new "renaissance" men and women, the so-called chief information officers, or CIOs.

Unfortunately, as the mythology has grown, so have the misconceptions, as a startling new survey from Big Eight accounting firm Coopers & Lybrand and *Datamation* shows. "Confusion over just what constitutes a CIO is rampant," says John Highbarger, the Coopers & Lybrand partner responsible for Information Systems and for the survey of 400 top IS executives across the nation. "Fifty-nine percent of respondents thought of themselves as CIOs, yet only 27 percent reported directly to the top of the company." A direct pipeline to the top, he notes, is a prerequisite for the job. As planner of long-term strategies uniting the business and technology sides of the house, the CIO must be part of the inner circle of top officers to have any influence at all.

Even more confusing is the title. "You'd have to look long and hard through corporate America to find executives that are actually called chief information officer," says Highbarger, who found only two in the survey. He adds that merely 14 percent of respondents felt the title actually described what they do. Small- and medium-size companies in the U.S. don't yet have CIOs. "If the function exists at all," says Gwen Peterson, manager of Dataquest's CIO Advisory service, "it is part of the CEO's job. If not, outside consultants are used."

Academia is similarly bereft of CIOs, according to Prof. James Wetherbe, director of the University of Minnesota's MIS Research Center. "The operational and administrative sides of the campus computing environment have separate IS chiefs," he says. "There's no top IS executive responsible for the whole show."

Where CIOs do exist, it seems, is in large U.S. companies, especially in the information intensive services sector: banks, insurance firms, and airlines. Those who function as CIOs usually sit behind such recognizable nameplates as senior VP, VP of information services, or information resources manager.

In the mythology, these men and women are seen as a new order, the information elite. The old maxim that "information is power" has led many to speculate that CIOs will take the top corporate spot in the 1990s by virtue of their control of information. The Coopers & Lybrand/*Datamation* survey debunks this notion. Only 7 percent of respondents believe they will ever secure the top position in their company. They believe the chances for advancement elsewhere are even slimmer. Only 3 percent expect eventually to become CEO of another company.

"The idea that we're some elite that controls information is totally false and misleading," says John Hammitt, who was VP, Information Management for Pillsbury Co., who functions as the food giant's CIO.[2] "We don't control information, we help make it flow." Hammitt is one of many who believe that the CIO moniker is not only misleading, but harmful. "It's pompous and self-serving, and obscures the issues," he says. "It confers on us an authority we don't have. We're servants of the corporation," he says of the job. "We empower others to succeed."

In most corporations, the CIO function has a toothless quality to it. As architect of the information system, the CIO sits on many corporate planning groups, but he or she has no direct control over any of the line organizations. The University of Minnesota's Wetherbe says he knows of several cases in which the title created such animosity that its bearers gave it up happily. "It turned them into targets."

Wetherbe's view mirrors much of corporate reality for these executives, whatever their title. Rather than rallying around the CIO, most business managers seem to be doing their best to make the CIO's life a misery. The CIO movement is only a decade old, yet already the first wave of managers functioning in the role have drifted into a nomadic existence; victims of power politics and unrealistic expectations, the C & L/*Datamation* survey shows.

"Few remain in their companies for more than four years," says Coopers & Lybrand's Highbarger, expressing surprise at the survey's findings. "Fifty-six percent of respondents—top IS executives, 59 percent of whom claim they are currently functioning as CIO—have been at their companies three years, and can be expected to move on soon."

INTRODUCTION

The managers of information technology organizations in most modern firms have considerable responsibility and authority. Power and responsibility flow from the line organization headed by the senior IT executives.

The pervasive deployment of information technology throughout the firm also presents IT executives with considerable staff responsibility and empowers them with additional authority. As the firm increases its dependence on IT for success, it also increases its dependence on the senior IT manager. Consequently, senior IT managers enjoy a position of opportunity—opportunity for great success or for substantial failure. To acheive success, IT managers must make sure their accomplishments are viewed positively by their superiors.

The premise of this text is that superior IT managers are made, not born, and that superb management skills are learned and developed, not the result of genetics.[3] Therefore, many significant management issues and trends have been discussed throughout the text. Management systems and practices designed to cope with these issues and trends dominate the course of this book. CIOs and senior IT managers are in a good position to appreciate the management issues, to follow the emerging trends, and to implement well-designed management systems.

However, do senior IT executives enjoy success commensurate with their responsibilities? Can well-trained individuals succeed in the job consistently? Or is the evolving nature of the technology and its influence on the organization such dominating factors that success is mostly a matter of chance? Given the difficulties of the position and the changing character of the responsibilities, what are the critical success factors of the CIO's position? These and other topics will be explored in this chapter.

CORPORATE CHALLENGES FOR THE SENIOR IT EXECUTIVE

The role of the CIO or senior IT executive in modern firms encapsulates the opportunities and pitfalls inherent in the technology, embraces the ebb and flow of managers' and employees' opinions and motivations, and symbolizes the dominant effect of information technology on business and industry worldwide. The role is an extremely difficult one. The title and position are alternately praised and cursed.

The dilemma of the CIO position is chronicled in a rash of articles. For example, writing in *Business Week*, Gordon Bock states, "There's a new breed of manager surfacing in the executive suite. Some members of this new information elite sit behind such recognizable nameplates as senior vice president, vice president for information services, or information resources manager. Others are beginning to get a higher-sounding title to reflect their new status: chief information officer, or CIO."[4] The position of senior IT executive is vital to the firm. As information technology shapes the future of many industries, chief executives recognize the increasing value of IT to their firms and see the need for sound leadership in this area. Thomas Friel, managing partner for information technology at Heidrick & Struggles, Inc. predicts that "in five years, virtually every major company will have a CIO who's a peer to the CEO."[5]

On the other hand, as Professor Wetherbe points out in the business vignette, "the title created such animosity that its bearers gave it up happily." But the title is not necessarily what makes the job risky for the incumbent—the task itself is hazardous. According to conventional wisdom, the turnover rate for senior IT executives significantly exceeds that of top executives generally.[6] Some failures can be attributed to individual performance deficiencies but others result from organizational consolidations, cost reduction exercises, or bad times for the industry or the firm.

The literature suggests that business firms have yet to sort out the CIO's fate. It also suggests that part of the CIO's identity problem may directly result from their behavior patterns in the firm. Corporate culture and corporate politics are powerful forces with which executives, including IT executives, must contend successfully. In addition to the executive's actual performance, the perceived performance in relationship to expectations may be the critical aspect affecting perceptions of the CIO position and title.

"I love and hate the idea of the CIO with equal vigor. The CIO is at once the would-be hero and potentially the most dangerous person you have on your payroll," says Tom Peters.[7] His remarks seem to characterize the feelings of many.

THE CHIEF INFORMATION OFFICER

Organizational Position

Senior IT executives, whether they are called CIOs, VPs of information systems, or information resource managers, have both line and staff responsibility for information technology within the firm. Because telecommunications and information processing technologies are unifying, telecommunications organizations and information processing organizations are also merging. The senior IT executive manages the merged organization. And because information technology is widely used throughout the firm, the senior IT executive has extensive staff responsibility. Just as the firm's chief financial officer is held responsible for the firm's expenditures, the firm's chief information officer is responsible for IT in the firm even though most IT resources are consumed outside the IT organization.

CIOs at most firms have responsibility for the traditional data processing operation and telecommunications services, and in some cases other responsibilities as well. For example, a survey of 137 health care CIOs revealed that 99 percent were responsible for information systems, 65 percent for telecommunications, 31 percent for management systems, 15 percent for medical records, and 10 percent for admitting services.[8] They believed their most important role was to integrate information systems, telecommunications, and management systems. Knowledge of hospital systems ranked fourth among top attributes needed by CIOs, according to these respondents, who placed more importance on leadership ability, vision and imagination, and business acumen.

The range of reporting positions for CIOs is large; fewer than 40 percent report to the CEO or to the firm's chief operating officer. Many, however, exert significant influence through committee appointments. In a survey of 300 chief information officers by Heidrick & Struggles, 40 percent were on a senior management committee, 7.7 percent reported to the CEO and/or president, and only 2 percent were members of the corporate board of directors.[9] Many report to executive or senior VPs, chief financial officers, VPs or division heads. Regardless of their reporting relationships, CIOs must be in close communication with the top echelon of executives. They must be an integral part of the executive management team if they are to be effective.

But being in close contact with the executive management team is not enough; the executive management team must recognize the importance and value of information technology. Top managers may fail to appreciate IT because they lack awareness of its strategic importance or because they see only the operational importance of computers. Senior executives may not view information as a resource or they may experience a credibility gap with the direction of IT. CIOs can overcome some of these difficulties by educating top management and by marketing IT accomplishments to management. It is especially effective to have users do the selling and promoting of IT's business image.[10]

Career Paths

The most popular route to the CIO position is through the information systems ranks, although some CIOs have combined their IS experience with that from other disciplines. Consulting and telecommunications backgrounds combined with IS experience is rather common. When telecommunications organizations are merged with IS, the former IS manager is most likely to head the consolidated organization. This occurs because the IS manager has experience developing or managing critical systems for the firm and is considered to have broad management skills. The telecommunications manager, on the other hand, is usually seen merely as someone who manages the phone system. Many other managers, however, move to the top IT position after having served in line or staff positions elsewhere in the company. Engineering, product development, or general management backgrounds are also common.

According to a Coopers & Lybrand survey, more than 75 percent of CEOs, COOs, and CFOs prefer that the chief information officer have both an IT and a general/line management background. Only 2 percent indicated that the CIO's background didn't matter.[11]

Firms are more likely to bring in CIOs from outside the IT organization when the IT function needs revamping. An IT organization that resists downsizing or decentralization or one that fails to respond rapidly to changing business conditions probably will be reorganized under a former line manager. In many cases it is easier to change the IT culture by replacing the CIO than by taking less drastic actions. Newly appointed CIOs from outside the IT fraternity usually bring with them a strong mandate for change.

Abrupt changes in the top IT position are common. A Touche Ross & Co. survey conducted in 1989 revealed that nearly one-third of the 568 CIOs contacted replaced individuals who had been dismissed or demoted. Part of the reason for the turmoil at the top can be attributed to the relative newness of the IT profession and to frequent redefinitions of roles and corporate relationships.[12] The difficulty of recruiting, training, and retaining competent IT professionals who can implement complex IT systems also adds to the CIO's difficulties.

Successful senior IT executives enjoy the option of considerable mobility. The careers of some of the top CIOs seem to indicate that exceptional individuals are very much in demand and that technology skills and general business management skills are largely transferable across industries. However, the job of the person to whom CIOs report may be beyond their reach, unless they work to become qualified to assume the higher position.

Performance Measurement

Regardless of the title held by the senior IT executive, the firm's top executives have certain expectations of the position and of the IT organization. Satisfying these expectations is the CIO's highest priority. What is expected of CIOs? What must they do to be successful?

Performance is measured by CIOs' ability to bring value to the firm through the application of information technology to the firm's goals and objectives. They are expected to identify technological opportunities and provide leadership in attaining advantage for the firm through the use of technology. They must educate the management team on the opportunities and incorporate these opportunities into the firm's strategies and plans. At the same time, CIOs must develop planned courses of action that balance opportunities, expectations, and risks. CIOs are measured on their abilities to accomplish these tasks.

CHALLENGES WITHIN THE ORGANIZATION

Businesses evolve as they change their business practices or adopt new practices to improve their competitiveness or to respond to changes in the external environment. Mergers, acquisitions, alliances, joint ventures, and new business formations are external indications of the evolution. Firms also evolve internally. Restructuring, downsizing, and internal reorganizations are taking place as firms seek to tighten up their internal operations and to improve their results. Support functions within the firm must be flexible enough to respond to these evolving corporate structures and to changing business requirements.

Organizational changes and new ways of doing business bear heavily on the senior IT manager position. New computing technology and advances in telecommunications systems and products are the tools IT executives must use to facilitate organizational transitions and to deal with competitive threats. Corporate executives are deeply concerned about competition. They must consider computer and telecommunications systems as vital elements in their competitive struggles. "To succeed fully in an era of rapidly changing technology, a corporation needs to understand the technology and to develop a strategic vision of how it can use technology to achieve something better than its competition."[13] The CIO is primarily responsible for making this happen.

As advances in information technology continue, new opportunities are created for alert CIOs and new pitfalls emerge for complacent ones. "[M]uch of the innovation in product development, marketing, sales, and service today is being fueled by advances in technology. The ability of IS to support technological adaptation is imperative if a company is to remain competitive in the world of tomorrow."[14] As firms explore these opportunities and introduce new technology, IT executives become change agents.

The need to change to improve the business is compelling for most firms in most industries. The CIO and the IT organization are also compelled by fundamental business forces to find new and better ways of doing business. Competitive pressures demand improved performance. CIOs have a mandate in most corporations to reexamine their methods of operation in search of improved productivity for the firm. The improvements may result from lowered costs, increased quality, or enhanced responsiveness. Top executives have high expectations that the IT organization can add substantial value to the firm's output.

Some employees, however, may not view favorably the changes required to keep the firm healthy and viable in response to competitive threats. IT managers responsible for these changes are perceived accordingly. In addition, the adoption of advanced technology may significantly alter the IT organization. Some IT professionals may resist these changes, and some may resent them.

THE CHIEF INFORMATION OFFICER'S ROLE

CIOs Manage Technology

CIOs are responsible for providing technological leadership to the firm in the area of information handling. The firm's executives depend on them to develop forecasts of technical direction and to provide assessments regarding the technology's significance to the firm. The CIO must assemble technology forecasts and develop assessments for use by the firm's senior officers in the strategic planning process. Because information technology advances so rapidly, the technology management task is difficult. Yet, because information technology is so important, technology management is a critical task.

The task is difficult because the pace of technological change is extremely rapid and the magnitudes of potential changes are great. For example, predictions are that over the next two decades computers will increase in processing power by two orders of magnitude with no increase in cost. Likewise, enormous bandwidth increases are occurring as utility companies, railroads, pipelines, and traditional communication companies install fiber-optic cables at breakneck speed. These trends, along with dozens of others ranging from new application development methodologies to advances in artificial intelligence to sophisticated optical storage devices, make technology forecasting both difficult and risky.

Technology forecasting is risky and difficult but it is critical for the firm. Decision making in the firm involves a vision of its future. In particular, at the senior or executive level, strategic decision making attempts to capitalize on future opportunities and avoid future pitfalls; strategic decision making also attempts to shape or define the future. Information technology has been critical in the strategies of firms in the airline, banking, and distribution industries where it has transformed both firms and the industries themselves. The increased pace of future technological innovation and the explosive deployment of information technology into all industries increase the critical nature of technology forecasting.

Technology forecasting is much more than extrapolating past trends into the future. For example, many forecasts indicate exponential increases in circuit density per chip or in bits stored per cubic centimeter. These forecasts lead to others that predict substantial computing power available at reasonable prices in the form of individual workstations. Making these predictions requires considerable understanding of the technology itself and of scientists' and technicians' ability to develop it. The ability to put the technological achievements into practice, and the capacity to manufacture the resulting products efficiently are also major factors. The forecasts are based on reasonable estimates of the difficulties involved in attaining these capabilities and on past progress in doing so.

The user of the technology faces another kind of problem when trying to estimate the technology's potential for the firm. This is because new technology leads to new uses for it, not just extrapolations of present uses. For example, experts predict that personal computers will someday have voice and handwriting recognition capability, extensive optical storage for text and voice, and artificial intelligence interfaces so the computer can be taught how the operator wants to use it.[15] In general, forecasting how capabilities like these can be valuable to the firm is a more difficult problem. There are many attempts to determine emerging technologies and to forecast new uses for existing and developing technological capability.

In one attempt to answer some of these questions, studies conducted by researchers at the University of Minnesota found that information management experts independently grouped technologies into five primary areas and rank-ordered them according to their perceived impact on organizations. The technologies in order of importance were found to be:

1. Human-machine interface technologies

2. Data and person-to-person communication technologies

3. System support or building block technologies

4. Limited impact technologies

5. Immature technologies for the 1990s

Topping the list of important technologies are those innovations designed to improve significantly the manner in which people relate to computing devices.[16] Very powerful individual workstations that employ natural language or voice input and output will permit workers to access the tremendous computing capacity and data stores predicted for the future. Straub and Wetherbe find the key human interface technologies to be speech recognition, voice input-output, and natural language interfaces supported on high-end workstations. The key communications technologies are voice mail, E-mail, fax, VSAT, and EDI. Powerful individual workstations will enable these technologies to be effective. ISDN, LANs, and desktop publishing were seen as important ingredients of the communications technology.

According to Straub and Wetherbe, technologies with indirect impacts during the 1990s include CASE, CD-ROM and optical storage devices, relational databases, and 4GLs/query languages. Indirect impact technologies are considered important because they support other technologies and thereby produce major second-order effects on the organization. Limited impact technologies were more controversial among the information management experts. Included in this group are stand-alone expert systems and artificial intelligence, mainframes and minicomputers, generalized decision support systems, computerized libraries, data extraction and conversion software, on-line database search techniques, PBXs, and calendaring software. ES and AI were most controversial within this group. Some experts thought these technologies will be critical while others have lost confidence because of past unfulfilled expectations for ES and AI.

Communication technologies, including data and personal interactions, will be of substantial importance, according to the study. Information from many sources and in large volumes will be easily available to the firm's knowledge workers. Communication technology and appropriate software (groupware) removes the barriers of time and distance and allows cooperative work groups to form, perform, and disband readily. Improved human-computer interfaces and expanded communication capability are the technological foundations for the information-based organization postulated by Drucker.

The job of the CIO and the executive management group is to evaluate future technology directions; select appropriate technology and products and develop strategies to capitalize on them; and establish plans to adopt new technology and introduce it to the business.

CIOs Introduce New Technology

Technology adoption and introduction begins after the firm's strategic plan has been accepted. The adoption and introduction process is formulated in tactical and operational plans developed by the functions for which the technology or products are intended. Implementation occurs as the plans are carried out by the individual units.

The adoption and acceptance of new ideas by an organization consists of many individual decisions to use the innovation or to adopt the new product. The process by which individuals become adopters is communication-based and consists of five steps. Individuals become aware of the innovation; they become interested and seek information about it; they evaluate the innovation in light of their needs; they give it a try; and, if conditions are favorable, they adopt the innovation. This process is called innovation diffusion.[17]

Individuals differ significantly in their propensity to accept new ideas or innovations. The people who are eager to accept a new idea become champions of the innovation and are called pioneers. Those who accept the innovation readily are the early adopters. The adoption process continues until the majority accepts the innovation. The last group to adopt is called laggards. The adoption process, represented as a distribution plotted over time, as shown in Figure 18.1.

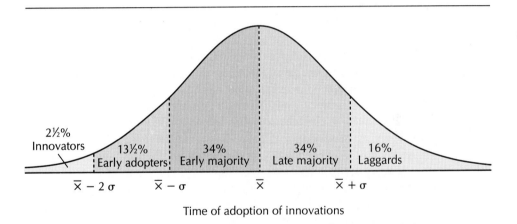

Redrawn with permission of the Free Press, a Division of MacMillian, Inc. from *Diffusion of Innovations*, Third Edition, by Everett M. Rogers. Copyright © 1962, 1971, 1983 by the Free Press.

FIGURE 18.1   Innovation Adoption versus Time

CIOs must understand these individual adoption differences because they are fundamental to the success of technology introduction. Perceptive managers search for pioneers and early adopters who are opinion leaders when introducing new ideas. Opinion leaders are effective in establishing awareness and developing interest. They pave the way for the majority to adopt the innovation.

Individuals are not necessarily pioneers for all new ideas. Some managers, for example, readily adopt and implement new organization structures but resist changes in personnel policies. Some managers accept changes in product strategy quite readily but resist personal computers in the executive suite. Skillful CIOs understand that these considerations apply to top executives as well as to nonmanagerial employees. Wise CIOs acknowledge these individual differences and utilize their knowledge to advantage when implementing innovations and new technology.

CIOs themselves have varying propensities to accept changes that enhance or inhibit their effectiveness. In today's fast-paced world, CIOs who cannot remain flexible and adaptable probably cannot lead effectively through the turbulence of alliances, restructuring, and new business processes. Although CIOs must be cautious and must recognize risk, they must not be viewed as laggards if they are to survive.

CIOs Facilitate Organizational Change

The adoption and use of new information technologies has brought steady change to the firm and to the IT organization. Microcomputers, networked systems, and strategic applications have altered work patterns, organizational structures, and competitive positions. Indeed, IT managers have justly earned the reputation for being change agents.

Today, most firms have installed large numbers of microcomputers and have implemented some form of end-user computing. Most have trained employees to use workstations for database query, statistical analysis, and many other applications. User productivity in these firms has improved as new applications ranging from word processing to executive support systems have been installed. End-user computing tools have allowed the firm's employees and managers to become more effective and more efficient.

Firms are now beginning to take the next step in the direction of distributed data processing. They are replacing their workstations with more powerful systems and connecting these systems through LANs to larger general-purpose computers located in the user area. This client-server architecture offers considerable potential to move mainframe work even closer to using departments. The process of downsizing mainframe-based systems is gaining momentum in business and industry.

But downsizing presents many problems for IT professionals and managers; it represents an organizational challenge for IT executives. As the firm migrates applications from the mainframe to decentralized processors, people, procedures, and data move with them. Skilled IT people migrate toward user organizations in support of the applications, and some IT professionals who can't make the transition to the new environment seek jobs elsewhere. Downsizing alters the function of the IT organization and changes its structure.

The new IT organization is more responsive to business needs and demands. A new IT structure is required to improve the firm's competitiveness. Structure is related to management style. Centralized companies with more autocratic management styles tend toward conservative competitive strategies; more aggressive companies tend to be less centralized. "IT structure is strongly related to competitive strategy, and specifically the degree of centralization of IT activities is significantly related to competitive strategies."[18] Highly competitive companies require flexible structures, especially in critical areas such as IT. CIOs and IT organizations must remain flexible and responsive to changing business conditions.

IT leadership and expertise are critical during the transition from a centralized mainframe operation to a distributed environment. After decentralization, IT professionals are needed to preserve the good features of the centralized environment such as data security and business controls. The assumption that downsizing will yield people savings in the IS shop may be invalid or, if people reductions take place, the reductions may reduce the users' capability.[19]

CIOs must recognize that some IT professionals may not welcome the plan to downsize because they believe their mainframe experience will not be very useful when the plan is implemented. They may elect to leave the organization. Others enjoy the opportunity to be part of a changing organization. They know their application skills will still be needed when the applications reside on LAN-based workstations. They also know that much of their knowledge of database operations, application controls, and business procedures will apply directly to the downsized environment. CIOs must remain alert to the people issues that arise during the process of downsizing.

Moving applications from the mainframe to LAN-based client-server architectures requires outstanding planning and implementation skills. Bankers Trust provides an example of good planning and implementation.[20] Their technologists created a flexible book of rules, standards, and policies to define an information architecture known as the Bankwide Technical Architecture. The architecture is part of their strategy to maintain the essential characteristics of the business during the bank's decentralization and growth in the increased globalization of the banking world.

The essential architectural concepts include standardized data processing hardware clusters, local and global communications networks, definition of the security environment, and defined integration interfaces.

The list of standards details over 100 specific items of software alone. According to Stanley Esikoff, vice president, "the Bankwide Technical Architecture was a response to the fact that to be entrepreneurial and decentralized, you would lose control if there wasn't something cohesive holding the community together." If the firm elects to downsize some of its mainframe operations, the CIO must ensure that the firm's information architecture supports the new strategic direction.

Sometimes the firm's strategic direction is to centralize. When the company adopts a centralized corporate structure, the IT organization must follow the corporate strategy. This was the case at Trailer Train Co. of Chicago. Trailer Train's corporate philosophy is to consolidate divisions that once were wholly owned subsidiaries and to centralize the corporate structure. The MIS organization replaced small computers in the subsidiaries with large central mainframe support and reduced costs in the process.[21]

Decentralization can sometimes be carried too far and the correct strategy is to centralize. This was the case for the city of Cambridge, Massachusetts, which had 50 separate agencies each responsible for its own data processing. A centralized system was installed, which resulted in a 11 percent reduction in the data processing budget over a three-year period. The payback took less than two years and annual savings of more than $800,000 are anticipated.[22]

There are other reasons to centralize. Some firms have had bad experiences with security or control problems, or have found that functional areas preferred not to manage computer operations. Some firms find that migrating back to the mainframe simplifies their operations and puts data processing back into the hands of data processing professionals. In some of these instances, the decision does not hinge directly on out-of-pocket costs, as the next business vignette shows.

The balance between centralized processing and downsized applications must be arrived at by the firm's top executives after considering many factors. The technology is generally available, but the decision to downsize is mostly driven by business factors, such as improved response to customers, higher quality customer service, and lower costs for the firm and its customers. In order to strike the proper balance, the CIO must have a careful blending of business, organizational, and technical skills.

"Most companies run their data processing operations the wrong way," says John Singleton, president of Security Pacific Automation Co. "DP is not about being centralized or decentralized. It's more that all managers have to be business executives first and technologists second."[24] A glance at Singleton's approach to IT business management is presented in the business vignette that opens Chapter 11.

A Business Vignette

Experience with Downsizing at John Hancock Insurance Co.[23]

John Hancock Insurance Company was faced with many of the same problems found in firms everywhere. The mainframe was overloaded resulting in poor response times for their important applications. In particular, one application utilized by nurses in their Managed Health Care department required rapid response. In addition, the development backlog prevented the programmers from getting to the application in a timely fashion.

The central MIS organization and the user department decided that the application was well suited for microcomputers tied to servers which are connected to larger systems. The agreement between the users and IS to proceed with a microcomputer solution was reached because both groups believed the downsized application was the only possible solution on the required schedule.

John Hancock hired a consulting firm to lead the project and provide the required skills. In order to gain experience, the IS organization worked closely with the consultants. The experience thus gained will be useful later when the system needs maintenance or enhancement.

The final system connecting microprocessors around the nation to the server and mainframe was extremely complex. As the system advanced in complexity the schedule lengthened too. John Hancock considers the application a successful venture into downsizing.

CIOs Must Find New Ways of Doing Business

Nothing is more fatal to the CIO's position and career than attempting to maintain the status quo. The old adage "if it ain't broke, don't fix it" definitely does not apply to the IT organization or to the CIO. CIOs must constantly seek ways to improve the organization's productivity and to create or sustain business advantage. They must control costs and find ways to improve cost effectiveness.

There are several ways in which effective CIOs have reduced costs and improved performance: using high performance and cost effective hardware; adopting alternative application acquisition methods such as joint development or purchased applications; and using highly effective disciplines in managing production operations and networks. Some firms find it cost effective to turn over some of their IT operations to third-party operators, a practice called outsourcing.

Outsourcing, is currently receiving considerable attention. The practice is not new. Computer service bureaus and subcontract application development have always been available. Recently, outsourcing has taken the spotlight as major firms such as Kodak and Merrill Lynch made major outsourcing decisions. Katherine Hudson, the very capable CIO for Kodak, negotiated a deal with IBM to manage Kodak's data center operation. EDS, partly owned by General Motors, performs the majority of GM's information processing. And DuWayne Peterson, general manager of information systems for Merrill Lynch, gave management of Merrill's network to MCI and IBM.

Many firms are considering the outsourcing alternative. Howard Anderson, managing director of the Yankee Group, believes that 2 percent of the Fortune 500 firms are engaged in outsourcing now and that by 1992, 20 percent will outsource at least part of their data processing or communications services. The economics of outsourcing remain controversial. American Standard claims 24 to 40 percent reduction in costs by outsourcing four of its MVS-based computer centers and a nationwide SNA network, yet Avon declined to outsource when seven outsourcing vendors failed to reduce costs by more than 5 percent. Avon had set the decision point at a 15 percent cost reduction. However, outsourcing will be a $50 billion business by 1992, Anderson believes.[25]

Outsourcing firms offer the advantage of economies of scale. By performing data processing for many firms, they are able to procure large processors and to keep these systems busy and fully loaded. Their economies extend to people and supporting functions too. For instance, because outsourcing firms have large production operations, they can afford to hire talented system-support programmers and network specialists. These highly skilled people are fully utilized in a large operation but may not be in a smaller one. Because of these economies, outsourcing should lead to cost reductions as compared to in-house operations.

Whether outsourcing develops as predicted depends on more than the economics of the deal. Other, major considerations play a significant part in the decision to outsource.

This decision turns on several noneconomic advantages and disadvantages. Outsourcing all or part of the production operation permits the IT organization to concentrate on important issues such as strategic application development instead of computer center or network management processes. IT can also concentrate on user support and business objectives. The technical considerations of hardware, systems software, and network management are the responsibility of the outsourcing firm. Advocates of outsourcing believe that computer centers are not the place to spend scarce technical resources. Even as the cost of hardware declines, the cost of acquiring, training, and retraining hardware, operating system, and network technicians is rising rapidly. More importantly, many of these technical skills are extremely difficult to acquire and retain.

Outsourcing has its critics as well. Though computer hardware and networks are considered mundane by some, others believe them to be vital to the firm's survival and prefer to retain firm control. Giving up data center and network operations to others, however capable they may be, may lead to long-term disadvantages. The issues center around quality, strategic direction, and control of production operations. Can the vendor provide long-term, high-quality service? How will the firm and the vendor work together on problems, changes, recovery, and other operational issues? The decision to outsource is not easily reversed. The firm may find itself locked into a computer architecture and management system ill-suited for its long-term goals and objectives.

Outsourcing is an alternative that must be considered by the firm's CIO for all or part of the production operation. The decision is primarily not a financial one in most instances, but financial considerations may be important in some cases. The decision to outsource is strategic in nature and probably depends on intangible factors such as the corporate culture as much as it does on strictly financial considerations.

SUCCESSFUL CIOs ARE GENERAL MANAGERS

Successful CIOs must demonstrate skills commensurate with the general management positions they hold. The general management role of the senior IT executive is demonstrated by Katherine Hudson's perspective on her job at Kodak, described in the next business vignette.

WHAT CIOs MUST DO FOR SUCCESS

CIOs are critically dependent on many people throughout the firm. In particular, the people at the top are extremely important to CIOs. They must gain the confidence of top executives by understanding the business from the executives' vantage point. The CIO must be a fully contributing member of the executive suite and must provide leverage to the firm's senior executives through leadership and vision. This leadership includes executive education and salesmanship. CIOs must not assume that their plan for the firm will be immediately adopted or even that it will be completely understood.

Kodak seems to be continually expanding its marketplace. With its purchase in 1988 of Sterling Drug Inc., it has become one of the world's largest providers of health-care products. A little-known fact, according to Katherine Hudson, vice president and director of corporate IS, is that Kodak has been in the health-care business since 1895, when it first began selling diagnostic X-ray film to hospitals and dental offices.

Hudson has been helping Kodak restructure and rethink many of its IS initiatives. Decentralization of many IS functions to business groups has occurred simultaneously with centralization of electronic messaging activities. The company actively supports standardization for all areas of computing and telecommunications, and is a founder of the Information and Technical Resources Corp., whose mandate is to move computer and telecommunications vendors toward an open architecture. And significantly, instead of "feeling like a first cousin once removed," corporate IS now reports directly to the office of the president.

Hudson sees her division's mission as aligning IS closer to the various users of its resources to make all workers more productive. For example, a LAN system being used in X-ray processor, camera, and electronic equipment plants is performing so well that inventory has been reduced by 90 percent, and on-time deliveries have increased to 98 percent.

Foreseeing IS requirements into the year 2000 is another aspect of Hudson's role at Kodak. To help her prognosticate, she established both an IS Executive Council, composed of 15 top IS managers, and several Centers of Excellence. These centers, located within the business units that best understand a given technology, are designed to develop that technology and transfer it out to other parts of the company. So far, centers have been established for artificial intelligence, data modeling, networking, and workstations.

Hudson, whose previous position at Kodak was managing and then dismantling the company's instant-camera business, believes that trying out such products as Kodak's new waterproof and panoramic single-use cameras can only help her get closer to customers. "We are focusing on customers and markets, not just on technology," she said.

The CIO must develop a vision for the firm's use of information technology that realistically adds substantial value to the firm. Reducing costs, saving the head count, and avoiding expenses are excellent activities, but the firm's executives expect much more. Executives expect IT to be a substantive contributor to the firm's value chain. They expect CIOs to contribute to bottom-line results and to have a shared vision for improving the firm's results; they expect the vision to attract advocates. The advocates and stakeholders must reside at all levels in the organization, not just at the top.

CIOs are people of action. They want to make things happen; they are impatient and intolerant of mediocrity, and they set high expectations for themselves and their organizations. At the same time, they have the patience to work with the situation at hand. They recognize that progress usually occurs in many small increments. They realize that effective executives continually strive to improve their operations across a broad front. They know that business success rarely comes in one grand stroke.

The corporate culture is a powerful force in nearly all organizations. The CIO must understand the culture and must learn to work within it. Realistically, the CIO cannot change the firm's accounting procedures or corporate personnel policies. If the firm is conservative in its behavior, radical infusions of high technology probably will not be acceptable. The firm expects the CIO to operate the IT function within the norms of the larger organization. Thus, the CIO should pay careful attention to the corporate culture.

CIOs must manage expectations within the firm regarding information technology. Unfulfilled expectations are one of the leading causes of failure for managers at all levels. Unfortunately, IT managers suffer more than most from this difficulty. The problem arises naturally because many communication channels to the firm's executives, managers, and employees are biased toward inflating the benefits of information technology while understating the costs or implementation difficulties. Additionally, lack of discipline within the IT organization itself frequently creates unrealistic expectations.

CIOs depend on many people for their success. Effective CIOs cultivate good human relations and encourage innovation. They withhold criticism of well-intentioned but flawed inventions, and they are free with praise for accomplishments. Exceptional CIOs are unselfish. They believe that there is almost no limit to the amount of good they can do if they don't mind who gets the credit.

There four kinds of people in most firms: those who make things happen, those who prevent things from happening, those who watch things happen, and those who don't know or don't care what's happening. Astute

CIOs can distinguish between them and have an approach for coping with all types. CIOs must make things happen and must develop supporters from among those of similar persuasion. CIOs should ignore those who always vote for the status quo and should concentrate on motivating the onlookers to get involved. For those who don't know or don't care what's happening, CIOs should prescribe education. With some effort, most people will buy into an idea that is good for the firm and, therefore, probably good for them.

CIOs must understand where the firm is positioned in its use of information technology and what ability it has to assimilate new technology. CIOs must balance the availability of new technology with the firm's need for it and with the firm's propensity to adopt it. Effective CIOs believe that the best place to work is where they are with what they have. They build on the capabilities currently in place. They understand the systems that run the firm, and they know the capabilities and limitations of the people who run the systems. Effective CIOs always try to improve the environment, but they use the present environment as the base for improvement within the culture.

SUMMARY

The position of chief information officer is precarious in both theory and practice. The CIO position, derived from the accepted CFO title, stands on shaky ground because information, unlike money, cannot be quantified or measured. Information can be created, used, and discarded without the CIO's knowledge or approval. Therefore, to identify an officer of the firm as the chief information officer is somewhat misleading.

In practice, the job of senior IT executive, regardless of the title, is large and important. But it is hazardous because of the nature of the work. The CIO is expected to be the technological leader of the firm and to provide the business direction for the selection and introduction of new technology. Information technology is critical to the firm, but its adoption and implementation frequently cause fundamental changes within the firm. Changes are difficult to manage and their success may be doubtful. When things go astray, the CIO is vulnerable.

CIOs usually report to executives at or near the top of the business and are part of the senior management team. CIOs usually emerge from within the IT organization but some were line or staff managers in other parts of the business. They are measured on their ability to bring value to the business through line and staff applications of information technology.

Chief information officers face many challenges as their firms alter strategic direction or adopt new forms for doing business. CIOs must assess technology changes and provide guidance to the firm on technology adoption and introduction. They must facilitate organizational changes within the firm and must find more effective ways of doing business for the firm and for the IT organization. They must play a strong general management role.

CIOs depend on many people for their success; they must be astute managers of people. CIOs are action oriented and must surround themselves with individuals with similar inclinations. They must study the persuasions of the firm's executives, but they must also comply with the corporate culture if they are to be effective.

Successful CIOs have learned the practice of management. They have learned to develop and manage IT management systems, and they have learned to manage themselves. People and technology are incredibly complex. After extensive study, considerable introspection, and prolonged experience wise managers maintain a respectable level of humility.

Review Questions

1. Why is the title of CIO surrounded by mythology and confusion? Does your university have a CIO?

2. Regardless of the title, why is the senior IT executive in a position of power and responsibility? Distinguish between line and staff responsibility for the CIO. Give examples of each type of responsibility.

3. Why is the CIO job so difficult? List six difficulties of the CIO position.

4. To whom do most CIOs report? What other avenues besides the direct reporting relationship do CIOs have to the executive management team?

5. Why do some CEOs fail to recognize the importance of information technology? What can the CIO do about this?

6. What are the career paths for most CIOs? Why is the CIO in a dead-end job in most firms? What, if anything, can the CIO do about this situation?

7. What kind of backgrounds do most CIOs have? What kind of backgrounds and experiences do most executives prefer in CIOs?

8. Under what circumstances is the firm most likely to import a CIO from outside the IT ranks? What do firms hope to accomplish by doing this?

9. How are CIOs measured? What can the CIO do to improve these measurements?

10. What are some of the changes going on, both externally to the firm and within it, that pose challenges for the CIO?

11. What is the principal driving force behind changes taking place in firms today? What signals do these changes give to the CIO?

12. What responsibilities do CIOs have regarding technology? Why is this a difficult task?

13. Why is technology forecasting risky, difficult, and important?

14. Why is forecasting the use of technology more difficult than forecasting the availability of technology?

15. What technologies are considered to be important for the near future?

16. Describe the process through which the results of technology forecasting are incorporated into the firm's business.

17. Describe the process of innovation diffusion. How would you use these ideas if you wanted to install new end-user applications?

18. What is meant by downsizing the IT organization? What are the advantages of downsizing to the firm? What are the pros and cons of downsizing to IT professionals?

19. What is meant by outsourcing? What is the motivation for most firms to consider outsourcing?

20. What are the disadvantages of outsourcing? Which disadvantage is most significant in your opinion?

21. What must CIOs do to be successful? What role does the corporate culture play in the CIO's plan for success?

22. What philosophical difficulties surround the title of chief information officer?

Discussion Questions

1. Discuss the conceptual difficulties with the CIO title and describe how the practical realities of the job are affected by the title.

2. Why do you think Tom Peters loves and hates the CIO idea with equal vigor?

3. Discuss the line and staff responsibilities that the CIO position generally has. How have these responsibilities changed over time? In large corporations the CIO may not have any line responsibility but may have considerable staff responsibility. Discuss what this means for the firm organizationally and what it means for the CIO.

4. Where do individuals gain experience to become CIOs? Sketch the path that might be taken by an MBA who desires to become a CIO. What experiences would you recommend for this individual?

5. Discuss the balance a CIO needs between technical skills and general or line management skills. How should a potential CIO obtain these skills?

6. CIOs must respond to changes initiated by the organization and they must take action which itself creates change. Discuss the interplay between these two activities.

7. If you were the CIO of a major firm, how would you obtain information on technology trends? Describe the process you would use to make sure the trends were evaluated by the correct people in the firm.

8. How do you believe the technology trends forecast by Straub and Wetherbe will impact businesses in the future? Give one example of how these trends might be put to practical use.

9. How might astute CIOs use the theory of innovation diffusion in their dealings with the firm's top executives? How would innovation diffusion apply during the development and installation of an executive information system? How might it apply if the firm wanted to install EDI between it and its suppliers?

10. The term *downsizing* was used in two different ways in this chapter. What is meant when the firm downsizes? What is meant by the term when applied to the IT organization? Under what conditions might both activities occur at the same time? What are the fundamental motivations for undertaking either type of downsizing?

11. Discuss the advantages and disadvantages to the firm of downsizing the IT operation.

12. Discuss the advantages and disadvantages of outsourcing the firm's external telecommunications system. Under what conditions would a firm outsource its telecommunications system and not its production operation?

13. What considerations are important in deciding to outsource the production operation? What industry factors might influence this decision? What strategic considerations are important? What role does the corporate culture play in these decisions?

14. Describe the aspects of Katherine Hudson's activities that are general management responsibilities.

15. Discuss the balance that must be achieved by the CIO among technology availability, the firm's need for it, and the firm's propensity to adopt it. What actions can the CIO take to affect the balance among these variables?

Assignments

1. "Conclusion: Effectiveness Must Be Learned" is the title of the final chapter in Peter Drucker's book *The Effective Executive*. Read this chapter and summarize what you learned for the class.

2. Conduct an interview with the senior information executive in a firm in your community. Determine from the interview what the job consists of and what the individual must do to be successful. How does this person balance technical skills with business skills in performing the job? Compare and contrast this person's view of the CIO title and job with that developed in the business vignette at the beginning of this chapter.

ENDNOTES

[1] Reprinted from *DATAMATION*, August 1, 1988. © 1988 by Cahners Publishing Company.

[2] John Hammitt is now IT chief at United Technologies Corp.

[3] Peter F. Drucker, *The Effective Executive* (New York: Harper & Row, 1986), 23. In speaking about executive effectiveness Drucker states that "effectiveness, in other words, is a habit; that is, a complex of practices. And practices can always be learned."

[4] Gordon Bock, "Management's Newest Star," Special Report, *Business Week*, October 13, 1986, 160. The article discusses the experiences of many practicing CIOs in large U.S. corporations.

[5] Jeffrey Rothfeder, "CIO Is Starting to Stand for 'Career is Over,'" *Business Week*, February 26, 1990, 78.

[6] Allen E. Alter, "A New Twist on Turnover," *CIO*, October, 1990, 49. Although conventional wisdom indicates high CIO turnover, there is little factual evidence to support these beliefs, according to this article.

[7] Tom Peters, "Peter's Principles," *CIO*, August 1989, 17.

[8] "Trends," *Computerworld*, March 5, 1990, 118.

[9] Rothfeder, "CIO Is Starting to Stand for 'Career is Over.'"

[10] Albert L. Lederer and Aubrey L. Mendelow, "Convincing Top Management of the Strategic Potential of Informations Systems," *MIS Quarterly* (December 1988): 525.

[11] Allan E. Alter, "Good News, Bad News...," *CIO*, January 1990, 18.

[12] Kathryn J. Hayley and Raymond W. Bolek, "CIO Survival," *CIO*, December 1989, 10.

[13] James Martin, "Strategic Information Systems: A Formula for Success," *PC Week*, November 28, 1988, 36.

[14] Jaque H. Passino and Dennis G. Severance, "The Changing Role of the Chief Information Officer," *Planning Review* (September-October 1988): 41.

[15] Jack Kuehler, "Putting New Technologies To Work," *Think*, no. 7, 1989, 2.

[16] Detmar W. Straub and James C. Wetherbe, "Information Technologies for the 1990s: An Organizational Impact Perspective," *Communications of the ACM* (November 1989): 1328.

[17] Everett M. Rogers, *Diffusion of Innovations* (New York: Free Press, 1962), 79.

[18] Hamid Tavakolian, "Linking the Information Technology Structure with Organizational Competitive Strategy: A Survey," *MIS* Quarterly (September 25, 1989): 309.

[19] Elisabeth Horwit, "Downsizing Quandry for IS Pros," *Computerworld*, March 5, 1990, 1.

[20] Emily Leinfuss, "Bank Decentralizes MIS," *MIS Week*, May 1, 1989, 1.

[21] Sally Cusack, "Keeping Centralization on Track," *Computerworld*, June 25, 1990, 67.

[22] Valerie A. Roman, "Time to Centralize," *MIS Week*, November 27, 1989, 17.

[23] Adapted from "Downsizing — A Company's Experience," *MIS Week*, September 11, 1989.

[24] Deborah Cooper, "High Five," *CIO*, December 1988, 55.

[25] Theresa Conlon, "Outsourcing Becomes a Controversial Policy," *MIS Week*, December 18, 1989, 1.

[26] Adapted from "Keeping IS in Focus," *CIO*, August 1989, 38. © *CIO Magazine*.

REFERENCES AND READINGS

Benjamin, R. I., C. Dickinson, and J. F. Rockart. "The Changing Role of the Corporate Information Systems Officer." *MIS Quarterly* (September 1985): 177.

Donovan, John J. "Beyond Chief Information Officer to Network Manager." *Harvard Business Review* (September-October 1988): 134.

Feigenbaum, E., et al. *The Rise of the Expert Company: How Visionary Companies Are Using Computers to Make Huge Profits*. New York: Times Books, 1988.

Gantz, John. "Telecommunications Management: Who's in Charge?" *Telecommunications Products + Technology*, October 1985, 17.

Gantz, John. "The Growing Power of the Telecomm Manager," *Telecommunications Products + Technology*, October 1986, 33-54.

Mahnke, John. "Downsizing: A Company's Experience." *MIS Week,* September 11, 1989, 16.

Orazine, Ron. "Why MIS Managers Are Becoming Network Experts." *Telecommunications*, January 1988, 57.

Owen, Darrell E. "Information Systems Organizations Keeping Pace with the Pressures." *Sloan Management Review* (Spring 1986): 59.

Rockart, John F. "The Changing Role of the Information Systems Executive: A Critical Success Factors Perspective." *Sloan Management Review* (Fall 1982): 3.

Wacker, William N. "The Search for Higher Beings." *CIO Magazine*, May 1988, 21.

Index

A

Accounting, **433**
  financial, **433**
  managerial, **433**
Accounting for IT resources
  reasons for, 432-35
Action orientation, a CIO CSF, 545
Administrative actions
  in planning, 111
Advanced Micro Devices, 153
Airlines, 47-50
  American, 47-50, 354
  Arkia Israeli, 66n2.13
  Delta, 48
  Northwest, 49
  TWA, 49
  United, 49
Akers, John F., risks due to technology, 148
Alexander, Michael, 394n13.1
Allen, Leilani E., 395n13.9
Alliances
  at Corning Inc., 493
  for application development, 272
  impact on value chain, 497
  importance of information technology, 497
  international networks support, 416
  reasons for, 497
Allstate Insurance Company, 191, 222
Alter, Allen E., 551n18.6, 551n18.11
Altman, Edward N., 122n4.11, 336n11.8
American Airlines, 47-50, 354
American Hospital Supply Corporation, 53,
  66n2.22, 496
American National Standards Institute (ANSI),
  53, 171
American Standard, 542
Ameritech, 158
Analog computers, 151n5.6
Analog signals, **160**
Analysis of computer center workload, 330
Analytic Systems Automatic Purchasing, 53
Analyzing measurements, 383
Andersen Consulting, 24, 34n1.28
Anderson, Lane K., 216n7.13
Anthony, Robert N., 33n1.13
Apollo Computer Co., 176
Apple Computer Company, 155-56, 175, 176
Application controls, **469**-75
Application development
  additional alternatives, 271
  an IT critical success factor, 220
  examples of difficulties, 221
  extended cost recovery, 447
  fourth-generation languages, 255
  funding by phases, 446
  phased approach, 224, 244
  problems with, 221-23
  programs for sale, 254
  project management, 224-44
  reasons for difficulties, 221-22, 244
  reasons for failure, 236
  relation to strategy and planning, 245
  subcontracting of, 265
  tools and techniques, 255
  use of alliances, 253
Application development and maintenance,
  funding for, 446
Application development, (see also project
  management)
Application management, a model for, see Figure
  17.2, 510
Application management, attacks issues, 512
Application owner, control responsibilities of, 467
Application portfolio
  alternatives for managing, 196
  cost of enhancement, 198
  management of, 203-11
  migration to new hardware, 198
  source of problems, 344
Application prioritization
  attacks IT issues, 212
  identify strategic systems, 211
  input to IT plans, 210
  results of, 209
  revealing alternatives, 210
  some important factors, 209
  value of disciplined process, 212
Application program audits, **475**
Application programming backlog, **195**
Application project management, 224-28
Application resources, 193
Application user, control responsibilities of, 467
Applications
  as depreciating assets, 194
  costs and benefits of, 208
  depreciation and obsolescence, 193
  difficulties with, 191
  in IT planning, 107
  inflexibility to change, 223
  Maintenance and enhancement, 193
  results of operational analysis, 207
  results of strategic analysis, 207
Arco Coal, 258
Aron, Joel D., 250n8.9, 250n8.13
Arthur Anderson and Company, 21
Artis, H. Pat, 394n13.3
ASAP, **53**
Assembly language, **256**
Asset classification, methods for, 465
Asset control and protection, some
  considerations, 480-82
Asset identification and classification, 464
Asset management, a theme of this text

Asset protection 461
  as related to end-user computing, 286
  sensitive programs, 481
  steps in accomplishing, 465
AT&T, 145, 157-158, 176, 161, 168, 184n6.1,
  423n14.7
AT&T American Transtech, 66n2.26
Auditability of application programs, 475-76
Audits, application program, **475**
Availability
  of IT service, 329
  of networks, 412
Avon, 37-39, 65n2.1, 542

B

Backlog, **195**
Bandwidth, **161**
Bankers Trust, downsizing at, 539-40
Barcodes, Federal Express use of, 55
Bardeen, John, 151n5.5
Batch systems management, **374**-76
Batch systems management, problem reporting,
  375
Baxter Health-Care Corporation, 53-55, 273
Baxter International, 147, 488n16.3
Baxter Travenol Laboratories, 54, 496
Bechtel Group, sells IS services, 254
Belitsos, Byron, 336n11.1
Bell Atlantic, 158
Bell Communications Research Corp., 418
Bell Laboratories, 147, 151n5.5
Bell South, 158
Bell Telephone Co., 184n6.1
Benson, Robert J., 122n4.5
Bock, Gordon, 529, 551n18.4
Bolek, Raymond W., 551n18.12
Booker, Ellis, 33n1.11
Borovits, Israel, 66n2.13, 394n13.2,
Bower, Marvin, 33n1.5
Bowman, B., 93n3.9
Boynton, Andrew C., 122n4.16
Brancheau, James C, 22, 34n1.25, 184n6.4,
  312n10.13
Brattain, Walter, 151n5.5
Briggs, George, 278n9.11
British Telecom, 400
Budget management, see Figure 17.6, 517
Budgetary control, **436**
Buse, Jeanne, 457n15.6
Business case, **225**
Business case,ingredients of, 226
Business conditions, as affecting system
  capacity, 388
Business controls, **117**
  efficiency and effectiveness of, 466
  keys to effectiveness, 482
  on application operation, 469-75
  principles of, 464-66
  reports on effectiveness, 483
  see also IT controls
  why important, see Table 16.1, 463
Business issues
  in strategy development, 87

Business management, 517-18
Business objectives
  application prioritization, 203
  in downsizing, 296
Business strategy, **77**
Business System Planning, **115**, 122n4.17
Business vignette
  A Measured Response, 429-31
  Achievements Lag Expectations, 69-70
  An IS Management Triumph, 317-19
  Avon's Executive Information System, 37-39
  CIO, Misfit or Misnomer?, 527-28
  Computer Runaways, 191-192
  Downsizing at John Hancock, 541
  Fire KOs Chicago Networks, 339-340
  IS Shops Form Alliances, 253-55
  Joint Venture, IBM and Baxter, 273
  Keeping IS in Focus, 544
  Lenders Maintains Information Flow, 371-72
  McGraw-Hill's Unified System, 97-98
  Met Rings in Remote Sites, 397-98
  New Applications Cause Difficulties, 242
  PCs Are Changing the Computer Industry,
    155-156
  PCs Replaced Mainframes at Echlin, 281-282
  Risk Management Succeeds, 364
  Software Failures at DoD, 246
  Software Helps Build Software, 219-220
  Spreading a Virus, 459-461
  The Age of Hierarchy is Over, 493-94
  Volvo's Net Gains, 180-181
  When Disaster Threatens, 356
  Who's on Second?, 127-28
Buss, Martin, 122n4.10,

C

Capacity analysis, **385**
Capacity management, **384**-89
  dependence on other disciplines, 387
  linkage to service levels, 389
  of networks, 413
  the objectives of, 385
  the process of, 384
Capacity planning, **386**, 387-388
  effect of cost center on, 445
  effect of end-user computing 395n13.9
  for networks, 413
Carlyle, Ralph, 34n1.18
Carnegie-Mellon University, 418
CASE, 219, **258**
  as used by Ramada, 219-20
  suppliers of CASE tools, 279
  use and adoption by firms, 223
CASE methodology, **258**, (see also CASE)
CASE tools, 258-61
  functions of, 259
  support for managers, 259-60
  types of, 259
Cash, James I., Jr., 65n2.3, 122n4.8
Cash Management Account, **50**
Caterpiller Inc., sells IS services, 254
CD-ROM, 135
Centel Corp., 159

Centralization
    at Cambridge, Mass., 540
    at Trailer Train Co., 540
    CIO's role in, 540
    IT role in, 540
    versus decentralization, 497
Centralized control, decentralized managment,
    496
Certo, Samuel C., 94n3.13
Challenges for CIOs
    advancing technology, 533
    competitive thrusts, 533
    internal restructuring, 533
    need for business improvements, 534
Champy, James, 34n1.26
Change management, **349**
    for networks, 410
    for office automation, 303
    reporting results, 353
    the goal of 353
    the process of, 351
    the scope of, see Figure 12.5, 350-51
Change request
    a sample form, 369
    see Table 12.7, 352
Changes in business, adapting to, 500
Chao, Ben, 277n9.4
Chargeback, (see also IT cost recovery)
Chargeback systems, **436**
    goals of, see Table 15.3, 439-40
    impact on incentives, 440
    relation to client behavior, 450
    relation to IT effectiveness, 438
    value of, 451
Chase Lincoln Bank, 394n13.4
Chrysler Corp., use of computer-aided design,
    66n2.11
Ciba-Geigy, 399
CIO, 7, **529**
    a general manager, 543
    a summary of their situation, 546
    ability to measure system effectiveness, 430
    balance technology with firm's propensity for,
        546
    career paths, 531-32
    challenges for, 533
    must find new ways of doing business, 541-42
    organizational position, 530
    origin of the term, 33n1.6
    performance measurement of, 532
    reporting relationships, 531
    responsibilities of, 531
    role in centralization and decentralization, 540
    role of, 534-36
CIO, concerns with title and position, 530
CIO position, the dilemma of, 529
CIO propensity to accept change, 538
CIO's use of accounting systems, 452
CIOs facilitate organizational change, 538-40
CIOs manage technology, 534-36
CMA, **50**
Cobbin, W. Frank, Jr., 66n2.26
Collier, Andrew, 423n14.7
Common carriers, **165**

Communications technology, 144
Competition
    forces governing, 44
    Porter's model, 44
Computer Associates International, 278n9.11
Computer processing, controls on, 473
Computer-aided software engineering, see CASE
COMSAT, 159
Con Edison, 260
Configuration management, **408**-09
    scope of, see Table 14.2, 409
    see Figure 17.5, 516
Conlon, Theresa, 552n18.25
Contingency plans, **358**
Control, **117**, 462
    budgetary, 436
    in plan execution, 509
Control Data Corp., 176
Control points in application, 470
Control processes related to plans, 117
Control responsibilities, 467-68
Controls
    application processing, 469
    in production operations, 476
    on computer processing, see Table 16.4, 473
    on data output and distribution, 474-75
    on data storage and retrieval, see Table 16.5,
        474
    on input data, see Table 16.3, 472
    on networks, 476
    on transaction origination, see Table 16.2, 471
    why important, 462
Controls, (see also IT controls)
Cooper, Deborah, 552n18.24
Corman, Lawrence S., 311n10.9
Corning, Inc., 147, 493-94, 524n17.2
Corporate culture, **5**, 495
    impact on IT cost recovery, 438
    need for CIO to accommodate, 545
    related to SLAs, 333
Corporate funding for development, 447
Corporation for Open Systems (COS), **176**
Cosmos, **55**
    strategic advantage to Federal Express, 55
Cost center, **443**
    characteristics of, see Table 15.5, 444
    relation to capacity planning, 445
Cost recovery, see IT cost recovery
Costs and benefits of applications, 208
Cougar, J. Daniel, 311n10.8
Coursey, David, 151n5.3, 367n12.8
COVIA, 49-50, 59
Critical applications, see crucial applications
Critical success factors, **27**
    application development, 220
    for CIOs, 543-46
    for IT managers, 27-29
    in planning, 99, 114
    people management skills, 506
    portfolio management, 212
    production operations, 320, 515
Crucial applications, **358**
    environmental factors for, 359
    strategies for, 359-62

Cusack, Sally, 552n18.21
Customer satisfaction, daily measures of
  336n11.11

D

Data center
  physical security, 480
  operations, see production operations
Data encryption, **478**
Data output and distribution, controls on, 474-75
Data resources, 194
  as an issue, 22
  relation to application architecture, 211
  relation to application prioritization, 211
  the management of, 211
Data security, **478**
Data sets, controls for sensitive data, 482
Data storage and retrieval, controls on, 474
Databases
  as part of information systems, 194
  as related to end-user computing, 286
  for capacity planning, 387
  relation to downsizing, 295
Davenport, Thomas H., 34n1.27, 525n17.15
Davies, Andrew, 488n16.5
Davis, Gordon, 33n1.9, 93n3.9
Deal, Terrence E., 33n1.5
Decision support systems, **43**
Delivery services, 55-56
Delta Airlines, 48
Department of Defense, 246, 260
Depompa, Barbara, 151n5.9
Depreciation
  of applications, 193
  of databases, 194
Development manager, control responsibilities of,
  468
Digital computers
  basis for advances in, see Table 5.1, 129
  trends in innovation, 129
  evolution of, 137
Digital Equipment Corp., 176, 260, 262
Digital Network Architecture (DNA), **174**, 175
Digital signals, **161**
Digitizing analog signals, 186
Disaster recovery, after Hinsdale fire, 340
Disaster recovery planning, common carriers,
  368n12.14
Disaster recovery services, 360
Disciplines, **321**
  applicability of, 342
  impact of end-user computing on, 365
  network control and security, 479
  network management by MCI and IBM, 405
  network management, see Figure 14.1, 406
  of production operations, 321-323, 512-515
  relation to network management, 405
Disciplines, (see individual items, e.g., change
  management)
Distributed systems, in disaster recovery, 360
Doll, William J., 312n10.18
Downsizing, **293**
  application implications for, 296

at Bankers Trust, 539
attributes of, 294-95
impact on IT personnel, 539
pros and cons of, see Table 10.4, 294
security considerations, 480
Drucker, Peter F., 9, 10, 33n1.1, 524n17.10,
  551n18.3
Drury, D. H., firms' use of steering committees,
  122n4.14
Duffy, Carolyn, 250n8.2
Dyson, Ester, 66n2.15

E

Eastman-Kodak, 401, 542, 544
Echlin Corp., 281-82, 311n10.1
Electronic data interchange (EDI)
  executives expectations of, 404
  future for, 536
  use by Baxter Health-care, 54
  use by Wal-Mart, 9
Elliott, Dennis W., 423n14.6
Emergency Planning, **357**
Emery, James C., 67n2.31
End-user computing, 44
  a case discussion, 308-10
End-user computing, (see EUC)
Environment
  for crucial applications, 359
  in strategy, 71-72
  planning, see Figure 4.3, 116-117
EUC, **54**
  accounting considerations, 449
  expenditures on, 283, 287
  financial considerations, 287
  impact on firms, 284-85
  reasons for adopting, see Table 10.1, 283-284
  the issues of, 285-88
  the issues of, see Table 10.2, 285
European Economic Community, 400
Evolution of digital computers, see Figure 5.2, 137
Executive information systems
  benefits at Avon, 39
  expenditures on, 40
Expectations, 4
  congruence with performance, 333
  executives of controlled operations, 464
  executives of information technology, 63, 69
  in portfolio management, 202, 212
  in production operations, 321, 513
  IT support to business goals, 87
  management of by CIOs, 545
  of chargeback systems by users, 452
  of IT accounting systems by executives, 452
  of IT by executives, 40
  of IT managers, 5
  of networks by managers, 404-05
  of office automation, 301
  of productivity in application development, 222
  performance appraisal, 6
  users of networks, 403
Expert systems
  in network management, 415
  future use of, 536

F

Fallon, James, 424n14.20
Farhoomand, Ali, 488n16.6
Fault management, **406**, (see also Problem management)
FBI, 198
Federal Express, 55-56
  growth of, 66n2.25
Feedback mechanisms for managers, see Figure 4.4, 118
Feuche, Mike, 277n9.5, 311n10.1, 311n10.2
Financial and business controls, 516
Financial considerations
  in planning, 110
  in strategy development, 89
  of end-user computing, 287
Fire, Hinsdale, Ill., 339-40
First Boston Corp., 253-55
Fischer, Barbara S., 312n10.24
Fischer, Paul M., 250n8.8
Floating point operations per second, flops, **139**
Flower, David A., 457n15.8
Forbes, "Speed to Market", 66n2.10
Fourth-generation languages, **255**-58
  acceptance by programmers, 258
  characteristics of, 257
  productivity of, 256
  types of, 256
Frank, Werner G., 250n8.8
Frankfeldt, Chester, 311n10.4, 312n10.21
Friel, Thomas, 530
Fujitsu, 153
Functional strategy, elements of, 75, (see also IT strategy)
Funding
  application development, 446, 447
  maintenance, 446
  to encourage technology adoptions, 450
  IT via chargebacks, 437

G

G. D. Searle, 399
Galileo, **50**
Gartner Group, 17, 18
General Accounting Office, 216n7.6
General Electric, 176
Gibson, Cyrus F., 19, 34n1.24
Glass, Robert L., 216n7.8
Gluck, Frederick W., 93n3.6
Goal congruence
  IT and the firm, 24
  people and organizations, 503
Goals and objectives
  in strategy statement, 80
  planning aids in achieving, 117
Goff, Leslie, 215n7.4, 457n15.4
Gold, B., 122n4.15
Gordon, Carl L., 250n8.5
Gorry and Scott-Morton model, 10, see Table 1.3, 12
Gorry, G. A., 33n1.14
Gralla, Preston, 278n9.12
Gray, Daniel H., 93n3.4, 121n4.4

Green, James H., 312n10.23
Greene, Harold H., U.S. District Court Judge, 158
Greene, Richard, 456n15.3
Gremillion, Lee L., 312n10.11
GTE, 159

H

H. J. Heinz Company, 399
Hammer, Michael, 423n14.5, 524n17.8, 525n17.15
Hammitt, John, 551n18.2
Hannaford Bros., saves money with minis, 311n10.2
Harbert, Tammi, 423n14.9
Hardware monitors, 382
Hartford Insurance Co., 255
Hayley, Kathryn, 337n11.14, 551n18.12
Helliwell, John, 184n6.7
Henderson, John C., 93n3.5, 312n10.14
Heschel, Michael, 496
Hewlett-Packard Corp., 58
Hindin, Eric, 184n6.7
Hise, Richard T., 94n3.12
Hitachi Ltd., 153
Horgan, John, 186
Horizon, time
  in strategy development, **83**
  in planning, **101**
Horwit, Elisabeth, 552n18.19
Hot site providers, 368n12.12
Houghton, James R., 493
Howe, Douglas J., 395n13.8
Hudson, Katherine, 542-544
Huff, Sid L, 311n10.6
Humphrey, John H., 184n6.3

I

IBM, 115, 147, 176, 260, 262
IBM, analysis of risk, 250n8.12
Iida, Jeanne, 33n1.2, 423n14.11, 424n14.19
Illinois Bell, 339
Index Group, Inc., 21
Index Technology Corp., 219
Infomart, 371-72
Information, **10**
  as related to management position, 11-13
Information architecture, **22**
  as an IT issue, 22
  database role in, 211
  dependence on application management, 511
  relation to downsizing, 295
Information center, **291**
Information center, functions of, 291-92
Information requirements of phase reviews, 231-34
Information technology
  complexity of, 16
  impact on business, 57
  impact on firm's direction, 42
  impact on people and organizations, 19
  leadership via innovation, 41
  pervasiveness of, 17
  reshapes business processes, 24

Innovation
  diffusion of, see Figure 18.1, 537
  in digital technology, 128
  in recording technology, 134-35
  in semiconductor technology, 131-33
  in systems, 136-41
Innovation diffusion, **537**
Input data, controls on, 472
Integrated services digital network, see ISDN
Intel, 127-28, 151n5.2, 153
Interior Department, 221
Internal rate of return, 227
International considerations of networks, 400,
  415-17
International Data Corporation, 40
International networks, strategic considerations,
  416
International Standards Organization (ISO), 171,
  406
Internationalization, through restructuring, 496
Investments in IT, 41
  by Baxter Health-care, 55
  by Merrill Lynch, 52
  opportunities for strategic systems, 56
  return on, 429
Invisible backlog, see backlog
IS human resources
  as an issue, 22
  management of, 514-19
ISDN, **166**-68
  classes of service, 166
  impact on equipment market, 168
Issues
  of IT management, **21**-26, see Tables 1.5
    and 1.6
  leadership gap on, see Table 1.7, 26
  related to critical success factors, 28
IT and firm strategy misalignment, reasons for,
  509
IT controls, need for, 461
IT cost recovery
  additional considerations, 445
  benefits of, see Table 15.2, 437
  for networks, 449
  for production operations, 447
  see also, chargeback
  some compromises, 451
IT effectiveness, relation to chargeback systems,
  438
IT interaction with users, 518
IT management, 14
  current issues, 21-26
  impact of environmental changes, 18
  model for the study of, 29, see Figure 1.5, 30
  need for business controls, 18
  relation to strategic considerations, 19
  the difficulties of, 15
IT management system, the result of adopting, 519
IT managers
  demand for, 4
IT managers, control responsibilities of, 468
IT planning
  a model for, 105-13
  an integrated approach, 115, see Figure 4.3, 116
  Business System Planning, 115

inclusion of SLA information, 334
operational plans, **100**, **103**
planning horizon, 99, see Figure 4.1, 101
schedules, 103, see Figure 4.2, 104
some approaches to, 113
strategic plans, **100**, **102**
tactical plans, **100**, **102**
IT planning, (see also Planning)
IT resource accountability
  accomplishments of, 437
  alternatives, 441-45
  evolution of, 451
  objectives of, 435
  relation to planning, 435
IT resource accounting
  difficulties of, 432
IT resources
  in planning, 109
IT role and contribution, relation to IT
    accounting, 453
IT strategies
  as a guide to action, 85
  assemblage of, see figure 3.3, 78
  elements of, see Table 3.1, 71
  issues covered, see Table 3.5, 87
  maintenance of, 72
  maintenance process, 86
  relationship to plans, 73-74
  requirements of, 78
  types of, 74-78
IT strategies (see also Strategy statement)
IT strategy development, 71
Ives, Blake, 34n1.17
Izzo, Joseph E., 34n1.19

J

Jaedicke, Robert K., 216n7.13
John Hancock Insurance Co., 541
Joint ventures, see Alliances

K

Kador, John, 277n9.3, 367n12.7, 395n13.7
Kaufman, Stephen P., 93n3.6
Kawasaki, 258
Keefe, Patricia, 423n14.1
Keen, Peter G. W., 66n2.12, 122n4.12, 311n10.7
Keider, Stephen P., 250n8.11
Kellner, Mark A., 423n14.3
Kennedy, Allen A., 33n.1.5
Kidder, Peabody Inc., 253-255
Killkirk, John, 66n2.11
King, William R., 80, 94n3.11
Konsynski, Benn R., 215n7.4
Koontz, Harold, 33n1.8
KPMG Peat Marwick, 311n10.3
Kraft, Dennis J. H., 66n2.14
Kronke, David, 33n1.13
Kuehler, Jack, 65n2.5, 552n18.15
Kull, David, 423n14.15
Kulula, Kathryn, 65n2.2
Kuzela, Lad, 67n2.28

L

Lablans, Peter, 424n14.21
Lang, Rolf, 423n14.14
Lederer, Albert R., 122n4.6, 551n18.10
Legal considerations
  business controls, 463
  of strategic information systems, 59
Leinfuss, Emily, 423n14.7, 552n18.20
Leitheiser, Robert L., 312n10.15
Lenders, Jerry, 371-72, 390-91
Levine, Arnold S., 216n7.9
Levine, Ron, 367n12.9
Levitt, Theodore, 94n3.10
Life cycle, see SDLC
Local area networks, growth of, 177
Local area networks, LANs, **176**
Lotus 1-2-3, 222, 278n9.9
Lotus Development Corporation, 278n9.9
Ludlam, David A., 424n14.22

M

Machiavelli, A., comments on change, 500
MacMillan, Ian C., 67n2.32
Magnetic recording, **134**
Mahnke, John, 250n8.4, 312n10.20, 367n12.10,
  423n14.4
Mainelle, Michael R., 216n7.11
Mainframe computers, **136**
Maintenance costs
  see Table 7.1, 199
  for federal government, 216n7.6
Maintenance of programs, see Program
  maintenance
Management
  levels of, 3
  of batch systems, 374-76
  of networks, 515
  of on-line systems, 376
  of phase reviews, 235
Management by Results at Security Pacific, 317
Management information reporting, 389-90
Management information systems, 44
Management issues
  leadership gap in, 25, see Table 1.7, 26
  see Table 1.5, 23
  see Table 1.6, 24
Management issues, 21-26
Management processes, the collection of, 506-19
Management reporting for networks, 414
Management skills
  a learned characteristic, 529
  regarding people, 503-05
Management systems, for networks, 414-15
Managing development alternatives, 273
Managing systems, 374
Mangurian, Glenn E., 423n14.5, 524n17.8
Marsh, Robert E., 337n11.12
Martin, Barbara H., 311n10.6
Martin, James, 277n9.2, 551n18.13
Maryland National Bank, 317
Maslow, Abraham, 502, 525n17.12

Mason, Janet, 250n8.1
Matsushita, 153
McCaw Cellular Communications Inc., 401
McDaniel, Stephen W., 94n3.12
McDonnell Douglas Corp., 58, 401
McFarlan, Warren, 65n2.3
McGovern, Laura Cooper, 184n6.9
McGowan, William G., 424n14.18
McGraw-Hill, 97-98, 106, 121n4.2
McGregor, Douglas, 525n17.13
MCI, 159, 423n14.3
McKenney, James L., 65n2.3
McLean, Ephraim R., 122n4.11, 311n10.5,
  336n11.8
McNurlin, Barbara C., 122n4.18
Measurements of satisfaction, 331
Measuring systems performance, 382
Meeks, Flemming, 184n6.5
Mendelow, Aubrey L., 551n18.10
Mentzberg, Henry, 93n3.3
Merrill Lynch, 4, 50-52, 399,405,423n14.3
  cash management account, 50
  investments in IT, 52
Merrills, Roy, 66n2.10
Metropolitan Life Insurance Co., 397-98,
  423n14.2
Metsisto, Tauno J., 525n17.15
Microcomputers, **136**
Microcomputers, growth of, 138
Miles, J. B., 184n6.6
Miller, David R., 216n7.11
Milymuka, Kathleen, 312n10.19
Minicomputers, **136**
MIPS, **136**
MIT, 147
Mitsubishi, 153
Moad, Jeff, 34n1.18
Model
  for application management, see Figure 17.2,
    510
  for application project management, see
    Table 8.2, 225
  for controlling to plans, 118
  for IT planning, 105-13
  for IT planning and control, 508
  for IT strategy development, 78-82
  for portfolio management, 203-11
  for study of IT management, see Figure 1.5, 30
  Gorry and Scott-Morton, 10
  integrated strategic influences, 62
  open system interconnect, 173
  Porter's of competition, 44
  production operations management, see
    Figure 17.4, 514
Modem, **161**, (see also Telecommunications
  systems)
Morale, achieving high, 505-06
Moore, Carl L., 216n7.13
Motorola, 153
Munro, Malcolm C., 311n10.6
Murphy, Michael, 488n16.6
Musmanno, Thomas E., 52

N

National Bureau of Standards, 176
Naumann, J. D., 184n6.4
NEC Corp., 153
Necco, Charles R., 250n8.5
Neiderman, Fred, 22, 34n1.25
Net present value, **227**
Network architecture, (see also OSI, 172)
Network architecture, vendor, **174**-76
Network capacity assessment and planning, 413
Network controls and security, 476-79
Network management, **403**
  accounting and cost recovery, 449
  expectations of users, 403, 404
  Merrill Lynch and MCI, 405
  relation to on-line management, 403
  relation to system management, 515
  reporting results, 414
  scope of, 403
  systems for, 414-15
  the importance of, 399
Network managers
  international considerations, 417
  responsibilities of, 417-18
  training for, 418
Network performance assessment, 412
Network problem sources, see Table 14.3, 410
Network problems and changes, 410
Network security
  considerations of, see Table 16.7, 479
  steps in accomplishing, 477-79
Network service levels, **407**-08
Network technicians, security considerations, 481
Networks
  as part of information system, 399
  as strategic systems, 399
  international, 400, 415-17
  reasons for growth, 402
  used for competitive advantage, 401
Networks facilitate restructuring, 401
Neumann, Seev, 66n2.13
NeXT, 134, 262
Nolan, Richard, 19, 33n1.16, 34n1.24, 113, 299
Northern Telecom, 159
Northwest Airlines, 49
NTT, 424n14.19
Nucor, 9
Numeris, brings ISDN to France, 424n14.19
NYNEX, 158
Nyquist's theorem, **186**

O

O'Donnell, Cyril, 33n1.8
Object paradigm, **261**
Object-oriented design, 261
Object-oriented programming, 261
Obsolescence of applications, 193
Office automation, 297
  accounting considerations, 449
  advantages of, 298
  implementation considerations, 300
  managing change, 303-04
  planning for, 299

Office automation systems, **44**
Office of Management and Budget, 355
Office systems
  cost justification measures of, 429
OLDE & Co., 52
Olson, Margrethe, 33n1.9, 34n1.17
On-line systems management, **376**-79
  problem reporting, 379
  scope of 377
Open systems interconnect (OSI), **172**
Operating systems, **142**-44
  advances in, see Table 5.5, 143
  current trends, 144
  evolution of, 142
  security considerations, 480-81
Operational plans and controls, 103
Optical storage devices, **135**
Order Entry Systems, 53-55
Organization
  alignment of IS, 22
  information based, **3**
  IT support for, see Table 1.1, 7
  IT within the firm, 6
  of IT, 7 and see Table 1.2, 8
Organizational change
  downsizing, 538
  facilitated by CIOs, 538-40
  facilitated by networks, 401
  with end-user computing, 289
Organizational concerns
  in strategy development, 88
Organizational effectiveness, relation to
  accounting, 453
Organizational learning as an issue, 22
Organizational restructuring, impact on IT
  professionals, 496
Organizational transitions, 495-96
Organizations
  changes in due to technology, 495
  not for profit 34n1.30
OSI, **172**, see Figure 6.6, 173
OSI layer functions, 187
Oum, Tae H., 66n2.14
Outsourcing, **541**-543
  a strategic consideration, 543
  current examples of, 542
  some advantages of, 542

P

Pacific Telesis, 158
Palmer, Scott D., 216n7.6
Parallel processors, **139**
Parker, Marilyn, 122n4.5
Passino, Jaque H., 552n18.14
Passwords, **477**
Patent on CMA, 51
Payback method of return on investment, **227**
Pelton, Charles, 524n17.6
People
  and information technology, 501
  satisfaction needs, 502
  the enabling resource, 500

People management, **502**
  a critical success factor, 506
  attitudes and beliefs, 503-05
  effectiveness, 503
People management skills, 502-06
Pepper, Jon, 37
Performance assessment, of networks, **412**
Performance management, **379**-84
  analyzing measurements, 383
  measurements, 382
  planning, 380
  process of, see Table 13.3, 379
  reporting results, 383
  system tuning, 384
Performance measurement, techniques of,
    394n13.5
Performance monitors, **382**
Performance of computer systems, 380
Period support, funding for applications, **446**
Personnel considerations
  in downsizing, 539
  in end-user computing, 288
  in IT organizations, 501-02
  in planning, 110
  in recovery planning, 362
  in strategy development, 89
  with office automation, 301
Pesci, M. L., 525n17.14
Peter, J. Paul, 94n3.13
Peters, Tom, 9, 33n1.11, 530, 551n18.7
Peterson, DuWayne, 4
Peterson, James L., 151n5.17
Phase 1 review, management information
    requirements of, 231
Phase 1,activities of, 230
Phase 2 review, management information
    requirements of, 231
Phase 2,activities of, 231
Phase 3 review, management information
    requirements of, 232
Phase 3,activities of, 232
Phase 4 review, management information
    requirements of, 233
Phase 4,activities of, 232
Phase 5 review, management information
    requirements of, 233
Phase 5,activities of, 232
Phase 6,activities of, 233
Phase review, **228**
  contents of, 230
  documentation of, 235
  for large projects, 234-35
  ingredients of, see Table 8.3, 229
  issue management, 235
  objectives of, 230
  participants in, 234
  process of, 228-35
  relation to development risk, 234
  timing of, 230
Phase review, see individual phases by number
Pitts, John, 423n14.8
Plan reviews, 111
Planning
  a critical success factor, 28, 99, 114
  capacity of systems, 386, 387-88

for computer center operations, 375
for contingencies, 358
for disaster recovery, 355, 362
for emergencies, 357
office automation, 299
in production operations, 108
in recovery management, 355, 362
systems performance, 380
technology, 112
Planning and control, see Figure 17.1, 508
Planning, (see also IT planning)
Polack, Alexander J., 277n9.6
Policy considerations
  for office automation, 302
  with end-user computing, 292-93
Porter, Michael E., 44, 65n2.8, 67n2.27, 524n17.7
Portfolio alternatives, see Figure 7.1, 197
Portfolio asset management, 509-12
Portfolio expenditures, see Figure 7.2, 200
Portfolio management
  a process for, **203**
  selection of alternatives, 274
Poste Telegraphe et Telephone (PTT), 185n6.11
Price Waterhouse, 434
Prioritization
  a disciplined method for applications, 203-11
  of resources, some common approaches, 202
Private branch exchange (PBX), **178**
Problem, **342**
Problem log, a sample form, 369
Problem management, **342**
  for networks, 410
  implementation of, 346-48
  linked to change management, 350
  objectives of, see Table 12.1, 343
  scope of, see Table 12.2, 343-45
  the process of, 345, 348
  the tools of, see Table 12.3, 345
Problem management reports, **349**
Problem reports
  an example form, 368
  see Table 12.4, 346
  in batch systems, 375
Problem status meetings, **346**
Problems and changes, with networks, 410
Production operations, **17**
  a critical success factor, 515
  growth of, 18
  in IT planning, 107, see Table 4.2, 108
  internal systems management, 514
  performance to customers, 512
  systems management, 374-79
  tactical and operational concerns, 320
  the disciplined approach, 321
Productivity
  a critical success factor, 28
  from project management, 244
  of fourth-generation languages, 256
  of programmers using CASE, 260
Profit center, **441**
  advantages of, see Table 15.4, 442
  disadvantages of, 443
Program controls
  during maintenance, 482
  during operation, 469-75

Program maintenance, **201**
Program maintenance, auditability considerations, 476
Programmer workbenches, **258**
Programmers, maintenance, 201
Programming backlog, **195**
  causes of, 195
  need to prioritize 196
Programs controlling assets, protection for, 481
Project management
  business case, **225**
  elements of, 225
  phase review management, 235
  phase review process, **228**
  resource allocation and control, 236
  risk analysis, **237**-42
  risk reduction actions, 243
Project staffing
  at Avon, 38
  during development, 236
Prototyping, **262**
  advantages of, 262, 264
  an example of use with SDLC, 264
  at Avon, 38
  economics of, 265
  in office automation, 300-01
  the process of, 263
  used with SDLC, 264
Pruitt, James B., 423n14.15
PTT, **185n6.11**
Pulse amplitude modulation, **169-70**, 186
Pulse code modulation, 169-70, 186
Purchased applications, 266
  advantages, see Table 9.5, 267-68
  disadvantages , see Table 9.6, 269-71
  trends toward, 266
Puttre, Michael, 423n14.17
Pyburn, Philipp J., 312n10.11

Q

Quality
  expectations of executives, 222
  of old programs, 198-99
  purchased applications, 268
  using CASE, 260
  with fourth-generation languages, 258

R

Rackoff, Nick, 67n2.30, 122n4.9, 216n7.14
Ramada, Inc., 219-220
Rappaport, Alfred, 456n15.3
RBOCs, **158**
Recording technology, advances in 133-35
Recovery management, **354**
  difficulties for managers, 354
  Federal requirements for, 355
  for networks, 361, 411
Recovery plans, 362
  testing of, 363
Reduced instruction set computing, RISC, 140

Reliability of IT service, 329
Repairs to programs, see program maintenance
Reporting
  results of disciplined processes, 389-390
  for network processes, 414
Reporting, see individual disciplines, e.g., change management
Reservation systems, 47-50
  Apollo, 49
  DatasII, 48
  Galileo, 50
  PARS, 49
  Sabre, 47
  value of, 50
Reservation systems, see also Airlines
Response time, **328**
  measurements of 336n11.10
  in network SLAs, 407-08
  in user satisfaction, 331
Return on investment
  conflicts among alternative methods, 227
  for application development, 227
  intangible considerations, 228
  types of calculations 227
Riggs, J. Robert, 457n15.12
Risk analysis
  in change management, 352
  in subcontract development, 266
Risk indicators for application development, 237-41
Risks
  as part of strategies, 82
  caused by changing technology, 148
  dealt with by recovery management, 355
  in application development 228
  in application development, see Table 8.10, 238
  in application development, see Table 8.11, 239
  mitigated by change management, 352
  of end-user computing, 285
  reduction actions in application development, 243
  reduction due to phase review information, 234
  with common carrier networks, 411
  with office automation, 302
Roberts, Johnnie L., 121n4.3
Rockart, John F., 27, 34n1.29, 113
Rogers, Everett M., 537, 552n18.17
Roman, Valerie A., 552n18.22
Rothfeder, Jeffrey, 215n7.1, 551n18.5
Rothstein, Phillip, 368n12.15
Ryan, Alan J., 66n2.17
Ryerson, Wayne, 423n14.8
Rymer, John, 93n3.2

S

S. New England Tel., 159
S. W. Bell, 158
Sabre, **47**
Sarup, Deepak, 488n16.5
Satisfaction analysis, **204**
  factors in, 204
  of applications, 204-06
  results of, 205

Satisfaction surveys, 332
Satisfaction, measurements of, 331
Sawyer, George C., 93n3.7
Schedules
    for strategy development, see Figure 3.4, 83-84
    in SLAs, 327
Scheduling batch systems, 375
Schendler, Brenton R., 66n2.21, 250n8.3
Scott-Morton, Michael, 33n1.14, 524n17.3
Sculley, John, 155
Second sourcing, **127**
Security considerations
    in downsizing, 480
    networks, 479
    systems and networks, 476-79
    with critical programs, 481
    with network technicians, 481
    with operating systems, 480
    with systems programmers, 481
Security of systems and networks, 476-79
Security Pacific, 272, 317-19, 336n11.2
Security-General, 400
Semiconductors
    industry structure, 130
    major suppliers, 153
    performance factors, 133
    technology trends, see Table 5.2, 131
Separation of duties, **466**
    with crucial applications, 481-82
Service bureaus, **272**
Service bureaus, importance to disaster recovery, 361
Service offerings, telecommunications, 164-65
Service representatives of IT, **518**
Service representatives, responsibilities of, 518
Service-level agreement, **323-25**
    at Security Pacific, 318
    example of contents, 337
Service-level agreements, see also SLAs
Sethi, Vijay, 122n4.6
Severance, Dennis G., 552n18.14
SGS-Thomson, 153
Shannon, Claude E., 186
Sherkow, Alan, 394n13.3
Shockley, William, 151n5.5
Sifonis, John C., 93n3.5,
Silberschatz, Abraham, 151n5.17
Singleton, John P., 122n4.11, 317, 336n11.8
SLAs, **323-25**
    for networks, see Table 14.1, 407-08
    process for establishing, 324
    relationship to expectations, 321
    schedule and availability of service, 327
    timing of service, 328
    unusual service requirements, 333
    what they include, see Table 11.1, 325
    when to develop and modify, 326
    workload forecasts in, 329
SLAs related to corporate culture, 333
Slutsker, Gary, 151n5.10
Smart, John R., 423n14.10
SMARTVEST, **52**
Smock, Gary S., 184n6.3
Software costs, capitalization of, 434
Software monitors, **382**

Souvran Financial Corp., 260
Span of communications, **498-99**
Span of control, 14
Spending on IT, trends of, 434
Sprague, Ralph H., 65n2.7, 122n4.18
Staff responsibility, as a source of authority, 529
Stages of growth, **19-21**
    in office automation, 299
    in planning, 113
    uses for, 19
Stalk, George, Jr., 65, 67
Stand-alone strategy, **76**
Standards
    data encryption, 478
    network and communication, 171
Steering committee, **111, 518**
    firms' use of, 122n4.14
    relation to capacity planning, 388
    role of, 518
    use in application prioritization, 204
    use in planning, 111
Stock Brokerage, 50-53
Strassmann, Paul A., 457n15.5
Strategic and operational factors, of applications, 206-07
Strategic information systems, **43**
    AT&T American Transtech, 66n2.26
    changes over time, 62
    common characteristics, 59
    external factors 60
    financial implications, 59
    impact of organizations, 59
    internal operations, 60
    legal considerations, 59
    leverage of, 56
    networks as, 399
    origins of, 60
    some cautions, 61
    the search for, 61
    thrusts employed, 56-59
Strategic issues
    relation to other issues, 42
    why important, 40-42
Strategic management, **90**
Strategic planning, **102**
    as an IT issue, 22
    at Security Pacific, 318
Strategic planning, (see also IT planning)
Strategic systems
    funding for, 447
    in action 47-56
Strategic thrusts, **45-47**
    characteristics of, 46-47
    international networks, 416
    time as one, 47
Strategic vision 3
    factors driving, see Table 2.1, 41
Strategies for contingencies, 359
Strategizing and planning
    a management process, 507-09
    dealing with issues, 507
Strategizing, see Strategic planning
Strategy development
    steps in, 84
    time horizon, 83

Strategy maintenance, 86
Strategy statement
  ingredients of, see Table 3.4, 81
  outline of, see table 3.3, 79
  process for development, 83
Straub, Detmar W., 536, 552n18.16
Structures
  (see also Organization)
Structures and information, 8-9
Structures, examples of flattening, 9-10
Subcontract development, **265**
  advantages, see Table 9.3, 265
  disadvantages, see Table 9.4, 266
Sullivan, Cornelius H., 115, 122n4.19, 312n10.22, 423n14.10
Sumner, Eric E., 524n17.9
Sun Data, 356
Sun Microsystems, 176, 242
SUNEXPLOR, 308-10
Supercomputers, **138**-41
Supplier of services, control responsibilities of, 468
Synott, William, coined CIO title, 33n1.6
System control points, see Figure 16.1, 470
System development
  Baxter Health-Care approach, 54
Systems development life cycle, **223**
  phases of, see Table 8.1, 224
  the need for, 223
System security, see Network security
System tuning, **384**
Systems architecture, disaster recovery considerations, 361
Systems Network Architecture (SNA), **174**, 175
Systems performance, measures of, 381-83
Systems programmers, security considerations, 481

T

T-1 telecommunications service, **168**-71
Tactical plans, **100**, **102**
Tavakolian, Hamid, 552n18.18
Technical issues
  in planning, 112, see Table 4.4, 113
  in strategy development, 88
Technology
  object-oriented, **261**
Technology adoption
  CIO's role in, 537-38
  new programming tools, 261
Technology forecasting, 534-36
  by some experts, 535-36
  by technology users, 535
  in application development, 511
  in planning, 105
  role of CIO and executives, 536
Technology introduction
  adoption process, 537
  effect of chargeback on, 450
  need for controls, 464
  role of CIO, 536-38
Technology introduction, (see also Innovation diffusion)
Technology shapes organizations, 495-500

Technology trends, 145
Technology trends
  meaning for management, 145
  meaning for organizations, 146
Technology trends, (see also Trends)
Teeter, Lorie, 424n14.19
Telecommunications
  as related to end-user computing, 286
  equipment suppliers, 159
  growth of, 157
  impact on internationalization, 498
  impact on IT structure, 499
  importance to alliances, 498
  ISDN service, 166-68
  major U. S. companies, see Table 6.1, 158
  potential for competitive advantage, 59
  service offerings, 164-65
  services industry, 158
  systems, 159-64
  traffic growth, 524n17.9
  T-1 service, 168-71
Telecommunications components
  cluster controller, **161**
  communications controller, **163**
  line adapters, **161**
  lines or links, **162**
  satellites as links, 163
Telecommunications, in affecting disaster recovery, 360
Telecommunication services, pricing of, 165
Telecommunications systems, 159-64
Terplan, Kornel, 184n6.8, 423n14.13, 423n14.16
Testing disaster recovery plans, 363
Texaco, 69
Texas Instruments, 153
Third-generation languages, **256**
Thyfault, Mary E., 151n5.11
Tichy, Noel, comments on restructuring, 14
Time
  an asset, 57
  a factor in competition, 58, 66n2.10, 66n2.11
  a strategic thrust, 47, 66n2.10
  importance to Hewlett-Packard, 58
  just-in-time manufacturing, 58
Time horizon, see Horizon
Torkzadeh, Gholamreza, 312n10.18
Toshiba, 153
Trailer Train Co., centralization at, 540
Transaction origination, controls on, 471
Transaction processing systems, **43**
Transform Logic Corp., 219
Transistors, see semiconductors
Transmission Control Protocol/Internet Protocol (TCP/IP), **175**
Treacy, M. E., 312n10.14
Trends
  in application resources, 200
  in CASE tools, 260
  in digital computers, 137
  in object technology, 262
  in programming, 141
  in semiconductors, 131
  in systems architecture, 141
  in technology, 145
  in technology, see Table 5.6, 145

of IT spending, 434
  toward distributed data processing, 293
  toward purchased applications, 266
Tretheway, Michael, 66n2.14
Trust and confidence, how managers can
    increase, 506
Trust and confidence in managers, 505-06
Tsai, Nancy W., 250n8.5
Turnaround time, **328**
TWA, 49
Tymnet, 400, 423n14.7

U

U. S. West, 159
Ullrich, Walter A., 67n2.30, 122n4.9, 216n7.14
UNISYS, 176
United Airlines, 49
United Telecom, 159
Unocal, 500
User interface, for on-line systems, 378
User satisfaction surveys, 331
Users
  control responsibilities, 467
  interaction with IT organization, 518
  involvement in application development, 224-44
  needs in IT planning, 108
  role in portfolio management, 203-11
  role in strategy review, 85
  spending propensity for IT services, 388
Users, see also EUC and SLA

V

Vacca, John, 337n11.13
Value, additions to the firm, a CIO, 545
Value chain, **497**
  alliances, effect on 497
  IT contribution to the firm's, 545
Van Schaik, Edward A., 122n4.11, 336n11.3
Vincent, David R., 336n11.5
Vitale, Michael R., 65n2.3

Volvo, use of international networks, 147, 180-81
VSAT, very small aperture terminals, 536

W

Wal-Mart, 9
Walleck, A. Steven, 93n3.6
Wang Laboratories, 176
Wenk, Dennis, 457n15.11
Western Electric, 184n6.1
Wetherbe James C., 22, 34n1.25, 93n3.9,
    312n10.13, 312n10.15, 457n15.7, 530, 536,
    552n18.16
Weyerhaeuser Corp., 254
Wheeler, Richard K., 394n13.4, 395n13.11
Wide area networks, **177**
Williamson, Mickey, 423n14.9
Wiseman, Charles, 45, 65n2.6, 65n2.9, 67n2.30,
    67n2.32, 122n4.9, 216n7.14
Witzel, C. N., 336n11.7
Woodman, Lynda, 311n10.7
Workbenches, programmer, **258**
Workload analysis in computer operations, 330
Workload forecasts, **329**
Workstation store, **289**
  the functions of, 289-91

X

Xerox Corp., 261

Y

Young, John, CEO of Hewlett-Packard Corp., 58
Yourdan, Edward, 216n7.12, 277n9.8

Z

Zalud, Bill, 368n12.11
Zimmerman, Joel S., 312n10.10
Ziska, Edward A., 336n11.9
Zmud, Robert W., 122n4.16